David Bohm's Critique of Modern Physics

Chris Talbot
Editor

David Bohm's Critique of Modern Physics

Letters to Jeffrey Bub, 1966-1969

Springer

Editor
Chris Talbot
Abingdon, Oxfordshire, UK

ISBN 978-3-030-45539-2 ISBN 978-3-030-45537-8 (eBook)
https://doi.org/10.1007/978-3-030-45537-8

This Springer imprint is published by the registered company Springer Nature Switzerland AG
The registered company address is: Gewerbestrasse 11, 6330 Cham, Switzerland

Foreword

Beginning in October 1966, shortly after I began working as a post-doctoral research specialist in the Chemistry Department at the University of Minnesota, I began a correspondence with David Bohm that continued until August 1969, when I moved to New Haven to take up a position as Assistant Professor in the Physics and Philosophy Departments at Yale. It is, in some ways, a disturbing, even painful experience reading these letters again after so many years. One sees the gradual disintegration of our relationship as I began to develop my own ideas about the foundations of quantum mechanics. Bohm was something of a father figure for me, and there is probably an Oedipal element to my breaking the bond. I was Bohm's graduate student for three years at Birkbeck College, London University, from 1963 to 1965, and very much under the influence of his approach to physics in general and his ideas on the conceptual problems of quantum mechanics in particular at the beginning of the correspondence, less so at the end.

Unfortunately—or perhaps fortunately for the reader!—I did not keep my own letters to Bohm. But it's not hard to reconstruct the broad outlines from Bohm's often lengthy replies and the useful explanatory footnotes by Chris Talbot. I hope these comments provide some additional context.

When I arrived as a student at Birkbeck, Bohm was going through a book by Hodge[1] on harmonic integrals in his seminar, exploring ideas from algebraic topology as a new language for physics. I understood very little of the discussion and, since it was up to me to find a dissertation topic, I spent a lot of time exploring the stacks at the Senate House Library, a short walk from Birkbeck on Malet Street in Bloomsbury. There I discovered the measurement problem of quantum mechanics in articles by Henry Margenau [2] and decided that this was what I wanted to work on. Following Bohm's suggestion, I looked at a paper by Wiener and Siegel [3] on a 'differential-space' theory of quantum mechanics, which eventually led to my dissertation, parts of which appeared as jointly authored papers in

[1][Hodge, 1959].

[2][Margenau, 1963a, Margenau, 1963b, Margenau, 1958].

[3][Wiener and Siegel, 1955].

Reviews of Modern Physics in 1966.[4] The core idea, which Bohm proposed, was to modify the unitary dynamics of quantum mechanics by adding a nonlinear term that produced the notorious 'collapse' of the quantum state in a measurement to an outcome determined by additional 'hidden variables.' As such, the theory was an early 'dynamical collapse' theory, a somewhat primitive precursor of the later Ghirardi-Rimini-Weber theory [5] and its variants, very different from Bohm's 1952 hidden variable theory. Surprisingly, Bohm never talked to me about the 1952 theory during my three years at Birkbeck, and there are only some brief references to the theory in the correspondence.

During my first year at the University of Minnesota, I worked with Alden Mead, a physical chemist who was interested in applying the idea of a fundamental length and time to the Mössbauer effect. Mead spent a sabbatical year at Birkbeck during my last year there and offered me a position as his assistant, largely on the strength of my derivation of the Born probabilities in the collapse theory. For Bohm, it was a mathematical detail he left to me, but it took me months to come up with an acceptable proof. Bohm's method of working through new ideas, which he produced with astonishing frequency, was entirely informal. He had a remarkable physical intuition and was dismissive of mathematical proof, which he felt obscured rather than illuminated informal thinking—a recurring theme in the letters. As I recall, he was never wrong.

After a year in the Chemistry Department, I moved across the mall to Ford Hall, where Herbert Feigl directed the Minnesota Centre for Philosophy of Science. Feigl arranged a two-year position for me as a research associate, presumably a step up since it came with a slightly increased salary. At the Centre, I met Bill Demopoulos and we began a long collaboration that continued until his death in 2017. Demopoulos was interested in Whitehead's process philosophy,[6] in which process and relations are primary, rather than objects with properties, and we saw echoes of Whitehead in Bohm's concept of 'implicate order' and his emphasis on relations, orders, and structures as relevant for physics. Our position shifted radically after Hilary Putnam visited the Centre and argued that the failure of the distributive law of classical logic was the source of conceptual puzzles in the two-slit experiment and other quantum phenomena. We began to move away from the ideas of Bohm and Whitehead and came to see quantum logic as the core idea in deflating the puzzles of quantum mechanics. Around that time we discovered the more sophisticated work of Kochen and Specker [7] in which the crucial non-classical feature of quantum mechanics is the impossibility of embedding the non-Boolean logical structure of quantum mechanics into a Boolean algebra.

Influenced by Kochen and Specker, my new way of thinking about the foundations of quantum mechanics was quite at odds with Bohm's approach, and as the correspondence continues one sees Bohm becoming increasingly frustrated, even

[4][Bohm and Bub, 1966a, Bohm and Bub, 1966b, Bohm and Bub, 1968].

[5][Ghirardi et. al., 1986].

[6][Whitehead, 1978].

[7][Kochen and Specker, 1967].

exasperated, with the lack of communication between us. In the postscript to the long letter of January 18, 1969 (p. 258, C136), he complains:

> What struck me is that you tacitly dismissed all this work, which represents my deepest feelings on the subject, as v. Neumann and his fellow mathematicians tacitly dismissed Bohr as "inessential" when developing their formalisms which were aimed at capturing what was essential in the informal situation regarding quantum theory. If you had at least discussed the papers carefully and shown why you regard them as irrelevant, this would have made sense.... I understand that your whole background has been such as to make you believe that changes in the informal language can only be arbitrary. With this background, it was natural and inevitable that you would dismiss these papers as irrelevant. But how are we going to communicate if you regard my deepest thoughts and feelings as arbitrary? Is it not necessary to begin by entertaining these thoughts, if your intention is communication?

On January 20, 1969 (p. 249, C136) he laments:

> I see again and again that our relevance judgements, are in certain ways, very different. It distresses me to see how far we are from communicating, even though we worked together and discussed so much together. How much further from communication I must be from others, with whom I never even had any common basis!

and a month later he ends the letter dated February 14, 1969 (p. 276, C136):

> My advice to you with regard to the measurement problem is very simple. It is just: "Drop it".

In the letters, there is a lot on measurement in quantum mechanics, the Daneri-Loinger-Prosperi theory I was working on, perception and the artist Charles Biederman (whom Bohm urged me to visit in Red Wing, Minnesota), Feyerabend and Popper, Bohr and the Copenhagen interpretation, Bohr's insistence on classical language for communication, wholeness, the difference between Bohr and Heisenberg, between Bohr and von Neumann, and throughout it all Bohm's attempt to get me to see the relevance of the informal as creative and my preoccupation with von Neumann and Kochen and Specker as a retreat inhibiting real creativity. For Bohm, Bohr's ideas were relevant but flawed, and Bohr's crucial insight was the 'wholeness' of the quantum description, while I had come to see 'classical' as 'Boolean,' and the quantum revolution as the transition to a non-Boolean theory. But quite apart from our intellectual differences, there was no way I could function in an academic environment as a Bohm clone.

The letters end in August 1969, when I moved to Yale. I got the job on the recommendation of Paul Feyerabend, who was at Yale briefly during the academic year 1969–1970 before returning to Berkeley. Feyerabend invited me to Berkeley after I sent him comments on a draft of his two-part critique of Popper on quantum mechanics [8] that Feigl showed me, and we continued to correspond afterwards. I saw Bohm occasionally and he visited me in Israel when I had a position at Tel Aviv University, but we never renewed our correspondence.

---

[8] [Feyerabend, 1968, Feyerabend, 1969].

Bohm's letters were handwritten and his handwriting is not easy to read. Chris Talbot has performed the Herculean task of transcribing and editing the correspondence. The letters should be a useful resource for philosophers of physics and historians of physics interested in Bohm's approach to the foundations of physics, but also for readers who simply want to understand what Bohm had to say on a range of topics, from physics to religion.

Jeffrey Bub
University of Maryland, USA

References

Bohm, D., & Bub, J. (1966a). A proposed solution of the measurement problem in quantum mechanics by a hidden variable theory. *Rev Mod Phys*, *38*(3), 453–469.

Bohm, D., & Bub, J. (1966b). A refutation of proof by jauch and piron that hidden variables can be excluded in quantum mechanics. *Rev Mod Phys*, *38*(3), 470–475.

Bohm, D., & Bub, J. (1968). On hidden variables—A reply to comments by jauch and piron and by gudder. *Rev Mod Phys*, *40*(1), 235–236.

Feyerabend, P. K. (1968). On a recent critique of complementarity: Part I. *Philos Sci*, *35*, 309–333.

Feyerabend, P. K. (1969). On a recent critique of complementarity: Part II. *Philoso Sci*, *36*, 82–105.

Ghirardi, G. C., Rimini, A., & Weber, T. (1986). Unified dynamics for microscopic and macroscopic systems. *Phys Rev*, *D34*, 470–491.

Hodge, W. V. D. (1959). *The Theory and Applications of Harmonic Integrals*. Cambridge University Press, Cambridge.

Kochen, S. & Specker, E. (1967). The problem of hidden variables in quantum mechanics. *J Math Mech*, 17:59–87.

Margenau, H. (1958). Philosophical problems concerning the meaning of measurement in physics. *Philoso Sci*, *25*, 23–33.

Margenau, H. (1963a). Measurements and quantum states: Part I. *Philoso Sci*, *30*, 1–16.

Margenau, H. (1963b). Measurements and quantum states: Part II. *Philo Sci*, *30*, 138–157.

Whitehead, A. N. (1978). *Process and Reality*. The Free Press, New York.

Wiener, N., & Siegel, A. (1955). The differential-space theory of quantum systems. *Nuovo Cimento*, 2:982–1003.

Acknowledgements

I would like to sincerely thank a number of people who helped to make this book possible. Professor Basil Hiley at Birkbeck College, University of London corresponded about the David Bohm revealed in the letters and gave copyright permission for use of Birkbeck's material. Emma Illingworth, the science librarian and archivist at Birkbeck, provided invaluable help in retrieving documents, and has always cheerfully found time for my questions despite her work overload. Angela Lahee of Springer has an interest in David Bohm and must be thanked for enabling publication of this volume.

My special gratitude goes to Prof. Jeffrey Bub of the University of Maryland, for writing the most valuable foreword to the book. It is not always easy to go over events in our distant past, but Jeff willingly provided material, information and advice which considerably enhanced the contents of this book. He also spotted a number of mistakes.

Thanks are also due to the historian of science Olival Freire Junior who, as he explains in his recent biography, has been working on Bohm history for three decades. Olival kindly read through my introduction, finding a serious omission and also noting mistakes. As usual though, the final responsibility for the introduction and the accuracy of the transcriptions and editing rests with the editor.

After moving to England, David Bohm developed an interest in art and his wife Saral was a keen amateur painter. It is therefore fitting that Australian artist Betty Davis of Alice Springs kindly allowed me to use a copy of her painting of David Bohm for the front cover.

Much of my time in recent years has been spent following up books and publications related to David Bohm and wider issues in physics and philosophy using the tremendous resources of the Bodleian Library, University of Oxford. Isabel McMann and the staff of the Radcliffe Science Library section of the Bodleian deserve particular thanks for their help and for providing an ideal environment in which to work. Sadly the RSL has recently been closed and moved to temporary premises. Hopefully it and the invaluable staff team will be soon be moved to more suitable accommodation.

I am fortunate in having the support of a loving family of three children, their partners and four grandchildren, to whom I express enormous gratitude. Many hours of sometimes tedious work in the preparation of the material in this book has only been possible through the support and encouragement of my wife, Ann. As a historian, she has a genuine appreciation of the importance of making these letters of David Bohm available.

Contents

Chapter 1
Introduction

I was inspired to transcribe and edit this collection of letters from David Bohm to Jeffrey Bub after reading drafts of Olival Freire Junior's recent biography of David Bohm.[1] Freire's book is an excellent contribution to the history of 20th century science. In writing it he used some material from these letters, struggling with Bohm's handwriting, to comment on the late 1960s. I was also pleased that Freire was able to use the transcriptions I had made of the Bohm correspondence for the 1950s period in Brazil and Israel.[2] I made some suggestions on Bohm's ideas which Olival was kind enough to take seriously.

There is a small but continuing interest in Bohm-related physics, including the "Implicate Order" approach which has been championed by Basil Hiley, Bohm's colleague for more than 30 years. Since Bohm's death in 1992, Hiley has produced more than 100 publications developing theoretical physics from his and Bohm's standpoint, rebutting numerous attempts to refute or dismiss it from the position of quantum mechanical "orthodoxy". The recently published "Emergent Quantum Mechanics, David Bohm Centennial Perspectives[3] exemplifies this continuing interest.[4]

As well as Freire's biography, we already have the first biography of Bohm by his friend David Peat,[5] written a little after Bohm's death, which, although containing less of the science, does give a colourful account of Bohm's life. In addition, giving some explanation of Bohm's involvement with the Indian thinker and teacher Jiddu Krishnamurti and his organisation, beginning in 1961 and continuing even after

[1] Freire Jr. (2019).

[2] Talbot (2017).

[3] https://www.mdpi.com/books/pdfview/book/1203.

[4] Note that Basil Hiley, now in his eighties, is joint author of no less than four papers in this collection.

[5] Peat (1996).

C. Talbot (ed.), *David Bohm's Critique of Modern Physics*,
https://doi.org/10.1007/978-3-030-45537-8_1

Krishnamurti's death in 1986, we have the recent informative book by David Edmund Moody.[6] Also most of the large number of books and articles written by Bohm, both scientific and philosophical as well as those relating to Krishnamurti are available in a good university library. Given all of this material, why is yet another book needed? This introduction is mainly an attempt to answer this question.

First me explain something about the material that is contained in this book. Bohm's correspondence with Jeffrey Bub is in the archives at Birkbeck College, University of London, in eight folders, numbered C130–C137, so I have kept them in the same order in the chapters here, headed with the same numbers. Altogether there are 594 pages in the original, mostly in Bohm's handwriting but a few are typewritten. Unfortunately Jeffrey Bub's replies were not preserved though Bohm often makes clear the point that Bub was making.

Since this correspondence the approach to physics followed by Jeffrey Bub and David Bohm diverged, as Bub explains in the foreword he has kindly written. I should give a brief outline of Jeffrey Bub's distinguished career for those not familiar with the world of modern physics. He received his PhD in mathematical physics from London University in 1966, where he studied physics with David Bohm at Birkbeck College and philosophy of science with Karl Popper and Imre Lakatos at the London School of Economics. He has published numerous articles in scientific and scholarly journals on the conceptual foundations of quantum mechanics and is the author of several books: *The Interpretation of Quantum Mechanics* (1974), *Interpreting the Quantum World* (1997), which won the prestigious Lakatos Award in 1998, *Bananaworld: Quantum Mechanics for Primates* (2016; revised paperback edition, 2018), and *Totally Random: Why Nobody Understands Quantum Mechanics* (2018; co-authored with Tanya Bub).[7] The latter, "An eccentric comic about the central mystery of quantum mechanics" is particularly recommended for non experts. Bub is currently a Distinguished University Professor Emeritus at the University of Maryland, College Park.

Jeffrey Bub was most helpful in providing additional material from his personal archives which I have added in the first three appendices. The first contains correspondence with the Italian physicist Angelo Loinger which relates to folder C130, Chap. 2, of the Bohm–Bub correspondence; the second is a letter from Bub to David Bohm relating to a paper by well-known physicist Leon Rosenfeld, also relating to C130 and the third contains details of the "game" referred to at the beginning of C131, Chap. 3. In addition there are appendices containing two previously unpublished articles written jointly by David Bohm and his postdoc researcher Donald Schumacher. They have been added because of their relevance to much of the correspondence.[8]

Olival Freire, quite correctly for a historian of science, concentrates in his biography on Bohm's scientific work and his relationship to scientists of his day, putting this

[6]Moody (2016).

[7]Bub (1974, 1997, 2016), Bub and Bub (2018).

[8]They are in the Birkbeck archives as B44 and B88. For more on Schumacher, including his severe mental illness which prevented him from continuing to work with Bohm, see Freire Jr. (2019), pp. 141–148 and Peat (1996), pp. 246–253.

into the context of the times and against the background of his philosophical ideas. However Bohm saw himself primarily as a philosophical thinker, taking philosophy in the broadest sense and not in the limited tradition of academic philosophy. One can see Bohm giving primary importance to metaphysics in the correspondence.[9] Referring to the Bellagio Conference in Theoretical Biology,[10] Bohm observes the way in which contributors talked at cross-purposes because of differences in their "deep tacit assumptions". "At a certain point, I plunged in and presented my own metaphysical notions. In doing this, I explained that each person inevitably has some kind of metaphysics, which is just a set of general and basic assumptions about reality as a whole." (Bohm was apparently pleased at the response to his intervention, reporting, perhaps somewhat naively, that everyone at the conference came out into the open with their own metaphysics.) I think this does leave room for an alternative approach, attempting to find the origins of Bohm's philosophical ideas, how he developed them, and how they relate to twentieth century thought. Perhaps one could say that this approach is more from the standpoint of the history of ideas, rather than from the history of science.

My first steps in this direction were an attempt to gain some understanding of "Causality and Chance in Modern Physics."[11] David Peat gives some explanation and made some references to Bohm's correspondence from Brazil and Israel, but it seemed to me that only by transcribing and editing all the correspondence could we gain a fuller understanding of the provenance of "Causality and Chance". I attempted to summarise Bohm's version of Marxist philosophy that the letters reveal.[12]

Moving into the 1960s we are presented with a far bigger task. Bohm's publications in this period discussing philosophical issues throw up difficult questions concerning what influences are at work. We consider this further below. But as with the 1950s we can hope to get a clearer understanding if all the important unpublished material in the Birkbeck archives is made available. Much of this is in Bohm's handwriting which is difficult and time-consuming to read. The publication here of the Bohm–Bub correspondence is a part of this project. A much bigger project, still ongoing, is the correspondence between Bohm and the artist Charles Biederman. Only the first letters of this correspondence, from March 6, 1960 to April 24, 1962 were edited and transcribed by the Finnish philosopher Paavo Pylkkänen.[13] In the Birkbeck archives (Folders C67–C92) there are letters continuing up to October, 1969 (3202 pages of letters written by Bohm and 674 pages of replies by Biederman). Note that in seven of the Bohm–Bub letters there is a request that Bub sends them on

---

[9]C132, pp. 147–148.

[10]The 2nd Symposium on Theoretical Biology, held August 3–12, 1967. The proceedings are published in Waddington (1969). Bohm's contributions were "Some remarks on the Notion of Order" (pp. 18–40), "Further Remarks on Order" (pp. 41–60), "Addendum on Order and Neo-Darwinism" (pp. 90–92), "Some Comments on Maynard-Smith's Contribution" (pp. 98–105).

[11]Bohm (1957). See also the important aftermath of this book in the Colston conference, Kožnjak (2017).

[12]Talbot (2017), especially pp. 23–37.

[13]Pylkkänen (1999).

to Biederman.[14] There are also a number of shorter collections of correspondence such as the letters to Bohm's brother-in-law, Yitzhak Woolfson. Some of these are published,[15] including an important introduction by Woolfson.

The importance of Bohm's correspondence can perhaps be emphasised by making parallels with the case of Wolfgang Pauli and Carl Gustav Jung.[16] I am not, of course, suggesting a direct comparison between Pauli, who belonged to an older generation of physicists, and Bohm and certainly not between the religion of Krishnamurti and the analytical psychology of Jung.[17]

Pauli is regarded as one of the foremost proponents of the "positivist" approach to orthodox quantum mechanics that was developed in the late 1920s. (The idea that there was a unified approach ignores the many differences between the leading figures, something which Bohm was well aware of as this correspondence with Bub shows, and which has been made clear in the work of Mara Beller Beller 1999.) It is remarkable that Pauli made a "metaphysical" turn from positivism to studying mysticism after 1930, to which his correspondence with Jung relates. Miller in his authoritative study comments that "Pauli told very few colleagues about his discussions with Jung. He feared their derision. Nevertheless his sessions with Jung convinced him that intuition rather than logical thought held the key to understanding the world around us."[18] Thus little was known about Pauli's ideas, even after his death in 1953, though Bohm apparently knew something about Pauli's interest in Jung,[19] (Bohm considers Jung's ideas in the correspondence[20] Ironically, considering that Pauli opposed Bohm's support for determinism in 1951, Bohm's essential objection to Jung is that his theory is "too crude and mechanical to account for perception, especially its more creative aspects.")

Without the Pauli–Jung correspondence and Pauli's correspondence as a whole becoming available to researchers, the studies of Miller and the others noted above, all of which are relatively recent, would not have been possible. Pauli's complete correspondence has been edited and published by Karl von Meyenn. It contains more

[14]These are C131: Jan 7, pp. 63–74, Feb 22, pp. 74–80, May 30, pp. 89–92, June 1, pp. 92–95 and June 2, pp. 95–98 and C133: Oct 20, pp. 172–177. All are in 1967. Note that there references to Biederman's ideas in C130 and especially C131.

[15]Nichol (2002), pp. 199–234.

[16]Lindorff (2004), Gieser (2005), Miller (2009) and Atmanspacher and Primas (2009).

[17]It was Pauli who, after Bohm replied to a series of objections, finally accepted that Bohm's two famous 1951 "hidden variable" papers were a valid alternative to standard quantum mechanics and sanctioned publication in Physical Review. Freire gives the details (Freire Jr. 2019, pp. 71–74) and also points out that Pauli, in a letter to Markus Fierz, "raised the stakes on philosophical grounds criticizing the expectations of recovery of determinism in physics. He observed that Catholics and Communists depended on determinism to buttress their eschatological faiths, the former in the heaven to come, the latter in paradise on earth." (pp. 82–3).

[18]Miller (2009).

[19]C131, p. 80.

[20]C131, pp. 80–89.

than 7000 pages in eight volumes, published between 1979 and 2005.[21] As in Pauli's case, making available Bohm's correspondence can hopefully encourage a deeper understanding of his ideas, which, like those of Pauli, are unusual considering the widespread disdain shown towards "metaphysics" in the environment of modern science.

Consider now Bohm's published material in the 1960s relating to philosophy and the philosophy of science, as opposed to publications directly relating to physics. In the early 1960s there are a few of Bohm's papers within the "official" philosophy of science tradition.[22] Note that in the first paper cited Bohm writes that "guided by different conceptions, one is led to seek different types of facts, some of which may be possible in a given field, and some not." He also stresses that "facts are made". These are ideas that can be seen in the Bohm–Bub letters.[23] The second, in a collection dedicated to Karl Popper, has a discussion of "understanding" which may be considered to be an earlier version of "creativity."

Some insight can be gained by comments from the philosopher of science Paul Feyerabend. In a letter to David Peat in 1993, answering questions on his experiences of working with Bohm at Bristol University in the late 1950s, Paul Feyerabend recalled that Bohm had discussed Gilbert Ryle, Ludwig Wittgenstein and Karl Popper:

> Dave who knew about all of them remarked that Popper relied on logic which was OK, but too rigid to aid scientific research and that 'ordinary language philosophy' was not really using ordinary language but an artificial lingo which was also too rigid. He agreed that there could be 'category mistakes' in the sense that predicates could be applied to inappropriate subjects (such as calling the number two blue) but that regarding category mistakes as boundaries to talk as Ryle suggested, was going too far: scientific research often went right through such 'mistakes'. Dave's knowledge of Ryle and Wittgenstein was not very detailed but he had an amazing ability of getting the whole picture from a few hints. After that he critized the whole picture, not just a few embroideries here and there as is the custom of many philosophers.[24]

However, after moving to Birkbeck, University of London, Bohm begins to publish his own distinctive philosophical ideas. These are to be found in (1) the inaugural lecture given at Birkbeck (1963),[25] (2) the contribution given to a physics conference in Kyoto (1965),[26] (3) the Appendix to the Special Theory of Relativity (1965),[27] (4) contributions given to the Bellagio conference on theoretical biology (1967)[28] and (5) "On Creativity" (1968).[29] Of these (2), (3) and (5) are readily available.

[21] Atmanspacher and Primas (2009), p. 3, n 3. An English translation of the Pauli–Jung letters is Meier (2001).

[22] Such as Bohm (1961) and Bohm (1964).

[23] Especially in C132.

[24] Letter to Peat dated 07/09/93, Birkbeck archives folder A21.

[25] Bohm (1963).

[26] Bohm (1965), also http://www5.bbk.ac.uk/lib/archive/bohm/BOHMB.149.pdf.

[27] Bohm (2006).

[28] See p. 3, n 10 above.

[29] Bohm (1968) also Nichol (1998), pp. 1–18, or http://classes.dma.ucla.edu/Fall07/9-1/pdfs/week1/OnCreativity.pdf.

Between them the five publications present the concepts of *order*, *structure* and *process*, *perception* and *creativity* that characterise Bohm's philosophical views in this period.[30]

Much of (1) is summarised in the first section of the more readily available paper (2). In the appendix to (1) we find an important comment on "process". Bohm wants to begin with "process", "the assumption that what is is movement itself." How does one then explain things which are at rest? "Such an explanation is carried out in terms of the notion of invariant repetitive, ordered and structured relationships that hold only relative to certain conditions, at certain levels, within specific contexts, and to limited degrees of approximation." Here "order" and "structure" are referring back to the discussion in the main text, on relativity, quantum theory and the latest developments in particle physics. The approach to process is reminiscent of "Causality and Chance", especially Chap. 5, but the latter is not referred to.

Reading (2) we see the basic definitions of "order", "structure" and a good deal on "function", especially "reflective function." Order is explained as being based on "similar differences" leading to "different similarities" with examples from geometrical curves and wave motion in physics. Structure is defined as "order of orders". Bohm clearly considers that these are concepts that are known to everyone from daily experience, so it is a matter of making them more precise. But note that he thinks they can be extended to "all of our perception, thinking, feeling and action", quite a wide generalisation. However in the long discussion on "function" he makes the qualification of "abstracting" from a "limited domain", and opposes ideas of "absolute and final truths"—again reminiscent of "Causality and Chance". The idea of "creative process" is briefly referred to but is expanded on in (4) and (5). "Reflective function" is clearly a key idea for Bohm, it is primarily "ontological" and is developed in the panpsychism evident in the first half of the Bohm–Bub correspondence. This paper seems to have been central to the work of the research group at Birkbeck.[31]

In (3), another widely available publication from the 1960s, Bohm summarises his study of Piaget on child development and the latest work by psychologists such as James J. Gibson. Note that Bohm cites J. R. Platt, Professor of Physics and Biophysics at the University of Chicago with whom Bohm had over 100 pages of correspondence in 1963, also unpublished (Birkbeck archives C51–C54). It is clear that though the psychology of perception is the area of investigation, "perception" is also being regarded as a philosophical concept. Already in the preface, Bohm has stated that "science is mainly a way of extending our perceptual contact with the world, rather than of accumulating knowledge about it."[32] Then extending the ideas of (2) we read that "in the process of perception we learn about the world mainly by being sensitive to what is invariant in the relationships between our own movements, activities,

---

[30] I have omitted Bohm (1962) as this is concerned with a topological approach in physics, and also "On the Relationships of Science and Art" (1968) in Nichol (1998), pp. 19–28, which would take us too far afield.

[31] See, for example Bub (1969) and many later publications of Hiley, for example Hiley (2011), also https://arxiv.org/abs/1211.2107.

[32] Bohm (2006), p. x.

probings, etc., and the resulting changes in what comes in through our sense organs," and that "the invariant is finally understood with the aid of various hypotheses, expressed in terms of higher levels of abstraction, which serve as a kind of "map," having an order, pattern, and structure similar to that of what is being observed."[33] As with (2) we find these ideas being extended in the correspondence with Bub.

At the Bellagio conference (4) we have already referred to Bohm's stress on the importance of metaphysics. Bohm goes over the material on order, structure and process,[34] stressing that "the notion of order is evidently more fundamental than other notions, such as, for example, that of relationships and classes, which is now generally regarded as basic in mathematics." Everyone has some "tacit" knowledge of order so "with words we can 'point to' certain essential features of this tacit knowledge." Which he does, repeating the exposition of "similar differences" and "different similarities". Note that he makes an extension of the "difference" conception to "constitutive differences" (determining the essence of what we are talking about, such as chords in the curves example) and "distinctive differences" (defining the relation between one order and another, such as between the chords of different curves). This extended definition is noted in the correspondence.[35]

Bohm stresses that order and structure will not be static but in a process. What is essential to process is "not merely that there is a change of order and structure, but that the differences are similar, so that the changes themselves are ordered". Process is thus an order of change. In biology Bohm considers that there are evolutionary processes with "the coming into being of new orders, along with an ordering of the changes of order in the whole process".[36] Previously he suggested that "the breaks or changes in order of a given process can themselves be the basis of a higher order of process."

Bohm considers the mechanistic view in which the constitutive order of the universe is that of fundamental particles moving in some kind of mechanical motion. His position is that natural processes can "contain a really creative movement, in which there appear new orders and orders of orders."[37] But while he considers there has been a tendency in physics to move away from a mechanistic view (he takes up several pages on statistical mechanics, quantum theory and quantum field theory) he is concerned that biology and psychology are moving closer to it.[38]

Apart from a discussion with biologist John Maynard Smith on neo-Darwinism[39] which needn't concern us here, note that in the section "Further remarks on order", especially "On the self-regulating hierarchy of process" and "On the separation of

---

[33] Bohm (2006), pp. 164 and 169.

[34] Waddington (1969), pp. 19–25. See p. 3, n 10 above.

[35] C132, pp. 115–120.

[36] Waddington (1969), pp. 25–26.

[37] Ibid, p. 28.

[38] Ibid, p. 34.

[39] Ibid, pp. 90–94.

the observer and the observed"[40] there is also material that occurs in the Bohm–Bub correspondence.

"On Creativity"[41] (5) integrates, if in a briefer and more popular form, much of the material in (1)–(4). Although perhaps it could be considered as primarily a paper on psychology that is misleading. In fact as noted above, perception, even "creative perception", is a philosophical concept for Bohm. The concepts of order and structure are explained again but note that a distinction is being made between perception which is mechanical where "the order, pattern and structure of what is perceived come from the record of past experiences and thinking" and creative perception. In the latter "one first becomes aware (generally non-verbally) of a new set of relevant differences, and one begins to feel out or otherwise to note a new set of similarities, which do not come *merely* from past knowledge, either in the same field or in a different field. This leads to a new order, which then gives rise to a hierarchy of new orders, that constitutes a set of new kinds of structure." It is not difficult to see Krishnamurti's idea of mechanical thought that he contrasts with "choiceless awareness."[42] Since "On Creativity" is readily available and relatively straightforward to read no more comment is needed.

The five publications I have considered could be supplemented by material from the Cambridge conference, held in July, 1968[43] and the Illinois symposium, held in March 1969.[44] However these two more recent publications are somewhat different in that Bohm has made a shift in his views to accommodate "communication", following his discussions with Schumacher[45] Also Bohm's ideas on Niels Bohr and quantum theory at Cambridge are part of a discussion that is ongoing throughout the correspondence, some of it quite technical, that would require much more consideration than we can give here.[46]

For Bohm's "Perception-Communication" view of the nature of science at Illinois, the material in the correspondence[47] taken together with the joint papers with Schumacher given in the Appendices should help give a clearer understanding of Bohm's standpoint. The connection of Bohm's ideas to Feyerabend's "pluralism" can be understood further by noting Bohm's sympathy with Feyerabend in earlier

---

[40]Ibid, pp. 51–59.

[41]Noted by Bohm in C131, p. 83.

[42]See Moody (2016), especially Chap. 5.

[43]C133, p. 184, n 6 and C136, p. 250, n 1.

[44]C136, p. 263, n 6 and C137, p. 301, n 1.

[45]No records are kept at Birkbeck and the preprints given in Appendices D and E are undated, but discussions with Schumacher are referred to from November 1967 (C133, p. 191) to February 1968 (C134, pp. 213–219).

[46]But note Freire Jr. (2019), p. 150.

[47]See C136, pp. 287–289, 291–296 and 298, C137 pp. 301–306 and 319–320.

letters.[48] Such an exposition cannot be given in this brief introduction and must be left to later work.[49]

Returning to consider the brief review of Bohm's philosophical material published in the 1960s, I have indicated that a number of key ideas also occur in the Bohm–Bub correspondence. I would suggest further that there is a wealth of material, especially in C130–C133 which expands on and, hopefully, helps to clarify the documents (1)–(5). Of course it is unfinished, not everything is clear, and there are ideas going in a highly speculative, Krishnamurtian direction. Bohm writes in a rather unstructured manner, sometimes repetitive, often using several words together to try to clarify or stress a point, even making up new words. Overall though, the correspondence should help to clarify the ideas of "structure-process" behind Bohm's (and Hiley's) view of the "Implicate Order" developed in the decades since. Also Bohm can be seen developing a type of panpsychism, "neither materialism nor idealism",[50] "the observer is the observed",[51] based on perception but with a definite ontology of structure-process, which again sheds some light on his later work.

However, even if the Bohm–Bub correspondence is of considerable help, there are still questions remaining for the student of Bohm's philosophical views in the 1960s. Even allowing for Bohm's originality and creativity,[52] the influences at work in his views remain something of a mystery. In (3) there are references to work on the psychology of perception and in (4) references to biologists and even a reference back to (2). But no other references and certainly no philosophical references are given in (1)–(5), not even to Bohm's own "Causality and Chance."

It seems that Biederman played a major role in the formation of Bohm's ideas on order and structure as well as creativity. In the correspondence he discusses Biederman's views on art, especially in relation to "creating new orders".[53] Surprisingly Bohm states that although he got some of his ideas from Whitehead,[54] his correspondence with Charles Biederman was a lot more important.[55] Some of this can be seen in the letters published by Pylkkänen. Bohm introduces the Hegelian concepts of identity and difference[56] but Biederman objects to the term "identity", preferring "similarity". There is a discussion lasting over a year with Bohm eventually agreeing to replace "sameness" with "similarity", or as Pylkkänen puts it "to drop "identity" from their set of concepts, as long as certain conditions are accepted."[57] But there is

---

[48] C132, pp. 148–149, C133, pp. 160–161, pp. 167–168 and p. 191. Bohm is reading and referring to Feyerabend (1965).

[49] Note however Freire Jr. (2019), pp. 148–150.

[50] C130, pp. 36–41.

[51] C132, pp. 104–112.

[52] In C131, p. 86, Bohm writes, "you cannot do valid work on creativity without yourself being in the creative state about which you wish to talk."

[53] C131, pp. 89–98.

[54] To our knowledge Bohm does not comment on Whitehead in the 1950s and 60s period.

[55] C131, p. 92.

[56] Pylkkänen (1999), p. 109.

[57] Ibid, p. 244. Pylkkänen's chapter summaries are very useful here.

no explicit mention of "order" or "structure". Note however that by the end of 1962, Bohm is writing to Yitzhak Woolfson on the idea of "structure-process". He writes: "Each structure has a kind of order, a set of sequences of elements that are naturally most immediately related, as well as breaks as variations in this order," and "The problem of structure is basic to my work in physics. In essence, I am trying to find the general principle of the process-structure that can abstract as time-space."[58]

Krishnamurti features throughout the Bohm–Biederman letters and clearly is a major influence on Bohm's thinking from 1962 onwards. Biederman derived ideas from the Polish-American philosopher, Alfred Korzybski,[59] of whom Bohm was quite critical. There are increasing differences of opinion regarding A.K. (Korzybski) and J.K. (Jiddu Krishnamurti) throughout the letters. Hopefully the Bohm–Biederman correspondence will help to explain Biederman's influence but will also tell us much more about Bohm's obvious enthusiasm for Krishnamurti in that period.

Another influence on Bohm is the 19th century philosopher G. W. F. Hegel. Bohm kept returning to a study of Hegel all his life, as reported in the 1986 interviews with his friend Maurice Wilkins where far more consideration is given to Hegel's ideas than to Krishnamurti and Biederman (though Bohm had clearly become disillusioned with Krishnamurti by then). Hegelian views—or rather the Marxist interpretation of Hegel—are central to "Causality and Chance" as I attempted to show.[60] The interviews with Wilkins certainly show Bohm's facility with Hegelian concepts.[61] An important aspect of Hegel's philosophy is the conception of the world and of thought as a process.[62] We may assume that Bohm's commitment to process philosophy comes from this background.

The dialectical opposites that are central to Hegelian philosophy were apparently frequently used by Bohm and appear in the Bohm–Bub correspondence (such as Necessity and Contingency, Form and Content, etc.). As if in a revelation that other people, especially physicists, do not have his familiarity with Hegel, Bohm notes that Bub regards "contingency and necessity as incredibly complex notions" compared to the von Neumann-style axiomatic approach to quantum theory, which Bub regards, to Bohm's chagrin, as "manageably simple".[63]

It is presumably because Hegel, Biederman and Krishnamurti are such major influences that Bohm gives no references or discussion of what influenced his ideas. We could assume that Bohm does not refer to Hegel and Marx because he wanted to

[58]Nichol (2002), p. 218 and p. 219.

[59]Korzybski's ideas, known as "general semantics", published in "Science and Sanity", Korzybski (1994) enjoyed some interest in the 1930s and 40s but are now rarely mentioned. A few references are given in the Bohm–Bub correspondence: C130, p. 48, C133, pp. 188–189, pp. 193–198.

[60]Note that in the 1993 letter to Peat, Feyerabend writes that at Bristol Bohm either "read Hegel's logic, or had just read it and like Lenin interpreted it materialistically."

[61]Wilkins (1986), parts VII, IX, X, and XI.

[62]See, for example, Beiser (2005).

[63]C136, p. 274.

distance himself his pro-communist past.[64] But also one should not forget the general antipathy towards Hegel amongst philosophers and historians of science.[65]

According to David Peat[66] Bohm's discussions with Krishnamurti "caused considerable consternation among his former colleagues in the United States," so that despite his enthusiasm for ideas that he thought held the key to understanding the world, Bohm became quite guarded on the subject. There was something of a change after 1980. As Freire notes, although the 1980 UK edition of "Wholeness and the Implicate Order"[67] contained no mention of Krishnamurti, later editions did.[68] Bohm then held a number of dialogues with Krishnamurti. The first, "The Ending of Time" was held in 1980, and published in 1985.[69] The reluctance to go public with the Bohm–Krishnamurti relationship seems to be not just Bohm's responsibility but also came from within the Krishnamurti organisation. As Moody explains, an earlier series of dialogues between Bohm and Krishnamurti held in 1975 was blocked from publication in 1977 by Mary Lutyens, Krishnamurti's official biographer.[70] There was something of a conflict between Bohm and Krishnamurti at that time, revealed in Bohm's correspondence with Fritz Willhelm. This correspondence, though in the Birkbeck archives, has also not been published. Willhelm was a physicist who worked for the Krishnamurti Foundation in the late 1970s. Conflicting interpretations of the seriousness of the conflict are given by David Peat[71] (anti-Krishnamurti) and David Moody[72] (pro-Krishnamurti).

Perhaps Bohm also thought that suggesting some of his views on the philosophy of physics were influenced by Biederman, an artist, would not be well received. If so he changed his mind by 1971, including a footnote, "This notion of order was first suggested to the author in a private communication by a well-known artist, C. Biederman," with a reference to Biederman's, "Art as the Evolution of Visual Knowledge"[73] in the first volume of the journal Foundations in Physics.[74]

Having explained some of the remaining difficulties of understanding Bohm's philosophical ideas in the 1960s, I now turn to two recent developments that could perhaps help to revive some interest in Bohm's ideas today.

---

[64] See Freire Jr. (2019), pp. 105–107.

[65] This applies to all of the "Romantic" tradition in philosophy. It is to Basil Hiley's credit that he has pointed out the influence of such philosophy on mathematicians who played a key role in the development of theoretical physics, Hermann Grassmann and William Rowan Hamilton (see Hiley (2011), also https://arxiv.org/abs/1211.2107).

[66] Peat (1996), p. 200.

[67] Bohm (1980).

[68] Freire Jr. (2019), pp. 175–6.

[69] Krishnamurti and Bohm (1985).

[70] Some were in fact published as the first part of Krishnamurti (1977).

[71] Peat (1996), Afterword.

[72] Moody (2016), Chap. 11.

[73] Biederman (1948).

[74] Bohm (1971).

The first concerns panpsychism. It was quite a revelation to read an article by an analytic philosopher, William Seager, referring to Bohm's panpsychism, included in the above collection "Emergent Quantum Mechanics, David Bohm Centennial Perspectives."[75] Seager argues that "advances in science serve not to eliminate metaphysical questions, but illuminate them and sometimes to reawaken metaphysical options that had faded from view." Pointing to the rebirth of interest in panpsychism, especially relating to "the problem of consciousness", he notes that "mental features are a fundamental and ubiquitous feature of the world" in the panpsychic viewpoint. This is an approach that "integrates mind and the physical world, which leaves the physical world causally complete, avoiding outside influences distorting the laws of nature, but nonetheless provides a role for mind in the world. We can see Bohm as a kind of pioneer for this rebirth." He gives many useful references[76] including several from Bohm's later writings as well as Hiley's. However the correspondence published here shows that Bohm was developing a form of panpsychism as early as the 1960s.

Freire notes[77] that the ideas of panpsychism introduced in the last chapter of "the Undivided Universe",[78] the "most daring conjecture of the entire book" could be expected to "dismay some readers". The suggestion that "participation goes on to a greater collective mind, and perhaps ultimately to some yet more comprehensive mind in principle capable of going indefinitely beyond even the human species as a whole," does seem to take us in a quasi-theological direction. It is thus very interesting to read Seager on this issue, positioning it within the traditions of academic philosophy.[79] Note that Seager has also written on what is called the "dual-aspect approach to the mind-matter problem" in relation to Pauli's views.[80]

All the recent works on panpsychism referred to by Seager are "Western" in orientation, even when giving a history of panpsychism.[81] But clearly there is a rich tradition of Indian philosophy, which despite difficulties of interpretation is now becoming better known.[82] Articles on panpsychism from an Indian standpoint are appearing in western journals.[83] In Bohm's writing there can be found the stereotypical view of the "Greek western tradition" on the one hand and the "Oriental mystical tradition" on the other, which was prevalent in the twentieth century. It did not help that Krishnamurti presented his ideas as his alone and gave no references. However it

[75] Seager (2018), also https://www.mdpi.com/1099-4300/20/7/493/htm.

[76] See also the entry in the Stanford Encyclopedia of Philosophy, of which Seager is a joint author, https://plato.stanford.edu/entries/panpsychism/.

[77] Freire Jr. (2019), p. 197.

[78] Bohm and Hiley (1993).

[79] Though there is fierce opposition to panpsychism among neuroscientists, e.g. "Conscious spoons, really? Pushing back against panpsychism," by Anil Seth, https://neurobanter.com/2018/02/01/conscious-spoons-really-pushing-back-against-panpsychism/.

[80] "A New Idea of Reality: Pauli on the Unity of Mind and Matter," Atmanspacher and Primas (2009), pp. 83–98.

[81] For example Skrbina (2005).

[82] See, for example Hamilton (2001).

[83] For example Vaidya and Bilimoria (2015).

is increasingly recognised that this rigid separation emerged from the Eurocentrism of colonial times[84] and has become unacceptable.

The second recent development concerns the various criticisms that have been made of the current situation in fundamental physics.[85] Without committing myself to support any of these critics, which I am hardly qualified to do anyway, there does seem to be something of an impasse, especially following the failure of the Large Hadron Collider at CERN, Geneva to confirm any of the predicted "supersymmetry" theories after the Higgs boson success in 2012. No doubt new theories, or variations on old theories will be developed. Even so it would seem not to be a bad idea to consider the critique that Bohm was making in the 1960s and which is reflected in the title of the present book.

Bohm's arguments for "pluralism" and for the importance of what he calls the "informal" as opposed to the formal mathematical approach would still seem to have some validity. The problem is that the material in the correspondence is too much rooted in the issues he was dealing with at the time, and are not especially well argued. For example, although Jeffrey Bub was praised for his reply to objections to Bohm's views at Illinois, it would seem to me that he was wise to concentrate on quantum theory. When Bohm used different theories of malaria[86] as an example he seemed to be on shaky ground, especially his "psycho-social" theory with its suggestion that (following Krishnamurti) "[t]he centrally relevant feature is that man has for thousands of years lived disharmoniously."[87] This could easily be interpreted as anti-science. Also the argument for the "informal" does seem to be carried too far, perhaps in the heat of the argument with Bub. In an email discussing Bohm's criticism of Basil Hiley for his addiction to formal deduction,[88] Hiley, who must surely be David Bohm's greatest champion, explained that Bohm was not correct in suggesting that Hiley did not "understand the irrelevance of these equations." Rather Hiley considered that exploring a topic did mean "putting it into some mathematical framework". In the process "one has to think deeply about the proposal." Note also that if the impression is given in the correspondence that Bohm was anti-mathematics this is not correct. He saw the possible importance of algebraic topology for physics in the 1960s and did his best to understand it.[89] Also he was one of the first to appreciate the importance of Clifford algebras.

So consider how Bohm presented the same core arguments but in the light of two decades of experience in "Science, Order and Creativity",[90] co-authored with David Peat. Firstly on Pluralism:

---

[84] See McEvilley (2002).

[85] Woit (2006), Baggott (2012), Hossenfelder (2018). Note Jeremy Butterfield's review of Hossenfelder at https://arxiv.org/abs/1902.03480.

[86] C136, pp. 292–296, C137, pp. 302–304 and p. 319.

[87] C137, p. 302.

[88] C137, p. 307.

[89] See the reference to the British mathematician W.V.D. Hodge in C136, p. 262.

[90] Bohm and Peat (2010).

The development of science can be seen in a Kuhnian sense, "until today it is taken as perfectly normal for revolution to succeed revolution, interspersed by periods of relative stability." But instead, would it not be possible that "creativity can operate at all times, not just during periods of scientific revolution? . . . this would imply that, at any given moment, there would be a number of alternative points of view and theories available in each particular area of science." Bohm and Peat give arguments to show that this viewpoint does not necessarily lead to a lack of objectivity, i.e. that social considerations or subjective preferences would dominate.[91]

Secondly on the importance of "informal" considerations:

"Today the general atmosphere is such that a physicist can do little more than state, and restate, a particular point of view. Various approaches are generally taken to be rivals, with each participant attempting to convince the others of the truth of a particular position, or at least that it deserves serious attention. Yet at the same time, there is a general tendency to regard the whole question of interpretation and the role of informal language as not being particularly important, and instead to focus upon the mathematics about which everyone agrees."[92]

No doubt Bohm and Peats' appeal for "the opening up of a free and creative communication in all areas of science" which "would constitute a tremendous extension of the scientific approach," and their conclusion that "[i]ts consequences for humanity would, in the long run, be of incalculable benefit" could be seen as utopian in the current climate. But that does not mean that Bohm's critique should not be taken seriously.

References

Atmanspacher, H., & Primas, H. (Eds.). (2009). *Recasting reality Wolfgang Pauli's philosophical ideas and contemporary science*. Berlin: Springer.

Baggott, J. (2012). *Farewell to reality*. New York: Pegasus Books.

Beiser, F. (2005). *Hegel*. Abingdon: Routledge.

Beller, M. (1999). *Quantum dialogue*. Chicago: University of Chicago Press.

Biederman, C. (1948). *Art as the evolution of visual knowledge*. Michigan: University of Michigan.

Bohm, D. (1957). *Causality and chance in modern physics* (1st ed.). Abingdon: Routledge and Kegan Paul. Second edition with new preface by Bohm (1984).

Bohm, D. (1961). On the relationship between methodology in scientific research and the content of scientific knowledge. *The British Journal for the Philosophy of Science*, *12*(46), 103–116.

Bohm, D. (1962). A proposed topological formulation of the quantum theory. In I. J. Good (Ed.), *The scientist speculates* (pp. 302–314). London: Heinemann.

Bohm, D. (1963). *Problems in the basic concept of physics an inaugural lecture delivered at Birkbeck College 13th February 1963*. [Printed for the Birkbeck College by J. W. Ruddock], London. Satyendranath Bose 70th Birthday Commemoration Volume, Part II, Calcutta, pp. 279–318 (1965).

[91] Ibid, pp. 44–49.

[92] Ibid, p. 78.

Bohm, D. (1964). On the problem of truth and understanding in science. In M. Bunge (Ed.), *The critical approach to science and philosophy. In honor of Karl R. Popper* (pp. 212–223). London: Collier-MacMillan.

Bohm, D. (1965). Space, time and the quantum theory understood in terms of discrete structural process. In Y. Tanikawa (Ed.), *Proceedings of the International Conference on Elementary Particles.* (pp. 252–287), Kyoto. Kyoto University, Publication Office, Progress of Theoretical Physics.

Bohm, D. (1968). Creativity. *Leonardo, 1*(2), 137–149.

Bohm, D. (1971). Quantum theory as an indication of a new order in physics. Part A. *Foundations of Physics, 1*(4), 359–381.

Bohm, D. (1980). *Wholeness and the implicate order*. Abingdon: Routledge and Kegan Paul.

Bohm, D. (2006). *The special theory of relativity*. Abingdon: Routledge. Reprint of Benjamin original (1965).

Bohm, D., & Hiley, B. (1993). *The undivided universe: An ontological intepretation of quantum theory*. Abingdon: Routledge and Kegan Paul.

Bohm, D., & Peat, D. (2010). *Science, order and creativity*. Abingdon: Routledge.

Bub, J. (1969). What is a hidden variable theory of quantum phenomena? *International Journal of Theoretical Physics, 2*(2), 101–123.

Bub, J. (1974). *The interpretation of quantum mechanics*. Dordrecht: D. Reidel.

Bub, J. (1997). *Interpreting the quantum world*. Cambridge: Cambridge University Press.

Bub, J. (2016). *Bananaworld: Quantum mechanics for primates*. Oxford: Oxford University Press.

Bub, J., & Bub, T. (2018). *Totally random: Why nobody understands quantum mechanics*. Princeton: Princeton University Press.

Feyerabend, P. K. (1965). Problems of empiricism. In R. G. Colodny (Ed.), *Beyond the edge of certainty* (pp. 145–260). Englewood Cliffs: Prentice-Hall.

Freire, O., Jr. (2019). *David Bohm, a life dedicated to understanding the quantum world*. Berlin: Springer

Gieser, S. (2005). *The innermost kernel: Depth psychology and quantum physics. Wolfgang Pauli's dialogue with C. G. Jung*. Berlin: Springer.

Hamilton, S. (2001). *Indian philosophy. A very short introduction*. Oxford: Oxford University Press.

Hiley, B. J. (2011). Process, distinction, groupoids and Clifford algebras: An alternative view of the quantum formalism. In B. Coecke (Ed.), *New structures for physics* (pp. 705–752). Berlin: Springer.

Hossenfelder, S. (2018). *Lost in math*. New York: Basic Books.

Korzybski, A. (1994). *Science and sanity* (5th ed.). New York: Institute of General Semantics.

Kožnjak, B. (2017). Kuhn meets Maslow: The psychology behind scientific revolutions. *Journal for General Philosophy of Science, 48*(2), 257–287.

Krishnamurti, J. (1977). *Truth and actuality*. London: Victor Gollancz Ltd.

Krishnamurti, J., & Bohm, D. (1985). *The ending of time*. London: Victor Gollancz Ltd.

Lindorff, D. (2004). *Pauli and Jung. The meeting of two great minds*. Wheaton: Quest Books.

McEvilley, T. (2002). *The shape of ancient thought: Comparative studies in Greek and Indian philosophies*. New York: Allworth Press.

Meier, C. A. (Ed.). (2001). *Atom and archetype, the Pauli/Jung letters, 1932–1958*. Princeton: Princeton University Press.

Miller, A. I. (2009). *137: Jung, Pauli, and the pursuit of a scientific obsession*. New York: W. W. Norton and Company.

Moody, D. E. (2016). *An uncommon collaboration: David Bohm and J. Krishnamurti*. Ojai: Alpha Centauri Press.

Nichol, L. (1998). *On creativity*. Abingdon: Routledge.

Nichol, L. (Ed.). (2002). *The essential David Bohm*. Abingdon: Routledge.

Peat, D. (1996). *Infinite potential*. New York: Basic Books.

Pylkkänen, P. (Ed.). (1999). *Bohm-Biederman correspondence*. Abingdon: Routledge.

Seager, W. (2018). The philosophical and scientific metaphysics of David Bohm. *Entropy, 20*(7), 493.

Skrbina, D. (2005). *Panpsychism in the West*. Cambridge: The MIT Press.
Talbot, C. (Ed.). (2017). *David Bohm: Causality and chance, letters to three women*. Berlin: Springer.
Vaidya, A., & Bilimoria, P. (2015). Advaita Vedanta and the mind extension hypothesis: Panpsychism and perception. *Journal of Consciousness Studies*, 22(7–8), 201–225.
Waddington, C. H. (Ed.). (1969). *Towards a theoretical biology 2*. Edinburgh: Edinburgh University Press.
Wilkins, M. (1986). *Interviews with David Bohm*. College Park, MD, USA: Niels Bohr Library & Archives, American Institute of Physics. (In 12 parts: https://www.aip.org/history-programs/niels-bohr-library/oral-histories/32977-1 to https://www.aip.org/history-programs/niels-bohr-library/oral-histories/32977-12).
Woit, P. (2006). *Not even wrong*. London: Jonathan Cape.

Chapter 2
Folder C130. Correspondence with Loinger, Rosenfeld, Schumacher and Bub. October–December 1966

Oct 3, 1966

Dear Jeff,

I was very glad to hear from you, and to learn that all is well. As for me, I am just getting back to work, with a number of new ideas.

I am having 25 reprints of your article sent to you.[1]

I have already written to Biederman, and he is very willing to see you, if you will visit him in Red Wing.[2]

Everyone here sends you and Mead[3] our best regards, also to your wife. Please let me hear what you are doing, and how you get on with Biederman.

Yours sincerely

David Bohm

This is my reply to a letter I received from Professor Loinger.[4] He also sent one to you here and this has been forwarded.

---

[1]Presumably (Bohm and Bub 1966b)—CT.

[2]In a letter from Biederman to Bohm dated Dec 1, 1966 (C91b in the Birkbeck archives) he reports that Bub and his wife visited on Thanksgiving Day, Nov 24, 1966. Biederman and his wife enjoyed their visit. In relation to Bohm, Bub "has a great admiration for what you are searching to do in science, in spite of those who oppose your intentions"—CT.

[3]Alden Mead, was a physical chemist at the University of Minnesota. He spent a sabbatical at Birkbeck College during Bub's last year there as a graduate student and offered him a post-doc position in the Chemistry Department after graduation, which was his first job (information from Jeffrey Bub)—CT.

[4]Angelo Loinger was one of a trio of Italian physicists, Adriana Daneri, Angelo Loinger, and Giovanni Maria Prosperi referred to in these letters as DLP. Their main publications were Daneri et al. (1962) also at Wheeler and Zurek (1983), pp. 657–679 and Daneri et al. (1966). See Freire (2015), Chaps. 4 and 5 for more details. See Appendix A for Bub's correspondence with Loinger—CT.

C. Talbot (ed.), *David Bohm's Critique of Modern Physics*,
https://doi.org/10.1007/978-3-030-45537-8_2

12th October, 1966.

Professor A. Loinger,
Universita Degli Studi di Pavia,
Istituto di Fisica Teorica,
Via Taramelli 4,
Pavia,
Italy.

Dear Professor Loinger,

Thank you very much for your recent letter. I agree with you that refinements of the conventional formation by hidden parameters is an unlikely way of discovering a new theory along these lines. The only purpose of our article[5] was to clarify the question of what it would mean to have a hidden variable theory. I would be inclined to think that you are right in saying that a generalized field theory in the Einsteinian sense is the probably real alternative to the Copenhagen point of view, though it is my view also that this field should refer to some sort of discrete space time, rather than to a continuum. When I discussed the question with Einstein in 1951, he agreed that there was no intrinsic reason why the underlying reality should not be discrete. However, he said that he knew of no mathematics that could handle such a theory. I am at present working on such mathematics, and a very early version of this work is enclosed with this letter.[6] Naturally, it has gone a lot further since then.

About your own paper, I am not sure that you have actually succeeded in definitively settling the measurement question in the present theory, at least not to my satisfaction. You have made an important step in clarifying the role of the measuring apparatus. In this connection, I would like to suggest that you have carried to its logical point of completion a notion that was implicit in what I wrote in Chapter 22 of my book, Quantum Theory (Prentice-Hall, N.Y., 1951).[7] However, with regard to the status of the present physical theory, I feel that the significance of your work would be clarified, by pointing out that it leaves certain questions unanswered, which are reasonable to ask in any physical theory. Here, I want to stress not the questions that arise merely in implementing Bohr's philosophy, but rather, those that arise by careful inspection of the physical facts themselves.

Now, the essential point about a classical probability concept is that when new information is obtained, the domain of possibilities is always narrowed down. But in quantum mechanics, it may be narrowed down in certain ways and extended in others. Thus, if we have a state of spin $\frac{1}{2}$, with $\sigma_2 = 1$ then if we measure σ_2, it is impossible to obtain $\sigma_2 = -1$. But after you measure the X component of the spin, it becomes possible to have $\sigma_2 = -1$. So measurement did not merely give information about σ_2; it created new possibilities for σ_2. Thus, as everybody agrees, is something not contained in the classical concept of probability.

---

[5]Bohm and Bub (1966a)—CT.

[6]Not in the Birkbeck archives, but see Bohm (1965), also http://www5.bbk.ac.uk/lib/archive/bohm/BOHMB.149.pdf—CT.

[7]Bohm (1989)—CT.

Now, you manage to deal with this question by considering the role of the observing apparatus. Generally speaking, after an interaction between a system I and its observing system II, but before anyone knows what the result of the interaction is, the wave function of I + II is

$$\Psi_{I+II} = \sum_i \varphi_i(y) c_i \psi_i(x) \tag{1}$$

where x is the coordinate of I and y of II, and where the initial wave function was

$$\Psi_{I+II} = \varphi_0(y) \sum_i c_i \psi_i(x) \tag{2}$$

Now, you show that because of the ergodicity of the apparatus system II, there is no interference between the terms of Eq. (1). Therefore, it is impossible by looking at system II to create new possibilities for I, because these latter depend on interference between the terms in Eq. (1). Therefore, the quantum mechanical rules for probability have now become more like classical rules.

Nevertheless, there is something left out of your account. This is that when a certain result is obtained in the measurement (e.g. $i = n$), then this result is reproducible. It is this behaviour which indicates the individual properties of a system, i.e., that certain properties of such a system are uniquely determined by the wave function. These include not only those operators for which the wave function is an eigenfunction, but also all properties that have probability unity. For example, one knows for certain that the electron will be found in the region in which its wave packet is appreciable, while its momentum will be found in the region where the fourier coefficient of this packet is appreciable.

This was all explained in our article, in connection with the diffraction experiment carried out with the cine camera, in which one electron entered the apparatus at a time. Each electron has its wave function, which is similar in shape to that of the others, but different in time of entering the system. Whether you say the wave function represents our knowledge or an objective reality beyond this, you must admit that when the wave function is given, then properties of this kind belong to the individual system, and are uniquely determined by the wave function. This is verified by the fact of reproducibility of a series of measurements of properties of this kind.

Now, as I take it, we both agree that the actual state of the system does not depend on our knowledge of the system. Therefore, after interaction of system I with the apparatus, II, but before anyone has looked at the apparatus, the wave function must be

$$\Psi_{I+II} = e^{i\phi} \varphi_n(y) \psi_n(x) \tag{3}$$

where ϕ is an arbitrary phase factor and n is the actual state of the system, which will be known later when someone looks at the apparatus.

Therefore, it is wrong to say that after interaction of I and II, but before anybody looks, the wave function is given by (1). If it were given by (1), then one would be forced to conclude that as soon as someone looked at the apparatus the wave function suddenly changed to III when he became conscious of the result of the measurement. For there is no doubt that if the result is to be reproducible, the wave function must be given by (3) after someone is conscious of the results of the measurement. So if we accept your claim that your treatment of the question is complete, we must either adopt the assumption that the system is influenced by our consciousness of it, or else we must say that both before and after an observer looks, the wave function is (1). But this latter assumption does not provide for the reproducibility of the measurement. Therefore, it is unacceptable. Since you also do not accept the influence of the observer's consciousness, you have not accounted for why the wave function is (3) and not (1), as soon as the interaction of I and II is complete, but even before anybody looks at the results of this interaction.

We see then that a simple inspection of the physical facts, along with an acceptance of the "realist" philosophy about the significance of measurements in physics, forces us to say that while the interaction between I and II is taking place, something more is happening than the destruction of interference between the terms of (1), because of ergodicity of the apparatus. This something more is that the wave function is not only undergoing a unitary transformation that turns it from (2) into (1). It is also undergoing a non unitary transformation that turns it from (1) into (3).

One is then immediately led to ask questions, such as: What is the equation of the process by which the wave function turns from (1) into (3)? Just when does this take place? (For until it does, the measurement is not complete.)

These are the sort of questions that our article shows how to answer, in terms of a theory that is admittedly not likely to prove to be on the correct lines. But we only wanted to make it clear what the nature of these questions is. Any theory of hidden variables would have to face similar questions, and I feel that your article would have been more clear and useful, if you could have brought this point out.

It may be that I have misunderstood your article. If so, please let me know in what way I am wrong.

With best regards, Yours sincerely,

D. Bohm.

Nov 15, 1966

Dear Jeff,

Thanks very much for your letter. I think your position is correct, as you will see from the enclosed copy of a letter to Loinger. By all means, write a detailed letter to him on the subject.[8]

---

[8] See Appendix A—CT.

I am glad to hear that you might get job in the philosophy dep't in Minnesota in Feigl's[9] dep't. I hope it comes through.

I shall be interested in hearing how your talk with Biederman goes.

It seems to me that the epistemological question raised by Bohr (see my letter to Loinger) is a crucial one. But his answer is on the wrong track. The key question is that of regarding physics as an extension of perception (as in the appendix to the Relativity book).[10] (But of course, it has little or nothing to do with Bertrand Russell). The question is: "What is really the immediate fact for physics? What does the physicist really perceive? Isn't classical physics largely a "descriptive fiction?" It is an exceptionally slippery question to discover just what the directly and immediately observed fact really is. Many philosophers have postulated fictions such as sense impressions. But these are as abstract as are the "atoms" that they are intended to replace. Somehow, man really perceives order, pattern, structure, potentiality along with actuality, totality and its partial aspects, etc., etc. Our mathematical terms are very far from these perceptions indeed. That is why physics seems to have degenerated into a purely formal one [of] practical manipulations of algorithms. Perhaps you can raise this question with Biederman, and ask him how art can help the physicist to be aware of how and what he actually perceives.

Best regards

D Bohm

Nov 24, 1966

Dear Jeff

I received your second letter. Meanwhile, I hope you received my letter to you, along with a copy of my answer to Loinger.

In my view, it is not possible to prove that Bohr's epistemology is compatible with "objective" probabilities in quantum theory, if only because the whole theory has been constructed in such a fashion that the interpretation of the "algorithms" depends essentially on what an observer is supposed to be able to communicate. As long as the equations are linear, you can't get out of the conclusion that the wave function spreads over a range of "quantum states", until another apparatus (III) observes (I+II). Any "objective" description would have to discuss the probability that the system I+II is in a certain state, S, (which we could call $P(S)$). But I+II is not in a state corresponding to eigenvalues of what is measured, unless the wave function is $\psi_i(I)\varphi_i(II)$. So there is no way to talk about the "objective" probabilities. We can only talk about probabilities of what III will find, when a further observation is made. I think that this is the essence of the argument against Loinger. No doubt,

---

[9]Herbert Feigl, Austrian philosopher and member of the Vienna circle, was for many years Professor of Philosophy at Minnesota, establishing the Minnesota Center for Philosophy of Science in 1957—CT.

[10]Bohm (1996)—CT.

Rosenfeld[11] has confused both what DLP say and what Bohr says. It would be much better if he had said nothing at all.

Could you write to Loinger a letter[12] doing the following?:

(a) Explain where DLP go wrong.

(b) Explain where Rosenfeld confuses DLP.

(c) Explain where Rosenfeld confuses Bohr.

How did you get on with Biederman.

Best regards

D Bohm

10th November, 1966.

Professor A. Loinger,
Universita Degli Studi di Pavila,
Istituto de Fisica Teorica,
Via Taramel1i 4, Pavia,
Italy.

Dear Professor Loinger,

To some extent, we have been arguing at cross purposes and not meeting. I have always meant to say that you have made a contribution to the more nearly complete expression of Bohr's point of view. What I have been trying to emphasize is that I am not satisfied with this point of view, and that it has inherent inadequacies, which are not really changed by your work.

When I wrote my book, Quantum Theory, my main objective was to try to understand what the subject is all about, and in particular, what Bohr is actually saying. After writing it and thinking the whole question over, I began to feel that, after all, I did not understand. This was perhaps the main reason why I started then to inquire into hidden variables.

Since that time, I have had several opportunities to discuss with Niels Bohr and with his assistant, Aage Petersen.[13] Although we never brought up my own book in a direct way, I did gather (largely from Petersen) that Bohr did not really like my approach in this book, and felt that I had not understood what the point of complementarity actually is. I also gathered (though it was never stated explicitly) that Bohr and Rosenfeld did not really see eye to eye on these questions. Indeed, I

[11] Léon Rosenfeld, long associated with the Niels Bohr Institute in Copenhagen, was then professor at Nordita (Nordic Institute for Theoretical Physics). For more on Rosenfeld including his earlier conflicts with Bohm see Freire (2015), Kožnjak (2017). For a letter from Bub to Bohm on Rosenfeld's paper, see Appendix B—CT.

[12] See Appendix A—CT.

[13] Petersen was Bohr's assistant from 1952 to 1962. His 1966 doctoral thesis was published as Petersen (1968)—CT.

had long suspected that Rosenfeld's belief that Bohr is a natural born dialectician was, to a considerable extent, merely Rosenfeld's own interpretation. Later, when I received a manuscript from Petersen (about a year ago) on Bohr's philosophy, this suspicion was confirmed.

Now as I see it, Bohr has been giving top priority to the role of language. As Petersen says, Bohr was fond of emphasizing that "Man lives suspended in language". Therefore, to Bohr, the principle of complementarity is primarily a linguistic question. As Kant raised the question of what conditions (such as space, time, causality, etc.) are the preconditions of all experience, so Bohr asked what is the precondition of the *precise* communicabil ity of descriptive information about nature. He came to the conclusion that the concepts of classical physics (position and momentum) are what determines this precondition. This conclusion is not primarily *physical*, but rather, *epistemological*. What it amounts to is that Bohr believes that *the language of classical physics is* in some way a part of the "human condition" (or perhaps one could say that it is an intrinsic aspect of human nature, and of man's interactions with his environment). Another closely related aspect of the "human condition" is the language of subject and object. That is, we always say that there is an observer who is looking at some object, that is being observed. As long as the interaction of the observer with what is observed is not subject to restrictions, then there is no limit to the precise communicability of information about momentum and position. But because of the quantum, subject and object can no longer be separated, and therefore precise information about the object is no longer possible.

Bohr then emphasizes that precise classical descriptions are to be replaced by the mathematical algorithm of quantum mechanics, which contains just the required limitation on precision of information about the object.

It can be seen that the essential assumption of Bohr is in the field of epistemology, and more particularly, in the field of linguistics. Bohr's belief that the language of separation of subject and object, along with the description of the object by classical concepts such as position and momentum, is inherent in the "human condition", means, in effect, that he puts certain aspects of the human mind (i.e., language) in the first place, in man's relationships with nature. This approach is not really compatible with the dialectical materialist point of view adopted by Rosenfeld.

I think that Rosenfeld does a disservice both to Bohr and to his own dialectical materialist views, by failing to recognize that Bohr is not a materialist (though he may perhaps be, in certain ways, a dialectician). Indeed, I learned in conversations with Petersen that in his early life, Bohr as strongly influenced by Kierkegaard. One can in fact see the influence of existentialism in the principle of complementarity. Man is somehow an individual and yet he must be in indivisible union with God. This unresolvable contradiction must cause man to be in a state of torment. For every time he tries to determine some aspect of his individuality, this comes into contradiction with another equally significant aspect. Somehow, man *exists* as a totality, but in his perceptions and actions, he can always only define one fragment of himself at the expense of another. Similarly, in physics, Bohr takes subject and object as being indissolubly united by the quantum that "connects" them. Yet, this unity can never be defined. When one aspect (position) is defined, this comes into contradiction with another aspect (momentum). Our language, according to Bohr, forces us to assert the separation of subject and object, while we know that they are united (as it also by implication forces us to assert the separation of God and man, though all religious people have said that they know the two are inseparably united).

Now, while I think it is useful to raise all these questions, I do not believe that Bohr's answers are pertinent. There is no evidence that the classical language of position and momentum is part of the "human condition". Rather, it evidently evolved historically for specific reasons. Indeed, in a typical cloud chamber, we never observe positions and momenta. We observe the order of a series of droplets (approximated as on a curved arc) and then we translate these into positions and momenta. So I would like to emphasize that man's direct perceptions are not limited by the need to be expressed in classical concepts of position and momentum.

To be sure, mathematics today is not able to say much about order, pattern, structure, etc. But this deficiency can ultimately be remedied. I am publishing an article on the subject, which I will send to you when reprints are available. In essence, I have reason to believe that both mathematics and physics can take the concepts of order, pattern and structure as basic. When this is done, quantum mechanics will be seen to fit naturally into this new language. The dichotomy of subject and object is thus removed. For order, pattern, and structure refer both to the order in the mind that perceives them and to the order in what is perceived. Indeed, the whole concept of an object is then merely an abstraction from an over-all structure. To each abstraction of an object, there must correspond an abstraction of a subject, which determines the "perspective" from which the object is "observed". But the basic reality is the total structure, in which subject and object are both contained as abstractible aspects.

So much for the Bohr point of view. Now, I understand that you are trying to base quantum theory on objective laws applied both to the measuring apparatus and to the observed system. Of course, you realise that this approach is diametrically opposed to Bohr's philosophy. I am sure that Bohr would have regarded it as being on a wrong track altogether, just as he probably thought of the approach in my Quantum Theory book. For this reason, I think that Rosenfeld is confusing the issue, by falling to distinguish clearly between his own point of view and Bohr's basically epistemological approach. In Bohr's point of view, it is absolutely essential that the ultimate observing apparatus be treated purely in terms of classical concepts, and that no attempt be made to represent an "objective" system in terms of the quantum mechanical algorithms.

The way Bohr would interpret your work is as follows. It is permissible to treat "object" (system I) and "observing apparatus" (system II) as a combined system, I+II, which is handled in terms of the quantum mechanical algorithms. But then, this presupposes an observing apparatus, III, which is "observing" system II. Apparatus III must be treated in terms of classical concepts only. The usual probability laws of quantum mechanics apply then to I+II. This means, of course, that the "reduction postulate" applies to the wave function, I+II, when the system, I+II, is "observed" by III.

As I indicated at the beginning of this letter, your treatment could be interpreted by Bohr as an extension and completion of his own notions. In effect, your theory provides a detailed explication of how it comes about that the "cut" between "observer" and "observed" can be moved around freely in the large scale domain. In your case, this "cut" is placed between apparatus II and apparatus III, which latter is assumed to belong to the purely classical domain, concerning the state of which human language allows the precise communication of detailed information. But of course, it is, according to Bohr, entirely inadmissible to discuss the quantum mechanical side (I+II) as if no human being were present. For the whole meaning of the quantum mechanical algorithm is that it refers to what a human being can communicate in precise terms about system III. To treat system I+II as if it existed objectively without III (which is taken to be, in essence, an extension of the observer's sense organs) is to deny the deep meaning and basic spirit of the principle of complementarity.

You ask my opinion of Tausk's paper[14] In my view, it is essentially right, though I think that Tausk does not pay enough attention to the significance of the ergodic properties of large scale systems in making possible a consistent theory, along Bohr's lines, in which the "cut" can be moved freely in the large scale domain. Dr. Bub has written to me that he also has analyzed Tausk's paper and found it to be essentially correct. I understand that he will be sending you some detailed comments on this subject.[15] Meanwhile, I would only emphasize that it is not enough to get the right probabilities for the properties of system I. If your treatment is to be "objective", it must also show that after I and II have interacted, but before system III has "observed" system II, the wave function is definitely on one of the "channels" of II, corresponding to non interfering results of possible measurements. As I see it, Tausk's treatment shows clearly that you have only provided for the "weak" reduction postulate – i.e., that after I+II have interacted, the "probabilities" are the same as if the system II were "objectively" in a certain channel. But where do these "probabilities" come from and to what do they refer? Basically, they are brought in only by tacitly assuming system III, which will "observe" I+II, and which can obtain certain well defined results with corresponding probabilities. But once you have presupposed system III, you have also brought in the reduction postulate. It is this postulate which provides for the fact that the wave function II is in one of the channels. As Tausk points out, if you don't assume the reduction postulate, then after I+II have interacted, the wave function of I+II still spreads out over all the channels. As long as this is the case, there is no way to provide for the reproducibility of a measurement of II. For after you obtain a certain result for II in any one measurement, the wave function still spreads out over all the channels, so that in the next measurement, you will in general get some other result. It is only because you have tacitly assumed the operation of system III, which is in essence an extension of the observer, that we can say that after the observation of II by III, the reduction postulate implies that II is in a part of its Hilbert space corresponding to a single one of its channels, and not in a part corresponding to its being in many channels at the same time.

[14]Tausk's paper is not available. See Freire (2015), Chap. 5 and Pessoa et al. (2008). See also Appendix A—CT.

[15]Appendix A—CT.

In your letter, you refer to "one parameter distributions". These are, as far as I can see, equivalent to von Neumann's notion of an indecomposable distribution - i.e., one that cannot be divided into sub ensembles. Now, I do not think that such a distribution is a meaningful mathematical concept. The very notion of a distribution, as defined in all mathematical theories, is inseparable from the idea that the various members of a distribution can be distinguished from each other, so that each member is an individual. This has no essential relationship to the question of determinism. Indeed, even in Brownian motion theory, which is not deterministic, there is a distribution over possible trajectories. Each trajectory is distinguishable from the others, and this distinction is made possible by a suitable set of parameters which specify the orbit completely, (e.g., Wiener's use of differential space). When you talk of a "distribution" over elements that are not distinguishable in any way at all, it seems to me that this is only a set of words that has been put together, and not a meaningful concept.

Indeed, in reality, there is a "hidden" distinction in the ordinary quantum theoretical probabilities. To each "first" wave function, Ψ, and to each observable, $\mathcal{O}$, there corresponds a distribution of "second" wave functions, ψ_i, with probability $|a_i|^2$ (where $\Psi = \sum_i a_i \psi_i$). These "second" wave functions are those which result from the reduction postulate, which asserts that the wave function "collapses" to ψ_i (in your theory, when system III observes system II). And this "collapse" is not compatible with an "objective" over-all treatment, (since ultimately system III must be observed by IV, and so on until we reach a conscious observer). Thus, we are back in Bohr's philosophy, which asserts that the content of physics is limited by linguistic considerations, built into the "human condition" and referring to what can be communicated precisely by an observer.

As Dr. Bub and I point out in our paper, there is a further distinction between individual members of an ensemble, not usually taken into account in the concept of a "one parameter" distribution. This is that (as we showed with the aid of a hypothetical cine camera experiment), each individual electron comes into the system at a different time. So, in a way, the "indecomposable" distributions have already been "decomposed" even without the help of hidden variables. In this point alone, the current quantum theory is confused, as to how it deals with the relationship between individual and ensemble. In my view, this confusion along with the other unsatisfactory features that I have described in this letter, can be dealt with in terms of the concept of hidden variables. There may well be other and better ways to do so, and I am trying to look into some of them. But something must be done about the problem, which cannot be solved in the framework of the current theory, not even with the aid of the theory that you and your colleagues have developed.

With best regards,

Yours sincerely,

D. Bohm.

Dec 2, 1966

Dear Jeff

Enclosed is a copy of a letter to Loinger. I don't think we will get much further with him.

How are you doing!

Sincerely yours

D Bohm

2nd December, 1966.

Professor A. Loinger,
Universita Degli Studi di Pavia,
Istituto de Fisica Teorica,
Via Taramelli 4,
Pavia,
Italy.

Dear Professor Loinger,

Thank you very much for your letter of November 23rd.

(1) With regard to Bohr's philosophy, I should be very doubtful that he would have agreed to the notion that the quantum algorithm directly represents an "objective" property of nature, in the way you wish to propose. Rather, he would have said that objective and precisely describable (therefore communicable) properties must refer only to classical variables, such as position and momentum. The quantum algorithm is then a kind of "metalanguage" which makes statements about the "objective" properties appearing in the classical "language". But to attempt to describe objective properties of nature directly through the quantum algorithm is surely contrary to Bohr's philosophy.

(2) I do not think that Bohr's concept was completely coincident with that of Von Neumann. Von Neumann was an outstanding mathematician, but his physical concepts were often best describable as "primitive" in their naive character. As far as I can tell his concept was that of "psycho-physical parallelism". That is, there is something appearing in the mind, which is parallel to what also exists physically. Bohr was much more subtle than this. He said that all precisely communicable "objective" properties must be described in classical language. At this level, the concept of objective reality is therefore not basically different from that used classically. The quantum mechanical algorithm is nothing more than a "metalanguage" making statements about what happens to "classical" quantities, such as position and momentum. These "classical" quantities generally belong to the directly observable aspects of the measuring apparatus.

If you treat the system II in terms of the quantum mechanical algorithm, then Bohr would say that this algorithm refers to predictions of the behaviour of system III, which is used to "observe" II. I don't see how you can possibly get out of this, without giving up the very essence of Bohr's philosophy. (This point has been explained very clearly in a recent preprint by Aage Petersen, also by a man named Schumacher in Cornell University, who has been in correspondence with Rosenfeld.)

I am sure that Rosenfeld has, as you say, an extraordinary esprit de finesse. Nevertheless, I do not think that he has really understood the essence of Bohr's point of view in spite of having lived for a long time with Bohr. Bohr's ideas are very hard to grasp. Indeed, they are so unclear that I think it would be good if those of us who are interested in them (including you and Rosenfeld and myself, along with others) could some day get together for a full and informal discussion.

In von Neumann's point of view, it is never clear just what is meant by the term "making a measurement". Sometimes it seems that he refers to the registration of the event on the mind of an observer and at other times, he refers to its registration in what he calls "classical observables". Neither of these notions coincides with Bohr's position. Indeed, as Bohr has explicitly said to me, "Nothing is ever really measured at all in quantum theory". Rather one observes the state of the apparatus and describes this state in classical terms, including the whole set-up of the "experimental conditions". One then applies the algorithm of quantum mechanics, to make statements about what can or will be observable later in the apparatus. One can see that the experimental conditions needed to measure conjugate variables precisely are not compatible with each other. This incompatibility is expressed perfectly by the failure of the corresponding operators in the algorithm to commute. As a result, the statements that one can deduce from the algorithms about these sets of experimental conditions have exactly the right degree of ambiguity to match the ambiguity of the classically describable specification of the experimental conditions themselves. This is the meaning of the uncertainty relationships. Physical experiments have nothing really to do with measurements of "objective" conditions at the quantum mechanical level. Rather, they have to do with statements about the behaviour of classically describable aspects of the world (usually of a piece of laboratory apparatus). Therefore, your whole programme is directly contrary to what Bohr wants to do.

(3) You say that practically, a superposition of macroscopically distinguishable states is equivalent to a mixture. It is just this word "practically" that I cannot accept. If we were satisfied to regard quantum mechanics as nothing more than an elaborate formula for an engineer's handbook, then your reasoning would be adequate. But surely, Bohr and von Neumann were not satisfied to look at it in this way, and I doubt that you are either. Rather, there is the implication that quantum mechanics is a logically coherent structure of physical and mathematical ideas. Whenever, in a purely logical argument, you identify something with something else that merely approximates some of its qualities, you have a contradiction. It doesn't matter how "small" the error is, it is still a contradiction. For logically, there is no such thing as a "small contradiction". Either the theory is logically coherent, or it is self-contradictory.

No matter how "small" a contradiction is, there is a theorem in mathematics that from it, one can derive any statement whatsoever (this includes both true and false statements). Of course, if you are careful to limit the domain of application of a contradictory theory, you can manage to get correct results in a limited field. For example, consider the contradictions, $X = 2X$, $Y = 2Y$. From these, you can derive correct results for any function of the ratio, X/Y, but not for functions such as $X + Y$. Similarly, your own contradictory theory will give correct results in a certain domain. If all that you want your theory to do is to state, in a different language what others have already stated more directly, then you can avoid coming into contact with the consequences of its contradictions. But if your wish is to apply your theory to the treatment of something that depends on the total logical structure of the theory, the latter cannot be relied upon.

What you have shown is that for "practical purposes", one can replace the pure ensemble of I+II by a mixed ensemble, which gives the same averages for system I as are obtained if one uses the usual quantum mechanical probability rules, which assume that II is "observed" by a classically described system, III. In other words, certain aspects of I+II can be treated, to an adequate practical degree of approximation by replacing the pure state of I+II by a suitable mixture. But logically, there is a gap in the argument. This gap is a contradiction. For it comes about because you identify something (a pure state) with something else that it is not (a mixed state). The "smallness" of the error in predicting the properties of I does not change the fact that this is a contradiction.

To avoid a contradiction, you must restrict yourself to stating the facts. These are that the probabilities for "finding" system I in a certain state are the same as if the ensemble for I+II were a mixture. But what is it that would "find" system I in such a state? Surely, it cannot be other than another system, III, IV, V, etc., which will function as a higher order observing apparatus. Without such a higher order observing apparatus, there is nothing that could "find" either I or I+II in a certain state. Therefore, the basic formulae from which you start would have no meaning. Moreover, it is these higher systems that guarantee the "reduction" of the wave packet and the reproducibility of the results, for system I. Without them, you never have anything else but a pure state for I+II, no matter what you do (unless, of course, you admit the logical contradiction, pure state = mixed state).

(4) Finally, let me say that von Neumann's systematic foundation of q.m. strikes me as being very excellent mathematics, but extremely naive and weak as an example of physical reasoning. As has been demonstrated clearly, his proof that there are no hidden variables is due to his not understanding the physical content of his mathematical assumptions. I also think he does not understand the physical meaning of probability distributions. He has a coherent set of mathematical equations. But his statements that these refer to "indecomposable distributions" are, in my view, only a set of words that has been put together, and not a meaningful concept.

With best regards,

Yours sincerely,

D. Bohm.

Dec 8, 1966

Dear Jeff,

Just received your letter of Dec 4. Meanwhile, I have already sent you another reply to Loinger.

I rather doubt we are going to get a lot further with DLP. I would only add that their point (3) is unclear (See Loinger's letter to you).[16] If the probability of "finding" is a probability in the sense of q.m., it means that it is the probability what apparatus III would find, if it looked at I + II. So unless you begin with apparatus III, the rest of what DLP [say] makes no sense. Apparatus III will always "reduce" the wave packet of I+II. The only contribution of DLP is to show that it doesn't matter whether you regard apparatus II as being on the classical or q.m. side of the "cut". This is also what I did in my quantum theory book, Chap 22. But Loinger does not accept this, nor do I think that he will ever do so.

Also, about Loinger's point (4), von Neumann's work is not clear, as is shown by his false proof of the impossibility of hidden variables. It is only a structure of mathematics. Its physical notions are often confused.

Their conclusion that practically a superposition of macroscopically distinguishable states is equivalent to a mixture is hardly a surprisingly new discovery. Many people have shown this, in different ways. DLP do it in yet another way.

Their point (7) is totally wrong. Without apparatus III, there is first of all no meaning to "finding" I+II in a certain state. And secondly, without apparatus III, I+II is still always in a pure state. It is, as I pointed out, a logical contradiction to equate it with a mixed state, however similar its "practical" consequences may be.

I shall be interested in hearing your further questions about Biederman and related topics. I do not like the word "caricature" either. The question is whether theories are a direct reflection of things as they are. In a way, they have to be just this. But in another way, they cannot be this. The interesting question is: "What, if anything, do theories reflect?" This leads to the question "What is knowledge, and its relationship to what is known?" I shall be sending you something on this soon.

Sincerely

David Bohm

Dec 10, 1966

Dear Jeff

I enclose several copies of my recent correspondence with Dr Schumacher, which will give you some idea of my notions on questions of epistemology. I will write you in more detail later.

[16]In Appendix A—CT.

Best regards

D Bohm

P.S. I would say that a theory is, in certain ways, like a work of art, created by scientists. Instead of reflecting particular objects or scenes, it reflects the general order, pattern, and structure of nature. New theories are like new works of art. Different painters on the same scene can do very different works. In a way, all may be true reflections. They are all related to each other in certain ways. Each is a better reflection in some ways, worse in others.

24th October, 1966.

Dr. D. L. Schumacher,
Centre for Radiophysics and Space Research,
Cornell University,
Clark Hall,
Ithaca,
N.Y. 14850,
U.S.A.

Dear Dr. Schumacher,

Thank you very much for your two papers,[17] which I have read with great interest. I expect to get in touch with Peter Szekeres[18] in the near future, in order to discuss the questions further with him.

Let me say first of all, that insofar as I understand your position on complementarity, I do not fully agree with it. In particular, I would doubt the validity of identifying the term "quantum" with Bohr's particular epistemology. I would rather let it refer to the general body of fact and mathematical theory, which is underlying modern atomic physics. For in my view, the epistemological question is still very confused, and this confusion is seriously impeding the development of the other sides of physical theory. In this regard, see the articles which I enclose.

I do agree with you, however, that to understand physics as an extension of perception is the root of the matter. I think, in this regard, that an absolute separation of subject and object, is false. I am sorry to hear (in your footnote 11)[19] that you gath-

[17]Presumably one of these is Schumacher (1967). The other may be a preprint as there seems to be no other publication by Schumacher from that period—CT.

[18]Peter Szekeres, a respected mathematical physicist, now at the University of Adelaide was then presumably a colleague of Schumacher at Cornell. Szekeres did his doctorate at Kings College, London in the early 1960s so would be well known to Bohm—CT.

[19]Schumacher's note: "There is a useful review of recent pertinent work on perception in the Appendix, "Physics and Perception," in *The Special Theory of Relativity* by D. Bohm (New York: Benjamin, 1965). (Earlier edition of Bohm 1996—CT) It may be helpful to examine some of the general remarks and conclusions made there. The notion of a separation of subject and object is discussed in other terms, and while it is rightly emphasized in the context that this notion is itself

ered the opposite from my book, Relativity.[20] When I used the term "inner show", I wanted to distinguish the content of consciousness from a broader reality, which is reflected in this consciousness. My view is that consciousness is, in some way, embedded in a broader reality, of a nature that is basically unknown and indeed, even unknowable, if one considers its totality. There is no separation between this broader reality and consciousness. But the contents of this consciousness are evidently reflections of a reality that transcends the contents of consciousness, while containing the latter as part of itself.

The "inner show" is "inner" only insofar as the events on which it is based are taking place "inside the body" and mostly in the brain. But brain, body and the whole environment are one total process. The "inner show" can reflect not only the external environment, but also, its own contents. The test of the correctness of this reflection comes when we use it as a guide to function and action. This does not commit us to the view that reality is either external to man or internal to him. Rather, it means that whatever the world is we can function in a way that is directed by the contents of consciousness, and as a rule, our functioning produces more or less the expected trend of results.

This is the same as to say that perception is learning in action and action in learning. That is, we act (function) we learn how the results of function differ from that implied by the contents of consciousness, and these contents then change themselves, so as to remove the discrepancy. In other words, we do something, and see what happens, learning from our "mistakes". With regard to your ideas on the role of epistemology in science, I find them very interesting. Certainly, we must take into account the fact that knowledge is the result of a process of learning, in which man is inseparably interacting with the milieu in which he lives. However, I am not at all sure that Bohr's particular suggestions as to how to do this are right. Especially, I am dubious of his emphasis on the supreme role of language. Thus, Bohr has often said "Man lives suspended in language".

Of course, language does undeniably have a certain kind of importance. It is the major means of communication between men. But I don't think that limitations in what we have thus far been able to communicate by language are ultimate limitations on what we can perceive and learn. No doubt, we are in the habit of allowing our perceptions to be limited by what fits into our current linguistic structures. But if I may be permitted to say so, I would like to suggest that this is nothing more than a bad habit. Such a limitation is by no means intrinsic in man's perceptual relationships with

learned together with the terms of a physical description which is not of an absolute character, the general conclusions given there tend to attribute absolute significance to the separation of subject and object. In particular, the reliance on the notions of a so-called "inner show," and of the entity to which it refers, essentially prejudges the separation of subject and object to be an absolute. (Cf. especially *ibid.*, p. 204, par. 1, and pp. 216–217.) In this connection, it is instructive to appreciate that Bohr, in writing on these questions, rarely used terms of this kind; if they were used, they were clearly not intended to bear the full weight of the arguments. Even though Bohr apparently intended to give absolute significance to the separation of subject and object, he did not do so by means of particular terms which might be mistaken to establish the absoluteness to this notion."—CT.

[20]Bohm (1996)—CT.

his total milieu. On the contrary, as man can learn the structure of his environment by sensitive and careful perception, he can also learn in a similar way how he is being limited by language, and thus develop new ways of expressing the new content that is disclosed in perception.

As I see it, Bohr is emphasizing certain epistemological conditions of communicability of precisely defined information, which, as it were, come before all questions of the actual content of this information. In this sense, he is using an argument similar in some ways to that of Kant, who emphasized the conditions of space, time, causality, etc., that are the necessary forms of experience, as distinct from the particular content of this experience. I regard it as a valid line of study to raise these questions. But before one answers them, one must also ask a similar question: "What are the conditions of communicability of knowledge about epistemology itself?" Bohr seems to accept tacitly the notion that whereas we must learn the actual content of our knowledge of nature, we are given some kind of direct and completely reliable intuitions about epistemology, that are free of epistemological confusion. Thus when he says certain things about the relationship of subject and object, the "cut" between them, etc., etc., he seems to accept all this as a self evident truth, which will be eternally valid, and is not open to serious questions. To me, however, it seems that we are infinitely more confused about just these epistemological questions than we are about the content of our scientific knowledge. As you yourself have remarked, the unity of subject and object is inseparable. What can it mean then to think a "cut" between them? This "cut" is an absolute contradiction of their unity. Of course, Bohr argues that language forces us into this cut. But I do not accept this limitation as inherent or necessary. Probably, it is only a result of certain habits of thought, expressed in terms of certain linguistic structures that have been common over the past few thousand years.

Whatever the truth may be about this point, I feel that we have hardly scratched the surface of the question, and that Bohr has therefore prematurely claimed to settle the issue definitively with the principle of complementarity.

In addition, there is the further unclear question of what it can mean to have knowledge about epistemology. Epistemology is by definition supposed to deal with the general structure of knowledge and the means by which we attain it and express it. But when we come to learn about epistemology, what we learn in this way is part of the total content of knowledge. If there are inherent structural limitations in the applicability or definability of the content of knowledge (e.g., inherent in man's mode of communication through language), then such limitations very probably exist on the content of knowledge about epistemology. Therefore, one cannot be sure that limitations deduced from the content of knowledge about epistemology are genuine.

On the other hand, one may suppose that precisely with regard to epistemology itself, man may have completely reliable knowledge which is not limited by epistemological considerations (such as the need to express it in language). But then, this admits the principle that some kinds of knowledge are not subject to epistemological or linguistic limitations of any kind. If this is true of some kinds of knowledge, where is the "cut" to be drawn between this kind of knowledge and the other kind that is thus limited?

Does the "conscious subject" make a "cut" between "himself" and his knowledge about epistemology? Is he sure that what he seems to know about epistemology is not limited by the division of subject and object, in much the same way that Bohr says about knowledge of the electron? Or does a man transcend this division of subject and object only when he is communicating what he has learned about epistemology?

The consideration of these questions suggests to me that ultimately, all knowledge is one total field, and that divisions between the content of knowledge, and its overall structure as studied in epistemology, have only relative and limited domains of valid application. If the content of knowledge is assumed to be subject to limitations arising from limitations on what can be communicated in a clearly definable way that is free of epistemological confusion, then there will be a contradiction, if we also assume that these limitations are themselves similarly knowable and communicable. If they are not thus knowable or communicable, then for all we know, they may not even exist or they may be illusory and delusory, so that it is fruitless even to think about them. On the other hand, there is no contradiction in supposing that knowledge is the outcome of a process of learning, and that whatever we learn in this process, we can also learn to communicate. To place a-priori limitations on what can be learned is, in effect, to try to limit the process of learning in arbitrary way, and thus to impede perception. But epistemology has a valid and useful role to play, if it concerns itself with learning about how we do in fact learn, rather than with trying to set linguistically determined limitations on what we can learn.

I should like here to indicate some of my own ideas on epistemology. These are developed more fully in an article to be published soon in the Progress of Theoretical Physics.[21] When I have reprints, I shall send one to you.

I think the key point is in relation to concepts, such as order (and disorder), pattern, structure, individuality, totality, etc. These are evidently, as you point out in your article, really aspects of epistemology. But it is my view that they are also aspects of the content of our physical knowledge. However, because of the present inadequate development of mathematics, one has no language for the precise expression of the meaning of terms like order and disorder at the level of physical theory. So one is content to regard them as "undefinable" epistemological categories. However, what I am working on is to show that such terms can have a precise definition at the level of physical theory. (In the reprints, some notion of how this is to be done is given.)

Now, Bohr's tacit assumption is that the content of physical knowledge has to be expressed in terms of concepts like position and momentum. He seems to feel that this language is inherent, necessary and absolutely inescapable for anything that is to be precisely communicable in physics. Thus when we come to questions like individuality of a quantum and the disorder in statistical mechanical processes, he concludes that these have to be considered only at the next level of knowledge, i.e. as epistemologically determined categories determined by the nature of our language.

[21] I.e. Bohm (1965), also http://www5.bbk.ac.uk/lib/archive/bohm/BOHMB.149.pdf—CT.

On the other hand, I would say that concepts of order (disorder), pattern, structure, individuality, totality, etc., are universal. They apply in man's immediate perceptual experience and they apply in physics. The notion that in physics we measure only momentum and position is false. This notion is merely due to wrong generalization of the inadequate language of classical physics. In fact, in typical measurements (e.g., in a cloud chamber) we measure the order, pattern, and structure in a series of photographs of droplets. We translate these into momenta of particles, but I suggest that this procedure is inappropriate at the micro-level. Rather, we need a new theory in which order, pattern, structure, etc., are the basic terms of thinking (the basic language). So all the way through, from the micro-level to the macro-level to our own thoughts and perceptions, we are concerned with the same kind of thing – i.e., order, pattern, structure, etc. (The "disorder" in thermodynamic processes is itself merely a change of order, from one kind to another.)

Since the one set of concepts runs through the whole field of experience, the division of subject and object is transcended. The order in our physical theories is a particular reflection of the order in our perceptions and the order in the actual structure of the world, including man and his interactions with it. Our ability to know the order of anything (including ourselves) is based on the fact that thought can accommodate itself to a limitless set of possible orders. What is to be known is order, pattern, structure, etc., and this knowledge is itself a similar but different kind of order, pattern, and structure. So one denies the Cartesian dichotomy between "extended substance" and "thinking substance". Extension in space is ordered extension, and thought reflects this order. It is only the order that can be known. Whatever the "substance" of space (or anything else) may be, it cannot be known.

The essential point is to get away from the assumption that current "physical language" is all that the human brain is capable of. Actually, it is the result of a certain historical development. And the new discoveries in physics have, my view, demonstrated the need for a new language in which order can be defined relatively precisely. Perhaps we could discuss this point in more detail later.

It is also to be noted that all perception is an ordered interaction with the world. It is because these interactions are carried out in an appropriate order, related to what is already known, that we can learn something new, by discovering what is out of this order (and what therefore calls for a new order on thought and knowledge.) So the concept of order applies to man, his body, his mind, to the world as a whole, and to man's perceptual interaction, in which he learns the order of things by assimilating them into the order of his actions, as he observes with his senses or probes with his instruments.

Yours sincerely,

D.Bohm

23rd November, 1966.

Dr. D. L. Schumacher,
Centre for Radiophysics and Space Research,
Cornell University,
Clark Hall,
Ithaca,
N.Y. 14850,
U.S.A.

Dear Dr. Schumacher,

I would like to supplement yesterday's letter. From my discussion with Dr. Szekeres, I can see that our points of view have certain peculiar similarities and dissimilarities, in so complex a way that we could hope to understand each other only in a direct discussion. Failing this, it may help if I write an article on my own views. For I believe that my ideas are very different from the usual ones. It is not my aim, for example, to eliminate the concept of randomness from quantum mechanics, by introducing hidden variables. What I hope for is to present this concept in a more natural way. In my opinion, in quantum theory, randomness (like space, time, causality, etc.) is brought in a very confused manner. I have what I feel to be a quite new view on the meaning of natural law, in terms of various kinds of order. One extreme is the order in which two points on a trajectory determine the order of development of a whole orbit. This is complete determinism. The opposite extreme is that of a series of events whose actual order cannot be determined in terms of less than the total series in question. This is complete randomness. At present, it is tacitly assumed that these are the only two ways of thinking open to the human mind in physical problems. In my view, we can conceive of an order that lies between randomness and determinism. This is the order that quantum theory requires. But to explain this in detail would need a long article.

You say that Bohr is unclear but consistent. I question whether this is really true. In my view, when something is unclear, it means that either I have failed to understand it or that it is inherently confused. After a long study, I have come to believe that what Bohr says is inherently confused. However, he has developed a very clever way of hiding from himself and others the fact that his ideas are confused. That is to say, his words have consistent "ring". But when you look into what they mean, you discover that his ideas are contradictory. This contradiction is not presented with clarity. Rather, Bohr thinks in terms of a mechanism that breaks his concepts into disjoint fragments, such as wave and particle, subject and object, position and momentum, etc. One is directed to look at only one of these fragments at a time. Then, one is directed to "jump" to the opposite fragment, and to believe that "somehow the fragments meet, but in a way that one cannot precisely describe". By keeping one's concepts purposely ambiguous, one can fail to see that the fragments do not meet. This is the classical form of confused thinking. Usually, when we find ourselves caught in such confusion, we try to get out of it. But Bohr has turned confusion into a new principle, which he calls "complementarity".

Finally, I would like to make a few remarks about your method of regarding the distinctions of subject and object as the basic theme of epistemology, adopted because you have not been able to make sense of the concept of reality. Firstly, let me suggest that the basic distinction is between reality and illusion, truth and falsity, fact and fantasy. As long as you are using the notion that basic concepts are undefinable. Why is one not free to take the above distinction as an undefined term? In my view, the words for basic notions like truth, fact, reality, etc. have mainly a denotative function. For example, if you and I are looking at the same tree, I can use words such as "Look at the color, the form, the structure of this tree !" I assume from the way you behave that you see more or less what I see. Similarity, truth, reality, fact, are aspects of the total functioning of the mind. By these words, I only intend to draw your attention to a vast dynamic movement, which is a fact. This movement cannot adequately be described, any more than words can really impart the color of a sunset to a man who has been blind from birth.

What I find missing from your views is an adequate attention to what is to be meant by the fact and by its distinction from what is not a fact. (If you don't like the word "reality", then I would instead emphasize the need to discuss and understand what is meant by "factuality".) I get the impression that you regard what people say as the essential fact of human existence. For example, you want to start by studying all that they have said about subject and object (or about randomness). In my view, this is not an adequate approach. Rather, it seems to me that what is necessary is to discover the real fact, the true fact, about subject and object (as well as about randomness). It is entirely possible, for example, that almost all that has been said about subject and object is entangled in hopeless confusion. Is it possible to see for oneself what is the fact about subject and object? Thus, eventually, what people say about the sunset and the tree is significant only to the extent that it indicates or denotes the deeper non-verbal fact of the sunset or the tree.

Now, I have been going rather deeply into inquiring into the fact about subject and object. My inquiries indicate to me that it is confused to regard these as fundamental categories of thinking. They do have a relative, limited significance in a certain rather superficial domain of experience. But I feel that to start, as you do, by trying to see all the rich meanings that people have attributed to these terms, will only entangle you in the general confusion that has been built up around these terms over thousands of years.

I would take consciousness (or preferably awareness) as a basic undefined term. When I talked to Szekeres, he agreed that this has to him a denotative significance. While I cannot explain the meaning of this word in terms of other words, it is a fact, both to me and to Szekeres, that this word denotes something, a dynamic process too vast to be described or thought about. I suspect that it has the same significance to you.

Awareness is neither subjective nor objective. Indeed, awareness of the subject and awareness of the object are in one and the same field. So awareness transcends the distinction of subject and object. As Piaget[22] has shown, young infants very probably do not make this distinction. It is learned in the first few years of life. The first step is probably to learn the ordered distinction of "inside" and "outside". When the infant realizes that his body is one object among many, he begins to distinguish what is going on "inside the body" from what is going on "outside the body". But in fact, this distinction can be drawn only in the field of thought, which is based on certain orders, such as inside-outside, before-after, etc. The feelings in the solar plexus, the intestines, the chest, the head are sensed to be inside, and labelled as belonging to "I" or "me". It seems then that there is an entity inside the body, the "subject", who is "doing the observing" and is thereby looking at something outside the body called "the object". Later, the child learns to treat the body as yet another object. And still later, the child begins to treat the feelings and thoughts going on inside him as still another object, which he calls "me" or "the self". But then, it is tacitly assumed that somewhere still deeper in him is the "inner self" that is the "subject" who is supposed to be "doing the observing" of the "self" that is being "looked at".

Now in my inquiries into this question, I have observed that this is such a confused account of what is actually happening that it is best described as illusion and delusion. You seem to believe that merely to recognize the possibility of moving the distinction between subject and object freely will remove the confusion. But in my view, there is no way to make the concept of subject and object free of confusion.

If one begins with the totality of awareness (interpreted denotativity) it is an evident fact that all that one knows about anything, (whether it be subject or object) is in this field. So this is the place to begin the study of epistemology or the nature of knowledge. To begin here does not commit one to solipsism. For I do not assume that all that one knows is all that there is. But it surely cannot be denied that all that one knows is what is potentially or actually in the field of awareness.

The distinction of subject and object is never perceived directly (as one can perceive the tree). Rather, it is always inferred. Indeed, the subject is never really perceived at all. Various sensations are perceived, in the head, the solar plexus, the body, etc. These are called "I", "me", "mine", or the "subject". But in reality, anything that could be denoted in this way is not a subject. Rather, it is just another object, that happens to be inside the body rather than outside. This is a peculiar feature of the so-called distinction of subject and object. In fact, it always turns out that whatever the word "subject" a supposed to denote, this always turns out to be just another object. Ultimately, the word "subject" can never have any denotative significance at all. The distinction of subject and object is in reality always a distinction between one object and another object. Implicit in this distinction is an unperceived and unperceivable subject, who is looking at both the so-called "subject" and the so-called "object".

At this point, one may become suspicious of a word that has no denotative significance at all. Is it not possible that the word "I" corresponds largely to a pure fantasy? Of course, insofar as it refers to "this body", "this brain", etc, it has a denotative significance. But insofar as it refers to an "observing subject", it has no such signifi-

[22]Bohm summarizes some of Piaget's views in the Appendix to Bohm (1996). He gives references to Piaget (1953), Piaget and Inhelder (1956)—CT.

cance. I want to suggest that the latter meaning of the word "I" is an illusion. Perhaps it is not entirely a coincidence that almost all human misery stems from confusion about the meaning of the word "I". For while it is generally regarded as representing the most important thing in the universe, it may in fact represent nothing at all. Surely, a person who mistook a mere nothing for something that is all important must become confused about almost everything. And it may well be that this is the explanation of the confused state of what is called "human history" over the past few thousands of years.

Can one not suggest that the trouble here is basically verbal? Thus we say "It is raining"; mainly because every verb seems to demand a subject, who carries out the corresponding action. But no matter how much you look, you can never find the "it" that would be "doing the raining". Rather, a process of raining is "going on". Similarly, let me suggest that a process of "awareness" is going on, but that there is no entity called "I" who is "doing the awarenessing". In this process, awareness of what is called the "subject" is inseparably intermingled with awareness of what is called the "object". And in awareness, there is no "cut" between subject and object, whether one regards it as freely movable or not. Rather, there is one field of awareness. However, confused linguistic habits and confused social relationships that have been shaped by these habits cause one to give a fundamental transcendental significance to a rather superficial distinction, that is at best a mere convenience, and at worst, a gigantic source of general befuddlement of the brain.

Now one may perhaps say that no matter how far one goes in any such discussion as this, there is always firstly the distinction between the content of what is said (which is the object of thought) and the subject who is "doing the thinking". I want to suggest that this distinction is unreal, illusory, and not a fact. To understand what is happening here, it is necessary to draw a distinction between verbal reflective processes of the brain, which are called "thought", and some deeper perception that they are reflecting upon. For example, a verbal description of a tree is at best a reflection of what one actually perceives or could perceive, which is an infinitely richer experience than anything that words can evoke. I would suggest that thought is a reflective process, sensed as "going on" in awareness, along with a more direct kind of perception. It is as if one were looking at something directly and also seeing its reflection in one or more mirrors. Only the trouble is that the "mirrors" of thought do not always function properly. Very often, they are so disturbed that what is reflected in them does not correspond to any direct perceptions at all (whether potential or actual). It is evidently necessary that each person shall be alert and aware of the distinction between thoughts that are thus false, illusory or delusory and those that give true reflections of some directly perceivable fact. (Or do you say that this distinction has no relevance?)

Now, the word "subject" does not seem to reflect anything that could be thus perceived. Insofar as it is distinguishable from "object", it is just another object, one among many. Thus, to assert that the subject is distinct from all objects, and yet not an object, is to become entangled in confusion. It could only be distinguished from an object, to the extent that it was another such object. Even if the distinction is freely movable, this conclusion is not altered.

Let me suggest that the paradox arises because everything that words can possibly refer to must be contained potentially or actually in awareness. This is the only way in which words can mean anything at all. So if the word "subject" means something, whatever it means must be part of the contents of the contents of awareness. The moment one draws a distinction between one part of the contents of awareness and another, one has tacitly brought in the concept of an object (E.g., one distinguishes the object in the "foreground" from what is called the "general background", which is another object, on which attention is not centered.) So whatever the word "subject" refers to, it can only imply the disjunction of the contents of awareness into still another object.

Now, I admit that the concept of the distinction of objects (one of which is the body and brain of this human being, called by the name of "me") has a certain relative and limited domain of validity. But in the total field of human knowledge (which is what is appropriately studied in epistemology), it cannot be taken as a basic starting point without introducing hopeless confusion. Rather, I would say that insofar as there really is a "subject", this cannot be other than the totality of awareness. But this "subject" contains all objects. For the distinction of objects is basically the result of thought (though this thought may in some way reflect what is beyond the field of thought). So there can be no unconfused distinction between "subject" and "object". If the word "subject" has any meaning at all, it refers to the total movement of awareness.

So insofar as the "subject" is anything at all, it is the totality of its "objects". Rather than being distinct from its object, it is its object. This is not solipsism. Rather, it is a denial that the notion of "object" has fundamental significance (which necessarily denies that the distinction of subject and object has this kind of significance either). There is a reality that transcends the contents of awareness, but there is no meaning to committing oneself to the view that this reality is an "object", distinct from the awareness in which it can be perceived.

Perhaps I could make this point more clear by distinguishing between the substance of awareness and the contents of awareness. All knowledge is the contents of awareness. But in direct perceptual experience, there is a kind of control with something that transcends more knowledge. Here I use words denotatively rather than as descriptions or explanations. That is, I assume that you see and experience what I mean directly. If you do not, then there is no communication between us on this point. If you do, then I would go on to say that for all we know, the substance of awareness and the substance of the whole of reality (the universe or whatever you want to call it) may be basically one. But knowledge, (the contents of awareness) is particular and different for each person. It is rather like fire. Basically all fire is one kind of movement. But each individual flame is different, in its own way. So while awareness and existence are deeply one and the same, what we know about it is particular and different, for each human being.

This view is neither materialism nor idealism. For the distinction between matter and mind is like the distinction of subject and object, a category of thought. I suggest that reality, whatever it may be, transcends this distinction.

In many ways, we may be in agreement on these points. You wish to unite science, mathematics, art, and other things in one broader field is similar to my own. But I feel that the principal barrier to this is the confused notion of subject and object.

With best regards,

Yours sincerely,

D. Bohm.

P.S. It might be useful to answer a question raised by Szekeres: "When we make a scientific theory, do we not imply an observer, a subject, who is ultimately looking at the whole universe?" In my view, we do not. Each theory is a structure of thought, containing many complex, inter-related orders of its component parts and various aspects. There is "going on" an awareness of this thought, along with an awareness of a more directly perceived kind (e.g., the results of an experiment). If there is in thought a set of structures and orders similar to those in immediate perception, then this similarity (and difference from similarity) can also be observed. So the test of a good theory is that it contains structures similar to those observed which are ordered similarly to those found more directly in perception. Insofar as the two orders are different, the theory is said to be "falsified". So thought is a kind of "mirror" of the general order and structure of what can be perceived more directly.

Of course, similarity and difference are basically undefined terms. Indeed, an understanding of their meaning is presupposed in every language. For unless a man can perceive similarity and difference in words and their usage, he cannot even learn a language.

In this connection, I would like to indicate what it means to say that a theoretical idea is new. I would not say that this means that it is unanalyzable. Indeed, it is only by some kind of analysis that one can establish that the idea in question is in fact new. For example, suppose I assert that the word "glub" stands for a new concept, and that the basic quality of this concept is "blub", but that the words "glub" and "blub" are unanalyzable. I seems to me that such a procedure opens the door to any kind of mumbo jumbo.

I would prefer to stress that a new idea implies a new kind of order in our concepts, this order being understood mainly non-verbally (as one sees the shape, form and color of the tree). If the order is really new, it is not reducible to an already known order. Far from implying that the concept is unanalyzable, this means that it is just by means of suitable analysis that one can demonstrate the novelty of this order. For example, the geometry of Euclidean space implies a certain (infinitely rich) set of related orders of points, patterns, structures, and objects that can be conceived as being in this space. Non-Euclidean geometry implies a new set of orders that cannot be reduced to those of Euclidean space. It is by analyzing the geometry with the aid of concepts such as curvature (the result of displacing a vector parallel to itself around a circuit) that one can see that the order of non-Euclidean geometry is really new, because it is not reducible to that of Euclidean geometry. One does not do as Bohr does, by asserting that the novelty of non-Euclidean geometry is nothing more than the imposition of a non-Euclidean "metalanguage" on the content of the Euclidean language. It can be thought of in this way, to be sure. But more fundamentally, it contains a basically new order. And by analysis, we can see that the old Euclidean order is a special case of the new non-Euclidean order.

You emphasize the concept of randomness. I would rather emphasize the concept of a creative process. This is first of all a process containing orders that are not completely reducible to any order that is known and specifiable. And secondly, it is a process in which the order of breaks in a lower order is the foundation of the next (new) higher order. What is commonly called randomness is a particular, limited, and very specialized manifestation of the creative character of universal order.

Consider for example, a series of coin throws. Unless it could have say 100 heads in a row, it would not be totally random. So first of all, a random sequence in a given field has the potentiality of all possible orders in that field. Secondly, no matter how many heads in a row one obtains, he cannot say anything at all about the next throw. This means that in a random sequence, the order of $N + 1$ terms cannot be determined by that of N terms or less. That is what it means to say that a random sequence is irreducible. None of this means that a random sequence is unanalyzable. Quite the opposite, it is only by means of various analyses that one can see whether a sequence is random or not.

So I cannot accept your idea that novelty has absolutely nothing whatsoever to do with analysis. A genuinely new order will, in my view, manifest itself both analytically and synthetically. I do not believe that analysis can exhaustively reduce a new order to an older kind of order. But analysis is able to demonstrate that a new order is in fact different from an older one.

DEC 14th 1966

[Date added – CT.]

Dear Jeff

You may find this correspondence with Rosenfeld[23] interesting.

Best regards

D Bohm

NORDITA NORDISK INSTITUT FOR TEORETISK ATOMFYSIK
Danmark . Finland . Island . Norge . Sverige

BLEGDAMSVEJ 17 . KOBENHAVN O . DANMARK

---

[23]On Rosenfeld see p. 22, n 11—CT.

6th December 1966

Professor D. Bohm
Department of Physics
Birkbeck College
Malet Street
London, W.C.1

Dear Bohm,

Loinger has showed me your letter to him of 10th November, containing startling considerations over Niels Bohr's philosophy. I have shown the letter to Aage Bohr,[24] who was as startled as myself. We do not, of course, take very seriously this game of putting labels with various "isms" upon Bohr. No more seriously, in fact, than he himself took it. In the hope of being helpful, however, I feel I ought to write to you about the facts of the case, because I suspect that you have been badly misinformed by Petersen,[25] or that you have perhaps misunderstood what Petersen told you. Moreover, Kierkegaard himself has been distorted out of recognition by the modern metaphysicians who got hold of him.

Kierkegaard appears on the very provincial scene of Danish theology as the advocate of an extreme irrational subjectivism and it is a strange thought, to say the least, to seek any relation between such a position and the objective rationalism to which Bohr consistently adhered and of which his writings, correctly understood, give such shining examples. In particular, the idea of mixing up God with Bohr's thinking appears most incongruous to those who have known Bohr intimately and who remember his familiar saying: "God is just a word of three letters". With regard to the alleged influence of Kierkegaard on Bohr, the truth of the matter is the following. Bohr did read some, at least, of Kierkegaard's books and rightly admired his wonderful stylistic virtuosity. But whenever he commented on this, he added that it was a pity that such a literary talent should have been wasted on such fatuous speculations. The actual philosophical influences that Bohr underwent in his youth are, on the one hand, that of Høffding who was a friend of his father's, and that of Georg Brandes, who was then the dominant personality on the Danish scene and especially enjoyed a great authority in liberal Jewish circles. Now Høffding, who has written a whole book about Kierkegaard, while paying due respect to his personality (he calls him a "Socratic personality") sharply rejects the contents of his philosophy. As to Georg Brandes, he actually started his literary career with a polemic contribution to a theological squabble about "Faith and Knowledge" which agitated people at the time, and in which he took sides against the philosophy professor Rasmus Nielsen, who defended an anti-rationalist attitude in the trend of Kierkegaard's. These are easily ascertainable facts, which could have prevented you from being led astray in your earnest efforts to understand Bohr's point of view.

[24]Bohr's son. Awarded the Nobel Prize in Physics in 1975 for his work detailing the structure of the atomic nucleus—CT.
[25]See p. 22, n 13—CT.

I have the impression that you are looking for non-existing complications. Bohr's attitude was extremely simple-minded and dominated by a straightforward, commonsense approach to the problems. He was quite averse to metaphysical subtleties. He used to dismiss these as "trivial", by pointing out that the true epistemological problems which confront the physicist cannot be dissociated from the most refined analysis of the natural phenomena, of which the philosophers, who loosely talk about "matter" and "mind", have not the slightest understanding.

Incidentally, I am afraid that you have also misunderstood what I have said about the relation of Bohr to materialism. I certainly have never played the game of putting a materialist label on him. I leave that to the high-priests of Marxist-Leninist theology. All I wanted to suggest is that Bohr's thinking, as well as Einstein's thinking in his creative period, occupies a natural place in a dialectic development initiated by the materialistic philosophy of the French school of physics and chemistry (Lavoisier and Laplace) at the end of the 18th century.

With best wishes

Yours sincerely,

L. Rosenfeld

13th December, 1966.

Professor L. Rosenfeld,
Nardita,
Blegdamsvej 17,
Copenhagen,
Denmark

Dear Rosenfeld,

I thought it would be helpful to add a bit to my answer to your letter.

Firstly, let me say that I am not trying to minimize the significance of Bohr's work, when I say he was influenced by existentialism. As a matter of fact, I remember Petersen telling me that Bohr had "gone overboard" on Kierkegaard when he was quite young, but that his brother had got him to reconsider his views on the subject. Of course, you are right to point out that later Bohr realized Kierkegaard's deficiencies very keenly. However, a youthful experience of this kind generally "leaves its mark" in the form of a profound influence on a person's thinking, which is generally unconscious and tacit. I feel that I detect this influence in the principle of complementarity, though, of course, I may well be entirely wrong.

As I see it, the essence of existentialism, both early and modern, is to emphasize man's existence as the primary thing, having top priority. That is, one must begin with the fact of existence, and then one must, in some way, discover or choose what one is. This creates a very serious problem. For when man searches his own nature to discover what he is, he finds that his nature is ambiguous, with many contrary and even contradictory potentialities. However, man cannot remain in a state in which all is potential. Some one of these potentialities must be realized at the expense of others. As he defines himself in one way, he loses the possibility of defining himself in other ways. In a way, this is similar to Bohr's view of the electron. The conditions needed to define its momentum (or wave-like potentialities) are incompatible with those needed to define the position (or particle-like potentialities). Of course, for the electron, the situation is determined by external conditions (e.g., the laboratory apparatus). For most men, it is also external conditions that determine which of his potentialities are realized. But for those who are philosophically inclined, this is not satisfactory. They may feel the need to make the choice, either on moral or religious grounds, or on grounds of inner necessity. The trouble is that no-one can ever find in himself sufficient grounds to determine this choice. So reflection on this subject is likely to produce a disturbing sense of "nausea" or "sickness unto death".

I cannot agree with your notion that existentialism is utter nonsense. In a way, it represents an important facet of what some existentialists call "the human condition". But it is, of course, one sided. So to go too far with it can lead to confusion.

It was only my intention to suggest that existentialist notions of this nature played a key role (perhaps largely unconscious) in shaping Niels Bohr's ideas on complementarity.

Secondly, let me repeat that Niels Bohr's ideas are very far from "common sense". This is in no way a derogatory comment on these ideas. In fact, Bohr raised profound questions, such as the relationship of subject and object, and the role of language in shaping our approach to external nature. Nowhere will you find this sort of thing being done by the "ordinary man" in his everyday routine of activities. None of the great scientists or philosophers was content with "common sense". Indeed, it was just because Bohr raised such deep questions that I was initially attracted to his ideas.

For about ten years, I studied Bohr's notions very earnestly. In this study, I found his style of expression a tremendous barrier. As someone once said to me "Bohr's statements are designed to cancel out in the first approximation with regard to their meaning, and in the second approximation with regard to their connotation. It is only in the third approximation that one can hope to find what Bohr wants to say". His style of talking was even more difficult than his style of writing. Sometimes it seemed to me that he was obsessed with the question of communication, just because he found it so difficult to communicate.

I wrote my book Quantum Theory[26] in the effort to understand the subject. But when I had finished, I saw that I did not understand it. Indeed, whatever in it that was clear was the result of an effort to think of the electron in a unitary way, and therefore profoundly opposed to Bohr's whole philosophy. The more I considered the subject, the more I saw that quantum theory was involved in great obscurity and confusion. Considered as an elaborate mathematical algorithm for computing experimental results, it was clear enough. But as a coherent logical structure, it is very deficient. For example, after I wrote my Relativity book,[27] I considered the

[26]Bohm (1989)—CT.

[27]Bohm (1996)—CT.

notion of trying to write a similar book on Quantum Theory. But I soon saw that I simply do not understand the latter topic well enough to do this, and that nowhere in the literature is there any basis for even beginning to clear up the confusion that seems to hover around the subject.

Recently, in a preprint by Peterson, I obtained new insights into Bohr's position. These insights were further extended by correspondence with Schumacher (in Cornell University). But what I learned has only served to convince me that Bohr's notions are not (at least for me) an acceptable answer to the very profound and valid questions that he was original enough to raise seriously, perhaps for the first time, in physics.

As I said in the previous letter, Schumacher's style of expression is similar to Bohr's, so that I find it difficult to understand just what he wants to say. Nevertheless, I do gather from him (as well as from Peterson) that Bohr emphasizes the question of language, with an attitude exemplified by his statement: "Man lives suspended in language". Bohr says (and Heisenberg agrees with him) that all precisely definable and therefore communicable concepts are, in effect, only variations in the classical concepts of position and momentum (space-time and causal descriptions). As long as these two concepts are simultaneously applicable, we can then obtain an objective description of phenomena, enabling us to communicate to each other what we know about them. But because of the quantum, this possibility breaks down in a very accurate treatment. Thus, we are left with an inherent ambiguity in the sharpness with which we can think about the nature of things. Within this ambiguity, experimental conditions determine which aspect (position or momentum) is to be sharply defined. The quantum mechanical algorithm is then a kind of "metalanguage" that makes statements concerning terms appearing in the primary (classical) "language", which latter is used for describing phenomena. Thus, the quantum mechanical algorithm does not directly reflect nature. Rather, it directly reflects the "language", with which we communicate what we know about phenomena.

To me, it seems that Bohr's views are a long step toward idealism. Of course, he recognizes the objective existence of the material world. But he seems to suppose that in man's dealing with the material world, he is limited by certain linguistic and conceptual structures, which are regarded (at least tacitly) as eternal and unchangeable features of the "human condition". So the form of man's perceptual contact with the world is limited permanently by his way of thinking and talking.

Now, in my view, there is nothing inevitable about classical concepts. They are merely the result of a (dialectical) historical development, so that they can change radically. As I indicated in my Relativity book, man learns to perceive the world, as he interacts with it, while being sensitive to the relationship between what he does and what he sees. Bohr also recognizes this, in a way. But he seems to say that man's direct perceptions will always have to be in the form of classical concepts (at least those that he will be able to communicate in a precisely definable way). It is only in the "meta-language" that man can think in new ways, while in the "language" that describes man's direct perceptual contact with phenomena he will never do other than think in classical terms.

In my view, even man's direct perceptual contacts with phenomena need not be in terms of classical concepts. An extreme case of this fact is the richness of response to the beauty of nature or a work of art. Of course, you may argue that in physics, this is irrelevant. But I do not think so. For example, when we look at a cloud chamber photograph, we do not really observe a particle with position and momentum. We actually observe a series of dots in a certain order (arranged roughly along a circular arc). The "particle" is projected into the photographs by our conditioning to classical ideas.

In my view, if we had a new kind of mathematics of discrete structural process, we would not have to think even of classical physics in this way. I have in fact been developing a new formulation of classical physics not based on the concepts of position and momentum, but rather, on those of distinction, relationship, order, pattern, and structure. In a way, what is needed is a more "artistic" conception of large-scale physics, one that is less technical and mechanical than the current one. We also need a new mathematics that allows us to express this new physical conception in precise terms. I have reason to believe that this can be developed. In a paper that I am preparing, these points will be explained in more detail.

Be that as it may, the key point is that both classical and quantum concepts are in need of fundamental changes. When this is done, the quantum algorithm will be able to refer directly to the phenomena of physics. But these phenomena will not be positions and momenta. Rather, they will be distinctions, relationships, orders, patterns, structures, rhythms, etc. etc.

So I do not wish to return to classical physics. My work on hidden variables was intended mainly to clarify the situation, which had been left rather confused by von Neumann's "proof" of the impossibility of hidden variables. I think that to entertain the notion of hidden variables calls attention to certain questions, which are necessarily overlooked in Bohr's point of view. But I am as dubious as anyone else that they are the right approach for a new theory.

I hope that this makes my position more clear than it was. I think it too much to hope that we will ever agree. But perhaps we can at least understand what it is that we disagree about.

Finally, will you please give my best regards to Aage Bohr, and assure him that I had the very highest esteem for his father. If I am deeply moved by the questions that moved Niels Bohr, this may cause me to be a bit sharp with his views at times. But I am sure he would have preferred such a response to the indifference with which physicists now so generally respond to these questions.

If you wish, you may show this letter to Aage Bohr.

With my best wishes,

D. Bohm.

16th December, 1966.

Dr. J. Bub,
Department of Chemistry,
University of Minnesota,
Minneapolis,
Minn. 55455,
U.S.A.

Dear Jeff,

This letter is intended to supplement the material I just sent you in the copies of letters to Schumacher, in Cornell University.

As I indicated in the material referred to above, one can regard a scientific theory as a certain kind of "artistic" creation, in the universe of discourse. Each theory is a particular reflection of nature, as perceived more directly, either with the senses, or with the senses aided by instruments. Now, to reflect is to abstract (in the sense used by Korzybski[28]). Thus, a mirror image abstracts the form, colour, and space relationships of an object, apart from its tactile qualities, its inertial qualities, its chemical and general physical properties, etc. In our own minds, the optical image abstracts from the total perceptual content of something, its "feeling", its tactile qualities, its functional properties, etc. etc. As Piaget shows, the infant tends to regard seeing, hearing, touching, as distinct experiences. Later, he learns to see what he hears, to feel what he sees, etc, so that it is all integrated. But one then overlooks that the unity of perception is based on various sensual abstractions.

Our descriptions go on to abstract from direct perception, and our inferences abstract from descriptions. We can also abstract further, to first describe our inferences and then form higher level inferences about inferences. Thus, we rise to the level of theory, which can go on to indefinitely high orders of abstraction.

The test of a theory is what Piaget calls the "circular reflex" – i.e., an action aimed not mainly at a specific function, but rather at learning. That is, guided by our theoretical reflective abstractions, we draw inferences about lower level aspects. We then do something to test these inferences and see what happens. If this latter is in accordance with our inferences, the theory is confirmed. Otherwise, it is falsified. Thus, we determine the domain of validity of a theory.

In certain ways, scientific theories are similar to what Biederman calls mimetic art. That is, they are designed to abstract from and reflect certain aspects of the perceived world. But the purpose of the reflection is different. In mimetic art, one tries to reflect on some concrete, individual perception, revealing the universal in the individual (e.g., Rembrandt). In science, one tries to reflect the general order and structure of things. The "test" for mimetic art is largely in its effect on the psyche, which is very subtle, but nonetheless, real. The test for science is through our functioning in nature (partly with the aid of instruments).

[28]Bohm took up ideas from Korzybski (1994), in which he was influenced by Charles Biederman. See Introduction, p. 10, n 59—CT.

What about the relationship of theories? This is also similar to that of successive mimetic artists. In a way, each artist must be keenly aware of how preceding artists perceived nature. Of course, he sees it in his own way. But as long as art was aimed largely at mimesis, there was a general trend of evolution toward mastering the problem of realistic presentation. Each (great) artist learned something new. He may have lost certain things that earlier artists had. But the general trend is that the later artists not only portrayed what earlier artists had not been able to do. They also could see what it was that previous artists had missed.

But, of course, the more and more complete portrayal of what the eye sees is a limited field of development. So artists had eventually to transcend this whole field. In a way, science is not thus limited, as nature is so vast and immense, so unbounded in all its manifold orders and structures, as revealed to the unlimited development of scientific instruments. On the other hand, this programme is also limited, in a certain other way. For even the indefinite probing into nature's structure raises the question of the kind of reflection of nature that man is creating. It is in no sense a movement toward complete and detailed reflection, nor is it even a steady approach to this as an unattainable limit. Rather, as I indicated earlier, it is an unending series of artistic "creations". Later creations are able to make abstractions about earlier creations.

It is like this. A given theory (e.g., Newton's) comes out. This makes inferential abstractions on the lower order descriptive facts about the field of mechanics. Later, one discovers that these inferences have limited domains of validity, and become confused when extended too far. So we have already begun to become descriptively conscious of the structure of inferences in Newtonian mechanics, as soon as we are led to criticize this structure. Then in Einstein's theory, we make inferences about our descriptions of the structure of Newtonian mechanics (e.g., that mass is not a constant at higher velocities, etc.). Indeed, certain key aspects of Newtonian mechanics are recovered as approximately valid inferences from the theory of relativity. But the situation is changed. For in Newtonian mechanics, they were the basic assumptions, determining the structure from which inferences are to be drawn. Now, in Relativity, they are merely particular inferences. (Of course, some day, Relativity must suffer the same fate.)

In this process, the older theory is not totally recovered as an inference from a later theory. Rather, only certain key aspects are recovered. For example, Schrodinger's equation contains Newton's law as an approximate descriptive inference. But it fails to infer the concept of a well defined particle state (position and momentum) even as an approximation. Einstein's theory does not allow us to infer the Newtonian concept of a rigid body. This is a serious deficiency, leading to the unsolved problem of how to describe extended structures in relativity.

So now, we are led to form newer abstractions, which recover more of the useful or correct aspects of older theories than was possible earlier. This process is generally ignored by modern scientists, who tend to suppose that they stand on "peaks" that are the unique culminating points of all earlier work. But this is, in my view, a delusion.

The factual test of a theory is a very complex story. Each theory contains distinctions, relationships, orders, patterns, and structures that are similar to those found in lower level abstractions. For example, in Newtonian mechanics, the theory distinguishes the positions and velocities of a particle, (these distinctions being made mathematically, in the universe of discourse). When we work with actual objects, we make similar distinctions, which we describe in terms of words and numbers (coordinates). We notice the similarity of our theoretical distinctions and our lower order described distinctions. It has been verified long ago that these theoretical distinctions are faithful reflections of the corresponding descriptive distinctions. But in the theory, there are inferential relationships in these distinctions (e.g., that in free space $X_2 - X_1 = v(t_2 - t_1)$), where v is a constant, so that

$$\frac{X_2 - X_1}{t_2 - t_1} = \frac{X_3 - X_4}{t_3 - t_4}$$

We test the theory by seeing whether our descriptive distinctions are in fact related in the same way as our theoretical distinctions.

Note that the relationship between theory and fact is not a correspondence between "theoretical" objects or entities and "real" objects or entities. Rather, it is a correspondence of the abstract qualities, such as distinction, relationship, order, pattern and structure. These abstract general qualities are common to every order of abstraction, from direct and immediate perception on to the most subtle kinds of thought. Even when we see the form of an object, we do this by seeing all the distinctions, relations, orders, patterns and structures (DROPS)[29] in the outline and boundary surfaces of the object in question. When we feel the object, we feel all the DROPS in its movement and in the tactile and kinesthetic sensations that are produced by it. The "object" that we are aware of is constructed in the brain, as a provisional representation of all the DROPS that have been established about that "object".

This fact is brought out by a simple experiment. Two photographs of a scene are taken, one through a red filter and one through a neutral filter. These photographs are projected in the same screen, the first with red light, and the second with white light. So on the screen, the illumination must be everywhere a set of varying shades of pink. But you actually see all sorts of colours, such as blue, green, yellow, etc. Evidently, what the brain picks up is not colour, considered as a sort of "substance" or fixed quality. Rather, it is sensitive to differences of colour, to their relationships, orders, patterns, and structures as influenced also by their form and space relationships. All this information about DROPS is integrated by the brain into a "construction", which is what we see.

Would we say that our perception of the screen in the above case is "wrong"? Of course not. The test of its rightness, is whether the distinctions, etc., faithfully reflect real distinctions in the total pattern and structure of light, as registered in other ways (e.g., by other people or by instruments), and translated into words or mathematics descriptively.

In testing a theory, we frequently accept a lower order theory as a fact. We then use this lower order theory descriptively. For example, we may accept thermodynamics, with all its inferences, as a fact. In terms of this theory, we calculate free energy,

[29]The acronym DROPS is used a number of times in this folder and the next—CT.

entropy, and other quantities, which would have no meaning, unless the DROPS of thermodynamics fairly faithfully reflected the DROPS of the descriptions of observations based on thermometers, calorimeters, etc. In statistical mechanics, we can compute what Gibbs called analogies to the entropy, free energy, etc. But with the aid of atomic theory, we can draw inferences about these, beyond those that can be drawn in thermodynamics. Thus, thermodynamics is being used descriptively. This use is twofold. First, there is a domain where statistical mechanical inferences never contradict thermodynamic inferences. Here, statistical mechanics merely enriches the total set of inferences. But then, in another domain (fluctuations in Brownian motion or near the critical points of materials) statistical mechanics leads to inferences contradicting those of thermodynamics. Here, thermodynamics is no longer being used as a description of experimental fact. Rather, one is describing the inferential structure of thermodynamics, and making inferences about it (e.g., that its inferential structure is different from that of statistical mechanics). So we are now at yet another level of abstraction, which is abstracting from the inferential structures of several theories, and relating these structures. We must also, of course, compare all this to the described structure of the corresponding experimental facts.

In all this, there is no such thing as a fact. Rather, there is an ordered process of making the fact (which could perhaps be called "facting"). The "fact" is never finished, nor does it remain fixed. The "fact" is the momentary result of the movement or process of "facting". Although the result has a certain significance, it is evident that the process of "facting" itself must have top priority in our attention. It is like a tree. The fruits have a certain importance. But if the tree is dead, or malnourished, there will be no fruits. Unfortunately, mankind has generally been far more interested in "results" than in the living dynamic process, out of which these have to emerge. But such emphasis on "results does not in general even produce these very effectively. Truly creative and original "results" have almost always been produced by those whose main interest was in the living movement of creation (which is hardly noticed in philosophical discussions about the nature of scientific research).

One can raise an interesting question by relating the fact for art and the fact for science. In early days, these two kinds of fact were established in very similar ways. But gradually, they began to separate. In science (especially in physics), there is a systematic effort to make the observation of the experimental fact as routine and mechanical as possible (so as to remove the "personal" factor). Creative perception occurs only at much higher levels of abstraction of this fact. In art, however, the essential creative act must occur directly at the level of immediate perception. This is not to say that high levels of abstract thought have no bearing on art. They are indeed very important. But their significance is only in the way in which they are able to bear on the DROPS of the level of immediate perception. On the other hand, the aim of scientific research is not to create "beautiful" experimental results, at the level of direct perception. Rather, it is to create a theory that faithfully reflects the descriptive fact, and that does so in a coherent, logical, orderly harmonious way, that will therefore also be sensed as "beautiful".

The factual test of a work of art is very subtle and difficult. Nevertheless, it is not an impossibility. One who wishes to test it must, in certain ways, be similar to an artist, in his creative approach to perception. If he is sensitively observant, his brain will register the DROPS, first non-verbally, and ultimately in words of description and inference. Another person, looking at the same work, can do the same. If they discuss, they can eventually arrive at a common understanding of what they see.

At first sight, it may seem that the scientist has a more reliable and simpler means of perception of the immediate fact. In certain ways, this is so. But in other ways, this "simplicity" and "reliability" are misleading, because they are purchased at the expense of sensitivity and flexibility of perception. Even in physics, we are probably "imprisoned" by our current mechanical modes of instrumentation, which channel research into certain narrowly defined directions. But if one goes on to biology, surely this limitation must become more serious. Will it not cause us to lose sight of the subtler qualities of life? And in psychology, it may well be catastrophic. For the mind is infinitely creative and dynamic. To try to "measure" its behaviour as one measures the location of a planet is evidently an absurdity. Will it not therefore be fruitful in psychology, in biology, and perhaps even in physics, to inquire into the question of whether there is perhaps a more sensitive perceptive, and artistic way of establishing the fact, at the level of immediate perception?

Of course, it follows from what has been said that the development of theories influences the structure of the fact. Once a theory has a certain domain of validity, its DROPS provide a framework, that tends to shape the development of instruments, which latter then provide facts within this framework. (E.g., to distinguish things according to the theoretical category of "spin" leads to devising instruments and experiments that produce distinctions and orderings of "particles" according to their spins.) Nevertheless, there is some limit to the possibility of such development. For theories whose DROPS do not cohere well with the kind of fact that they lead us to observe will eventually be given up. [Of course, people may be reluctant for psychological, religious, or other reasons (e.g., status) to drop theories of this kind. But this is only a failure to engage properly in the process of perception. It is like a man who refuses to look at something, that he does not wish to see.]

In spite of this mutual influence between forms of theory and forms of the fact, the distinction between what is a fact and what is not is at the basis of all perception. In any particular instance, there may not be a hard and fast distinction. Nevertheless, unless the basic intention in perception is to perceive what is a fact (rather, for example, than what would be pleasing and satisfying, even though not true), it has no real meaning. What is needed is that the inferences drawn from our theoretical notions shall apply on lower levels of abstraction, that are ultimately founded on direct perception. The partial shaping of these lower levels by our theoretical concepts does not really basically alter the case. Wherever one finds inferences that are not confirmed at more nearly descriptive levels, then the theory must alter, to accommodate this fact.

Even more, there must be a continual questioning, probing and testing, aimed at checking all possible kinds of inferences of our theories. It is here that pre-scientific ages generally failed. For example, to prepare a certain potion, they might give directions for what materials were required, how they were to be heated, etc. These treatments might include the saying of certain magical phrases. In earlier times, people were generally rather pragmatic. If the potion worked, they assumed that all steps involved in preparing it were necessary. They did not generally think of testing the inference of necessity by varying the steps involved in preparation in some orderly fashion, and seeing what happens when these variations are made. Thus, they were unlikely, for example, to learn that the magic words might perhaps be dropped, without really affecting the activity of the potion. It may well be said that the notion of testing all possible inferences of our theoretical ideas is one of the really revolutionary things that has developed in more modern times.

It is important to remember that the whole fact is not merely external to man. It is also internal. These two aspects of the fact are actually inseparable. Indeed, in science (as opposed to art), the external aspect of the fact is generally rather mechanical and superficial, because of our current modes of instrumentation. The subtler and more profound aspects of the scientific fact are contained in the hierarchy of descriptive and inferential structures of ideas that has been developed over the centuries.

But, of course, the fact is not just a set of words and thoughts. Rather, it is, most basically, the movement of "making" the fact (or "facting"). In this movement, the mind is directly and non verbally aware of the verbal descriptive and inferential structures that I have referred to above. It is here that creative perception of what is new can take place in science. When such perception leads to some new order and structure (whether in a large or in a small context), we say that there has been a "flash" of understanding. If you are somewhat attentive to what happens in such a case, you may notice that this flash takes place only when the mind "lets go" of previously held orders and structures, thus leaving an "empty space" or a "silent period", within which there is room for creation of a new set of DROPS. The creative act is in no sense a deductive inference from previous knowledge, nor is it any other kind of inference (e.g., inductive or associative). But from the new set of DROPS, new kinds of inference can be drawn, permitting the new understanding to be tested.

It is crucial to realize that the act of understanding is non verbal, and not a reflection of anything else. This act is creative, in the sense, for example, that it creates new kinds of verbal and intellectual reflections or abstractions, which then help shape the lower order perceptual structures in new ways. The act of understanding is a very high order of awareness contact with the fact.

The act of understanding is evidently then not analytic (i.e., not a deduction from previous premisses). Nor is it synthetic. For by synthesis, we generally refer to the coming together of various constituents to make some "compound" that has new qualities, not contained in the components. Thus, there is the dialectical view that a thesis develops, and that because this is partial and one-sided, it leads to its anti-thesis. The two then lead to a syn-thesis, with new qualities, which in one sense contains thesis and antithesis, while in another sense, it puts these aside. But in my view, this is not what happens in the act of understanding. Rather, the essential point is that thesis and antithesis are both inactivated and dropped, when the mind is "empty" or "silent". The mind is then working on a faster, finer, more subtle level, far beyond that of words, with their thoughts of thesis and antithesis. It is "feeling out", this way and that, being in direct contact with the fact, "sorting it out" with tremendous speed, and coming immediately to a new set of DROPS, in terms of which the fact is comprehended without contradiction.

As a by-product, the valid aspects of earlier ideas can be recovered from the new set of DROPS as approximations, while at the same time, the limits of validity of the older ideas reveal themselves clearly. But the new ideas are in no sense a "synthesis" of the older ideas. Such a "synthesis", if it existed, would have to be confused, since it would be in essence a mixture of two or more contradictory ideas. It is necessary to stress the creative character of understanding, in the sense that this leads, in a single act, to a new total structure, with new "component" elements. This act takes place "all at once". It is not a time process, in which the older "elements" are combined and synthesized, in the hope of arriving at something new, which is not analytically deducible from the original starting point. For while it is possible to combine chemical elements synthetically in this way, ideas cannot be thus combined. For when ideas contradict each other, any effort to relate them can lead only to confusion. They must be dropped altogether – i.e., the mind must "let go" of them, and leave itself open to the unknown, which is the source of all creative movement. We generally find it very difficult to do this, because we are so strongly conditioned to hold onto the known, as if it were life itself. Very probably, the key difference between genius and mediocrity lies, not in the degree of talent, but rather, in the ability of the brain to "let go" of the known, so that it will be open to the creative movement of the unknown. And if creation is really the ability to lead to new orders, it must have its source in the unknown. Otherwise, it would merely be the adaptation of old knowledge to new situations.

I hope that by now you have "digested" your reading and your talks with Biederman. I am awaiting your comments with great interest.

With best regards,

D Bohm

Dec 20, 1966

Dear Jeff

Enclosed is a letter to Schumacher, that is relevant to the letter I sent you yesterday.

You might find it interesting to get in contact with Mr. Schumacher, who is doing a thesis on Bohr's epistemology. I have already mentioned your name to him.

You might find it enjoyable to look up an old friend of mine, Professor Melba Phillips,[30] when you are in Chicago. She is at the Physics Dep't of the University of Chicago. I have mentioned your name to her.

Best Regards

D Bohm

---

[30] See Talbot (2017) for Bohm's letters to Melba Phillips in the 1950s—CT.

20th December, 1966.

Dr. D. L. Schumacher,
Centre for Radiophysics and Space Research,
Cornell University,
Clark Hall, Ithaca,
N.Y. 14850,
U.S.A.

Dear Schumacher,

This is a continuation of my letter of a few days ago. I also enclose a copy of a letter to a former student of mine, Dr. Bub, which is relevant to the topics that we are discussing.

With regard to denotation and connotation, of course you are right to say that both are necessary for communication. But I want to add something more. Denotation and connotation stand in a naturally ordered relationship. Denotation is the fundamental role of words, which must be given primary emphasis or top priority. Connotation is significant mainly insofar as it bears on denotation, i.e., insofar as it helps to make the process of denotation more flexible, more subtle, more sensitive, richer, etc. A little reflection shows that this must be so. For as I indicated in earlier letters, words in themselves have very little significance, unless they refer to something else that is non-verbal. Thus, the word "dinner" may call up the image of food, the taste of food, the feeling of eating it, etc. But it will not nourish you. Unfortunately, with regard to more subtle questions we have fallen into the habit of trying to live in a "world of words", and the images and feelings that they evoke. Thus, the whole of the "self" is evoked entirely by words. And whenever we use terms for which it is difficult to find the denotation (e.g., love, duty, honor, etc. etc.), there is a tendency to let their meanings be determined mechanically by manipulating words. For example, a child is told that it is his "duty" to love his parents. Therefore, he tries to manipulate his feelings so that they will seem to be the ones that conform to what is expected of him, while his natural and spontaneous feelings are covered up. In a way, he is then starved of real feelings, as he might be starved of real food, if he tried to satisfy himself with words, images, and their associated sensations of eating. So unless our more abstract words ultimately denote something real that is going on either inside or outside of us, they can lead to endless confusion.

Of course, you are right to emphasise that an inability to find denotation immediately does not necessarily justify us throwing out what is new and not understood as "mumbo jumbo" or "gibberish". Nevertheless, you cannot expose yourself to the tremendous totality of "new" things, indiscriminately. You did, for example, choose Stravinsky as the one to listen to, and presumably failed to put in equal effort in listening to others, whom you may have regarded as mediocre. I want to ask you what determined this distinction. Was it not perhaps a kind of perception at very deep levels that although you did not understand him, there was a new set of non trivial distinctions, relationships, orders, patterns and structures (DROPS) here, that was worth paying serious attention to? In a way, our creative responses begin very deep, and stir in us in ways that are hard to specify. But I suggest that from the very beginning, your brain was, at deep levels, able to see that Stravinsky had a new set of DROPS, while other relatively mediocre composers did not. One has, of course, to be very sensitive and "vulnerable" to those faint "stirrings", or else they will be "drowned out" in the noise of everyday life.

I feel that it would be better to emphasize the creative character of novelty, rather than its synthetic character. As suggested in the letter to Bub, synthesis implies the coming together of older elements to form "compounds" with new qualities. Real novelty implies what is original and creative, going far beyond mere synthesis.

Also, I would not say that novelty lies in the "choice" of axioms or assumptions. The word "choice" has wrong connotations here. Choice implies the existence of a well defined set of alternatives, one of which is decided upon by the "chooser". Basically, there is nothing new in this, as all the alternatives were already present from the beginning. Creation implies the coming into being of a genuinely new total structure. Therefore, there is no "choice" in creation. On the contrary, one has a sense of logical inevitability in it, that it could not have been otherwise.

In any case, freedom is not the ability to "choose". For merely to choose among alternatives is to be determined by one's conditioning, which causes one to prefer a particular possibility over the others. Rather, freedom is born in the deep intention not to be a follower of old, familiar, and apparently safe patterns. This latter tendency is the principal barrier which impedes the mind from moving in freedom.

I still feel that you are paying too little attention to the question of factuality. Of course, the fact is often very subtle and difficult to specify. Perhaps you are conditioned to think of facts as isolated bits of information. This is unfortunately a result of our modern way of doing technology. As I indicated in the letter to Bub, the fact is something very subtle, vast, flexible, and dynamic. But it is nonetheless real, as well as the keystone of proper perception and a healthy mental life. When one ceases to pay attention to it, the result is illusion and delusion, born of the wish to present in perception something that is pleasing, rather than what is true and factual. Consider for example, our typical response to flattery. We find it very difficult not to respond semi-automatically, with an image of the "self" as sublimely beautiful, powerful, wise, good, etc, etc. This image evokes a wonderful glow of pleasure and a sense of well being and euphoria. Behind these feelings is hidden a "censorship" process, making us insensitive and "dead" to all perceptual and other evidence that the image is false. Thus, the brain deludes itself. For example, it may infer that one who flatters us always speaks the truth, while one who criticises us is a liar. In this way, it seems to "protect" itself from perceiving the fact that the flatterer is saying what he does, because he wants to take advantage of us. (Seeing this would make us "feel bad".)

It is clear that having a simple factual attitude to what people say about us is quite different from collecting hard "bits" of statistical information or "opinion polls". It involves a very subtle kind of perception, which is aware, not only of what people say, but of how we are reacting and responding. This the broad fact, both external and internal to us. Indeed, as I indicate in the letter to Bub, these are inseparably related.

You said that our ideas should be considered to be "heuristic". Could one not better use the word "provisional"?

Whenever you talk of subject and object, I somehow fail to understand what you are driving at. It seems likely that one of us is confused about this question. Which of us is it?

You say that the division of subject and object transcends anything that can be expressed in the content of scientific knowledge, because it is among the conditions required before such knowledge can even be discussed meaningfully. I wonder whether this is really a fact (in the broad sense of the word). On the one hand, it is quite easy to discuss this division within the content of knowledge, if we treat the subject as a special object (i.e. the one containing the "instruments" of perception). But then, of course, we see that the dividing line is arbitrary and freely movable. This ambiguity suggests that such a division is at best a matter of convenience, and not something of deep epistemological significance. On the other hand, in a deep sense, there is no way at all to discuss the division of subject and object, in fact, the word "subject" has no denotative significance whatsoever. While I agree that connotation is important, I also emphasize that unless a word contributes at least indirectly to determining some ultimate non verbal denotation in the total perceptual process, it has no meaning, and is merely a source of confusion. Thus, no matter what you do, you cannot prevent the mind from tacitly and unconsciously creating a "display object" out of images, thoughts, and sensations, which it calls the "subject" or the "self". This "display object" is pure illusion – it is in reality nothing at all. But it is created, because the mind cannot make sense of a word that has no denotative significance at all. Therefore, it tries to make sense by creating an illusory denotative significance. And this leads inevitably to confusion.

If you are perceptive of the whole fact, inward and outward, you will see that our theoretical constructions are also parts of this fact (as explained in the letter to Bub). So nowhere in awareness is there a division which could be denoted by the words "subject" and "object". Nowhere does one find any indication that theoretical ideas belong to a "subject" who is distinct from the "objects" that are the contents of these ideas. Rather, theoretical ideas are abstractions of high order, that reflect the general structure of abstractions of lower order, going all the way down to immediate perception. As Piaget put it, the infant learns to see what he hears and hear what he sees. Later we learn to think what we perceive and to perceive what we think. It is all one process, without a "thinker" who would be "doing the thinking". So where does the division of subject and object come in? Why do you insist that this is a precondition for all scientific descriptions? To me, it seems to be the result of certain related linguistic and social structures that have developed historically.

I must say that I do not understand the first paragraph of your letter. When you say "object" and "apparatus" are non descriptive notions, I ask you how the physicist is to know whether a given thing is a piece of apparatus or not. For example, I go to the grocery shop and buy a piece of cheese. However, the manufacturer has decided to sell it under the label "Observing apparatus". How am I to know that I should not also take this cheese to a laboratory and use it to observe an electron? Surely,

Bohr has some tacit notions in the back of his mind "describing" what an observing apparatus is supposed to do. It is not totally undefined and indescribable.

With regard to randomness, I wonder if our words on this issue are not sources of confusion. We speak of disorder. But this is impossible. Whatever we see is in some order. E.g., the throws of coin can be ordered as HTHHTHT, etc. etc. One feature of a random distribution is that there is not some "simpler" principle from which this order can be deduced. A random order is not deducible from another order with a smaller number of basic elements.

Now, you start to talk of the universe as random. Of course, this does not mean that it has no order at all. It means only that you have not been able to deduce the visible order from some simpler set of principles. In reality, we may then say that the order of the universe is known only descriptively and not as an inference from higher level abstractions.

Of course, when you look at such an "unknown" order, the brain creatively starts to "play with" higher level abstractions, to see whether it cannot find some that permit certain features of the perceived order to be inferred. If it should find them, these will then be stored up in the "reservoir" of known orders. The brain will then be able to "see" these orders relatively easily, because the whole mechanism of recognition and storage of information will be able to operate.

But notice that not every order that the brain "plays with" will turn out to be suitable. Here is where the question of factuality comes in. Only certain orders will prove to have a broad domain of applicability. Of course, men may prefer to hold onto orders that please them or appeal to them for religious, political, or other reasons. Then, as happened with the Schoolmen in the time of Galileo, they may protect their sense of security and pleasure by simply not looking at the facts. But this is, of course, the beginning of the deterioration of the mind.

I sometimes have the impression that you want to play a "game" of manipulating undefined words, such as "epistemology" and "subject", to see whether it does not produce interesting results. Perhaps this procedure may work in mathematics. But in physics, it will lead to confusion. It is true that we must use undefined words. But unless they ultimately denote something non-verbal, either external to us, or in our own mental processes, these manipulations must lead to confusion, in the way that I have described. And I insist that the word "subject" cannot possibly denote anything at all, either external to us, or in what we are aware of as going on the mind. Therefore, all it can do is to lead the mind to construct an illusory "display object", which it tacitly takes to be the "subject". And this is, of course, pure confusion.

Coming back to analysis and synthesis again, Rosenfeld's example of the potion that had to be made in a magic dish has been, in my view, wrongly interpreted by you. As I indicated in the letter to Bub, what is needed is continually to test our ideas by means of critical analysis, to see whether inferences of necessity (tacit and explicit) are really valid or not. The primitive way of thinking that Rosenfeld describes is also characteristic of young children, who very frequently fail even to think of the possibility of such test. As Piaget has shown, this is because they so often regard whole situations as "unanalyzable totalities" that "go together". When the components of such a situation have in fact no necessary property of being together,

Piaget calls this mode of thinking "syncretism" – i.e. the tendency to regard actually separable elements as indissolubly connected. Childish and primitive reasoning is full of syncretism. Nor is the civilized adult really free of such survivals of childish and primitive thinking. On the other hand, the civilized adult also tends to carry analysis too far, as he tries analytically to treat things that are in fact indissoluble wholes. One and the same individual generally reveals both extremes, in different aspects of his thinking.

I feel that in the effort to correct the wrong application of analysis, you have "gone overboard" in the direction of synthesis. Thus, it is not really right to equate the primitive man's belief in the power of magic with the scientist's belief in the power of chemical analysis. It is true that the latter's belief is also frequently wrong as he extends analysis beyond its proper scope. But to correct his error, it is not appropriate to return to the primitive syncretic way of thinking.

It is nevertheless frequently useful to combine elements of thinking synthetically, provided that they do not contradict each other. But then, these combinations must be treated as provisional. In order to test the validity of the new results that come from such combinations, however, a suitable analysis is needed. It is only with the aid of such analysis that one can criticize the inference of necessary connection of the various elements. Without analysis, one is likely to be trapped in making arbitrary tacit assumptions of necessity of connection, of which one is not even aware.

So analysis and synthesis are two sides of every process of reasoning. It is impossible to reason, unless our premisses are first synthetically combined in new ways, and then analyzed, for their results to see if the inferences drawn from the premises make sense logically and cohere with the fact.

But genuinely original discoveries transcend both analysis and synthesis. Rather, they arise in the creative perception of new total structures, involving not only new relationships, but new sets of basic "elements" that are thus related. Here is where creation differs crucially from synthesis. For the latter is just the combination of old elements in new ways, to give rise to certain new properties or qualities.

Of course, every creative discovery has to be expressed in way that involves synthesis and analysis of terms in our language. Synthesis may be compared to the composer's ability to combine already known themes in new ways. Analysis would then be the faculty involved in criticizing these combinations, to see if they "make sense" musically. But creative perception is a much vaster, deeper process, going on mostly at non-verbal levels of the mind. It is out of this process that there "wells up" an unending series of "ideas" that a great composer like Beethoven or Mozart is very probably always "playing with". Similarly, those who worked with Einstein said that in one day he tried and discarded more ideas than the average physicist does in a year.

The key point is this: Synthesis involves combining known structures in new ways. But most new combinations are "nonsense". Surely it is not pure chance that determines the possibility of ideas that are not "nonsense". Rather, there is a creative process that "throws up" combinations that tend, in some measure, to make sense. (These mast of course later be tested by analysis.) In this process, there is not only new synthesis, but also a new set of "elements" leading into a new mode of analysis. Every creative step thus involves both synthesis and analysis.

Unfortunately, because of conditioning, most scientists are seldom able to get beyond the analytical stage. They are probably discouraged by the fact that their efforts at synthesis generally lead to nonsense. But they do not realize that useful synthesis is a by-product of a non-verbal creative movement of the mind, which is of an entirely different order from ordinary thinking and feeling. Indeed, the essential quality of what is called "genius" is very probably due to the fact that in certain individuals, this creative movement was, for fortuitous reasons, not "killed" by society in its effort to produce people who conform to certain standards or patterns, in their thinking and feeling. For evidently, if new ideas are continually surging up in the mind, society may find it hard to keep people "in line". So it generally does it's best to instill conformity to a pattern at a very early age, and with it, of course, the concomitant suppression of the creative process.

You say in your letter that it is wrong to try to analyze synthesis. I do not agree with you. Indeed, just as one can obtain synthetically formulated insights into analysis, one can also obtain analytically formulated insights into synthesis.

Analysis and synthesis are two complementary extremes that are both of the same general order. Therefore, they can reflect each other. I feel that you are confusing synthesis with creation, which is of an altogether different order of subtlety and dynamism. I would say that it is meaningless to analyze creation, because the real pre-condition for all fruitful analysis is the creative mind. This is also the pre-condition for all fruitful synthesis. So synthetic insights are also inadequate for understanding creation. Creation can be understood only with the aid of a yet higher order of creative perception. That is to say, creation may first be concerned with some particular field (such as physics, mathematics, art, music, etc.) But ultimately creation begins to work on itself, to give creative insights into the creative process. This in my view, is a still higher order of creation,

With best regards,

D. Bohm.

References

Bohm, D. (1965). Space, time and the quantum theory understood in terms of discrete structural process. In Y. Tanikawa (Ed.), *Proceedings of the International Conference on Elementary Particles* (pp. 252–287). Kyoto: Kyoto University, Publication Office, Progress of Theoretical Physics.

Bohm, D. (1989). *Quantum theory*. New York: Dover; Reprint of New Jersey Original (1951).

Bohm, D. (1996). *The special theory of relativity*. Abingdon, UK: Routledge.

Bohm, D., & Bub, J. (1966a). A proposed solution of the measurement problem in quantum mechanics by a hidden variable theory. *Reviews of Modern Physics*, *38*(3), 453–469.

Bohm, D., & Bub, J. (1966b). A refutation of proof by Jauch and Piron that hidden variables can be excluded in quantum mechanics. *Reviews of Modern Physics*, *38*(3), 470–475.

Daneri, A., Loinger, A., & Prosperi, G. (1962). Quantum theory of measurement and ergodicity conditions. *Nuclear Physics*, *33*, 297–319.

Daneri, A., Loinger, A., & Prosperi, G. (1966). Further remarks on relations between statistical mechanics and the quantum theory of measurement. *Nuovo Cimento B*, *44*(1), 119–128.

Freire, O., Jr. (2015). *The quantum dissidents: Rebuilding the foundations of quantum mechanics (1950–1990)*. Berlin, Heidelberg: Springer.

Korzybski, A. (1994). *Science and sanity* (5th ed.). New York: Institute of General Semantics.

Kožnjak, B. (2017). Kuhn meets Maslow: The psychology behind scientific revolutions. *Journal for General Philosophy of Science*, *48*(2), 257–287.

Pessoa, O., Jr., Freire, O., Jr., & De Greiff, A. (2008). The Tausk controversy on the foundations of quantum mechanics: Physics, philosophy, and politics. *Physics in Perspective*, *10*, 138–162.

Petersen, A. (1968). *Quantum physics and the philosophical tradition*. Cambridge, Massachusetts: The MIT Press.

Piaget, J. (1953). *The origin of intelligence in the child*. Abingdon, UK: Routledge and Kegan Paul.

Piaget, J., & Inhelder, B. (1956). *The child's conception of space*. Abingdon, UK: Routledge and Kegan Paul.

Schumacher, D. (1967). Time and physical language. In T. Gold (Ed.), *The nature of time* (pp. 196–213). New York: Cornell University Press.

Talbot, C. (Ed.). (2017). *David Bohm: Causality and chance, letters to three women*. Berlin, Heidelberg: Springer.

Wheeler, J. A., & Zurek, W. H. (1983). *Quantum theory and measurement*. Princeton: Princeton University Press.

Chapter 3
Folder C131. Perception and Panpsychism, Jung and Biederman. January–June 1967

Jan 7, 1967

Dear Jeff

Thanks very much for your long letter, which I read with great interest. This question of creativity is certainly a difficult one to be clear about. I am glad that it is now one of your main interests, as otherwise, there would be very little chance of your ever learning much about it. And it is, in my view, the key question, not only in science and art, but also, in the whole of life. But we shall have to come to this fact by stages.

Firstly, let us consider the mathematical "game" that you gave as a first example of DROPS.[1] Of course, you were right eventually to give this line of approach up, because creativity is not merely a game (though it includes a kind of "play" as a part of what it is). Even more significant, the example is drawn from too narrow and fragmentary a field. That is why the abstraction of DROPS seems so arbitrary. In a narrow field, there is always an almost unlimited set of DROPS that might fit the facts. As the field is broadened, this set begins to narrow down. And if we include the whole of life, all that is perceived at a given moment, then in a certain sense that I shall explain later, the arbitrariness is gone.

But first, let me emphasize that any set of DROPS is always provisional. It must be tested by carrying out inferences drawn from it, and relating these to the observed fact. This is not the same as conventionalism. It is not merely that it is convenient to abstract the known DROPS in a certain way. When this abstraction is carried out with deep insight, then it leads to concepts, permitting a wide range of correct inferences to be drawn, very often in totally unexpected ways. But deep insight depends on a psychological state of mind, in which the problem touches the whole of a person very

[1] See Appendix C—CT.

C. Talbot (ed.), *David Bohm's Critique of Modern Physics*,
https://doi.org/10.1007/978-3-030-45537-8_3

deeply. Thus Einstein's new insights into space and time did not come mainly from abstractions of DROPS in technical data. Rather, from deep reflection in a wide range of experience in everyday life and in the laboratory, Einstein began to feel that the speed of the observer should not be very significant, in determining the fundamental and universal DROPS that are relevant to the basic constitution and motion of matter. As I indicated in my Relativity book, this is merely an extension of the Copernican notion that the location of the observer also does not matter in this field. And if one goes still deeper, this is in turn an extension of the notion that generally speaking, the peculiarities of the individual observer have little or no bearing on the essential, fundamental and universal characteristics of reality.

Now, would you say that this latter notion is only a convenient convention? Surely, it is a very well confirmed fact that the peculiarities of individual observers have no deep significance. The longer we live, the more we see this fact exemplified in different ways. But just how do we become aware of this fact? Evidently, this happens in a way that is too subtle and deep to specify in much detail. Nevertheless, this is not merely a metaphysical notion (although, of course, it is a part of your general metaphysics).

Now, let us come to your second point. You say that what I said about perception, creative and destructive, seems metaphysical. Again, let me say that this is because one is starting from too narrow a field, i.e., science (or even science and art). It is necessary to begin with the whole fact, involving all of life, and then to focus down on various aspects of this fact, being aware that we have focused on an aspect, as we do so. Otherwise, we will mistake a part for the whole. And this part will seem arbitrary, because what really makes it inevitable will be just the rest of the whole, that we have forgotten about, as inferred to be "irrelevant to these questions".

Now, I say that this notion of "facting", "understanding" etc. is primary and comes before the result, is evident in every phase of life. It only takes a certain attention, awareness, interest, and sensitivity to see it. Without these qualities, no amount of philosophy has any meaning. It will all be just arbitrary "metaphysics".

You can begin, for example, with the fact about flattery. This is a malfunction that arises in human relationships. So how does one meet it properly? Evidently, one must be sensitive and aware, paying attention to what one is doing, and seeing what happens, as a result of what one does. Usually, we don't do this at all, and thus get caught in the trap, either of accepting flattery because it pleases us, or flattering someone else to obtain an advantage. But what is it that we must see? First of all, it is the difference between real joy and the false pleasure that comes when one has a pleasing image of the "self". Secondly, it is the relationship between this false pleasure and a general tendency to falsify perception, so as to increase this false pleasure. Thirdly, it is the order of cause and effect, in which failure to differentiate leads to a whole series of mistakes, each one piling up on top of the next. Then, it is the general pattern of one's life, in which one's drive towards status, security, the appearance of "success", gratification of various cravings, etc., becomes a major feature of the "self". And then, it is seen that this pattern is abstracted from a gigantic structure, involving the individual and society, which is almost totally wrong, self-defeating and poisonous. So each example of flattery is only a tiny aspect of a deep and gigantic structural process, in which all of us are submerged, for most of the time.

What I have just described is the act of learning. As we engage in relationships with nature, with other people, and with ideas, we always pay careful attention to what we do and to what happens, and especially, to the discrepancies between our apparent intentions and the actual results of our actions. Each discrepancy points to something wrong in our basic assumptions, which has to be corrected, with the aid of intelligent perception and insight. This is surely the proper way to do scientific research and what is not generally realized is that it is the proper way to approach the whole of life. Unfortunately, there is a common tacit assumption that we learn in order to live better. But the truth is the other way. The "good life" can only be an outcome of a spirit in which unending learning is taken to be the very essence of life. In each relationship, it is always necessary to be learning how we approach it with the wrong DROPS drawn from false generalization of inferences made in the past (either tacitly or explicitly). When the error in the DROPS is seen, then one discovers that there is a sensitive, intelligent creative perception, which reveals the new order of approach that is appropriate. This is always provisional. Yet, generally speaking, it tends to be right, for the most part. And where it is wrong, it is corrected by an extension of the process of learning.

Now, can you understand this by conventionalism or by traditional approaches? Is it only metaphysics? I would say that all of this is a fact, which any man can establish, if he is at all sensitive, observant, aware, attentive and interested. (If he is not any of these, then he may as well give up trying to learn about such deep questions.) It is a fact that perception of what is new and original is always a process of learning, in everyday life, in science, in art, and in every other field. Only routine and mechanical perception can be carried out by manipulating known DROPS and making "choices" among these. So we are in complete agreement on this point. Evidently, one cannot properly meet flattery, for example, by a "conventionalist", "traditionalist", or "metaphysical" approach. Rather, when one actually meets it, one comes into direct awareness contact with the mind, creatively "sorting out" the DROPS of the whole process. However, many people feel that such questions are "irrelevant" to science. In my view, this "specialist" approach is one of the most destructive poisons of our era. It takes only a little attention to see that real scientific progress involves the same kind of learning that is involved in properly meeting the fact of flattery.

Another principal difficulty of our era is overemphasis on the role of words (and formulae). The example that you gave of the "numbers game" is a case of this difficulty. Evidently the whole problem is mainly a verbal one. But as I indicated in the letter to Schumacher, verbal DROPS have little significance, unless they either reflect or "point to" a vast field of non-verbal DROPS, either external to us, or in the mind, or in both. It is in this vast field that the arbitrariness of our actions is seen to be removed. And this must include the psychological field as well. For after all, one's psychological attitudes act as "major premisses" in all our reasoning. For example, one is trained to do physics with the tacit premiss: "The purpose of this operation is to give me status, security, prestige, success, etc., etc." If one is at all observant, he can factually see how this premiss is in the long run, far more significant in determining our "choice" of hypothesis than are the experimental data or theoretical conclusions that have been drawn in the past. What one generally does not notice is this: When all the tacit psychological premisses are taken into account, along with the other premisses, then our DROPS are no longer arbitrary in any way at all. They only seem arbitrary, because we tacitly assign our psychological premisses to an imaginary entity called "I", "me", or "the subject" who is assumed to be "completely free" and to "stand above" the field about which he is thinking. We don't notice that all of thought is one field, whether this be the thought of the electron, the thought of getting a better job, or the thought of eating a satisfying dinner. It is in this field that the DROPS of our thought are determined, as anyone can verify directly and factually, if he will only pay fairly careful attention to what is actually happening, as he is in the process of thinking.

Evidently, then, what is called for is that everyone, scientists included, shall be always learning about the whole fact, not only what is external to man, but also about the DROPS with which man approaches the fact. When he sees how these are inappropriate, there is a sensitive creative intelligence that can "feel out" and "explore" new DROPS provisionally, until a set is found that is appropriate. This, I say, is a fact, that anyone can observe by paying proper attention to what happens when he faces a somewhat new situation.

In each instance, one's ideas are determined by the total field of thought, whatever it is at that moment. Careful observation of the DROPS of thought leads to their being changed, in the way described above. Clear thinking demands that our intention be to establish DROPS that are appropriate, in each case. Of course, man can never be perfect. Indeed, perfection would mean to be following a pattern, and therefore to be in a wrong order of mental operation. So we cannot reasonably expect that any man will ever be totally free of arbitrary "subjective" features, among the factors that determine his thinking. But an enormous improvement is evidently possible, in this regard.

Now, to come to Existentialism, about which the remarks of Jammer,[2] quoted by you, seemed very interesting. The Existentialists do see the fact that man cannot by thought alone really determine the whole of life. This is not only because thought is always concerned with parts, aspects, and fragments of totality. It is also because one does not know all the tacit features of the "self" that are actually the major premises, determining the general pattern and structure of this thought. But the Existentialists are totally wrong in saying that this fact calls for an act of "decision" or "choice" that seems to be irrational and unfounded in reason. For in reality, "irrational" decisions are determined mechanically by our conditioning, which leads to our perceiving through a tacit set of DROPS. Thus, a man who is sick of the dull deadly pattern of daily life may react by "going berserk" or by engaging in "wild orgies" of pleasure. But in doing this, he has merely surrendered his whole being to the residues of his childish conditioning. Far from being free, he is now a slave of even more mechanical impulses than are those that cause him to conform to the boring routine of everyday

[2]Jammer (1966), pp. 185–189 (information from Jeffrey Bub)—CT.

life. Or if he suddenly "chooses" a new religion or political party, he is still a slave of the conditioning that makes these seem attractive and right. In such a situation, no "choice" or "decision" has any real meaning.

Now, what the Existentialist is tacitly seeking is a meaningful kind of freedom. This is not arbitrariness of "disorder", nor is it the arbitrariness of an imposed pattern of order. Rather, true freedom means to realize "what has to be", because of the very order of external nature and human nature. But this is meaningless, unless one can first learn what is wrong or arbitrary about one's present state of order. So what the Existentialist leaves out is the need to be always learning how what is called the "self" is imposing arbitrary DROPS on the whole field of awareness. Then the Existentialist complains that it is all "arbitrary", not noticing that it is the very act of apparently "observing" from a separate "self" that has introduced the "arbitrariness" in the first place (Of course, it is not really arbitrary, but in fact determined by the tacit assumptions that are mistaken for inferences from the supposed character of the apparent "self"). As one learns how all this is taking place, then the mind is freed of arbitrary DROPS, because they cease to be confused with DROPS that have been seen to cohere with the whole fact. So fundamentally what is needed is to learn the difference between arbitrary DROPS imposed by past conditioning, and genuinely appropriate DROPS that cohere with the whole fact. This requires sensitivity, interest, attention, and awareness. (If any "philosopher" feels that he doesn't know what these are, then it would be best, in my view, if he left this field altogether.)

The existentialist is right in saying that no metaphysical framework is adequate to determine the order of what is. But he is wrong, in saying that one cannot always be learning about this order. The fact is that everybody, whether he likes it or not, cannot do other than have a metaphysical framework, either tacit or explicit, relating to this order. For metaphysics is just the name for one's most general structural assumptions about nature, society, the "self", the universe, God, or whatever else one feels to be deeply significant. All of one's thinking tends to be shaped by tacit metaphysical premisses. (E.g., the conventionalist's metaphysics is the assumption that our theories are only convenient conventions.) What we can do is first to be always observing and learning how our thinking is in fact conditioned by metaphysics.

But then, is all metaphysics necessarily destructive? I do not think so. Rather it seems to me that one can distinguish two kinds of metaphysics. Firstly, there are the general structural assumptions that we know we have, or that we are at least ready to learn about, and to change if they should not cohere with the fact. Then there are those that we don't know about, and don't want to learn about, because to do so might seem to threaten the order in which the "self" appears to be founded. These two kinds we may call "useful" metaphysics and "destructive" metaphysics. Many philosophers, reacting against destructive metaphysics, failed to distinguish these from the useful kind, and thus attempted to do the impossible – i.e., to get rid of all metaphysics.

Why do we need metaphysical assumptions? One reason is that (as demonstrated recently by science), interest and attention are determined by the thalamus and the reticular structure at the base of the brain. This structure does not respond directly to abstract thought. Rather, it responds mainly to feelings and images, which contain the essence of the DROPS of abstract thought. If one is to do creative work, it is necessary that the thalamus shall collaborate whole-heartedly and fully. That is why a good conceptual "model" that presents the DROPS of thought to the thalamus is so necessary. For if interest and attention has to be determined by laboriously working out all the inferences of abstract thought, one will not be able to deal with subtle and difficult questions, requiring rapid, accurate and penetrating perception.

In classical physics, one was at least able to present the general structure of thought to the thalamus in terms of mechanical "pictures", but in quantum theory, this possibility is largely gone. But even in classical physics, the general metaphysical "model" is very far from what is experienced "intuitively" in direct and immediate perception. Thus, if one looks at a stream, it is easy to see that it is constituted of a certain substance, "water", that one has known in other contexts. But no one can directly and "intuitively" appreciate the atomic constitution of the stream. So we have a situation where our metaphysics and our intuition fail to coalesce. This is very harmful, as it causes man to fragment, either into an "intuitive" type that rejects reasoning and logic, or into a "hard-headed" type that rejects the intuition and the deeper feelings that are our basic contact with reality. The wholeness of man demands a new kind of metaphysics that coalesces in its general structure represented in direct and immediate perception. But this in turn will require careful attention to what we actually see in direct and immediate perception. At present, we seldom distinguish what is actually seen from the thalamic "display" of the content of thought, in the form of images and feelings. This failure to make a proper distinction leads us to confuse the influences of thought with directly perceived reality – a comparison that is the ultimate source of all illusion and delusion. So to <u>learn</u> about our metaphysics and our direct and immediate perception has almost unimaginable potential significance. For if mankind were even <u>relatively</u> free of illusion and delusion, this would in itself <u>be</u> a revolution that would put all other known social change in the shade, (as would happen to a candle placed in the direct sunlight).

It is perhaps best to approach the metaphysical question indirectly and obliquely, because our "direct" approach would be determined by the DROPS of past conditioning, and therefore false. Indeed, we can usefully approach this question by considering your third point – i.e., the suggestion that man's constructive "architectural" activities are essential, to be taken into account, if one is to avoid a "metaphysical" (in the bad sense) version of creativity.

First of all the word "architecture" is a bit too specialised for the meaning that you wish to assign to it. Biederman once pointed out to me that the ancient Greeks had a word "Techne" meaning an indissoluble union of science and art, working in all of man's "constructive" activities. (Note how different this is from technology.)

But then, could we accept your assumption that inventive art and science are two aspects of what could be called "Pure Techne"? You must ask Biederman about the situation in art. But in science, can you imagine that even in an ideal society, there would no longer be such a thing as "Pure Science", whose culmination was taken to be discovery of new DROPS and which was regarded as crowned by factual proof that the inferences from these DROPS are correct? In such research, the aspect of utility for Techne would still be a by-product of the discoveries. In other words, scientific research is most deeply a part of the process of learning, which is intrinsic to human existence. Out of this learning, man may discover the emergence of a creative intention, in the field of Techne. But in my view, one should not say that one learns in order to create something. Rather, learning is the foundation of the life of the psyche, as breathing is of the life of the body. And as Beiderman has frequently pointed out, what one learns about nature and about man (including the mind) is the very "soil" in which creation can emerge, whether in Techne, in art, or in relationships in everyday life (e.g., parent and child). It is meaningless to try to create something new and original in a field that you do not understand. If you are an artist, you must learn about light, space, color, structure, and form. If your field is Techne, you must understand all this, and science (including sociology) as well. If your field is the raising of children, you must understand human nature, body and mind, in addition.

To sum up, learning is itself a creative activity of intelligent perception and it is the true foundation of all other orders of creativity. We do not learn in order to create any more than we breathe in order to create. Of course, our learning process is now diseased, suffering from various functional disorders, so that it is often painful (as happens with breathing when the lungs are diseased). But when the mind is healthy, learning is natural and joyful, as is breathing fresh pure mountain air.

One difficulty is that we seldom notice how we actually do learn from the DROPS of immediate perception. This has become so dull and automatic that we don't even notice how it takes place. In order to reveal what happens more clearly, I propose an experiment.

Get hold of a pencil or a long stick. Grasp it tightly and close your eyes. Start to "feel out" objects in your neighbourhood. Notice that the stick now seems to be an extension of "the observer". Get up and start walking, "tapping" your way along, like a blind man. Notice how the mind pays little attention to the stick as an object. Rather, one is interested almost entirely in DROPS of the tapping. It goes: Similar, similar, similar, different. Then similar, similar, similar, different. Each difference denotes an edge or change in shape or orientation of the surface that the stick is contacting. Notice now, how a vague conception or "image" of the room is slowly being "constructed" in the brain. It is very striking to "see" the objects emerging into consciousness. Some of them are proved to be "wrong" by further tapping, and these disappear. If you blindfold yourself and enter a room into which someone else has introduced objects unknown to you, it all grows much more interesting. Try it for an hour or so, and you will see that the "space" of the room has been pretty well constructed in the mind. Your muscles begin confidently to be guided by the "space", with the main "objects" in it. You see how action starts from this mental "space", and how the brain is always looking for discrepancies between the DROPS implied by this "space" and the ones that are actually experienced.

All of this is a plain fact. It is not explained conventionalism, traditionalism, or any form of metaphysics. You see clearly that the "space" is constructed in the mind. Yet it is not merely a convention. For it serves as an accurate "map", from which inferences of the DROPS observable with the stick can correctly be made. After all, a map has certain conventional features, but yet, it reflects DROPS that are not purely conventional.

It is not a big step to see that optical perception works in a similar way. However, the operations that built up our optical "space" in the mind took place mainly in the first few years of life. So we have forgotten all about it. Indeed, we habitually see in terms of our conditioned notions of space, not even realising that we are thus conditioned. And all scientific research is also evidently a basically similar process.

One can see in the "tapping experiment" that one is tacitly dividing the world into two parts. One of these is the body plus the stick. The other is the rest of the world. The first of these is the "subject", and the second is the "object". In the "subject", the brain is mainly interested not in the essential DROPS, but rather, in those that are, for it, "accidental" (e.g., the pattern of the tapping). It is these "accidental" DROPS that reflect the intrinsic DROPS of the "object". When the brain "constructs" the latter, the "subject" is largely tacit and implicit, being vaguely indicated only by the "feel" of the body and the stick.

One can generalize this. In the "subject", one is aware mainly of DROPS that are accidental to it, but that reflect something else called the "object". The division of subject and object is freely movable, because the brain can look at any object (e.g., a photograph), and treat its DROPS as reflective of something else (e.g., the person portrayed in the picture). Indeed, most generally, the "subject" is that aspect of awareness whose DROPS are mainly reflective of something else. This reflection is however active. That is, it includes the outgoing movements needed to produce the DROPS in question, as well as the incoming sense impressions that carry these DROPS as "accidental" features. (i.e., not necessary in the limited context of the body by itself).

Now, as you look at the world, you can treat the skin as the boundary between subject and object. But (remembering the "stick"), you can regard the light as an extension of the body, which is somehow "probing" the environment. You are not interested in the DROPS of "light itself," (e.g. Maxwell's equations), but only in how the DROPS of light reflect those of the "object". So the space is now seen as part of the "subject" while what you see in the space are the "objects" (as if the light were a set of "sticks" tapping against the objects). But now you can go further. Even the objects can be seen primarily through their DROPS. The whole universe is their "subject". But if the whole universe is "subject", is there a separate observer! Evidently not.

But now, the "Gestalt" can switch. Suppose we consider that the whole universe, including the body, is the "object". Once again, according to the point of view that I am proposing, all objects, from macro-objects to atoms to electrons – etc. are, most deeply, determined by their DROPS. There is no fundamental level. It is all DROPS. Therefore, in a deep sense, the "subjective" and the "objective" coalesce. True, their DROPS are different. But in any case, all DROPS are different. That is their most basic characteristic. In being different, subject and object are similar to everything else.

The main new point is that we cease to regard any particular set of DROPS as basic, whether these be atoms, or neutrinos, mind or matter. In a certain way, all DROPS are similar, in that they are DROPS. Therefore, when one is aware most strongly of the character of immediate and direct perception as DROPS, one is in direct and immediate contact with what may perhaps be called the fundamental "energy-substance" of the universe, in all of its manifestations and aspects. Everything is DROPS. With regard to what we are, it then follows that we are what everything is. It is as if every mental event were a direct contact with the essence of existence. Indeed, we can say that the totality of appearance (not merely its results, but also its process) is the essence of existence. For appearance in awareness is the creation of DROPS. But every form of existence is the creation of DROPS. So we can get a kind of direct "intuitive" feeling for our basic metaphysical notions. By assuming in our metaphysics "All is DROPS" we have removed the gulf between immediate intuitive perception and our basic metaphysical notions.

But to exist is not only to be DROPS. It is also to reflect other DROPS. Thus, the mind reflects what transcends awareness. In this regard, its reflection is limited and particular. That is, although, in a way, we are everything, we do not know everything.

Nevertheless, the mind can reflect certain relatively universal features of DROPS. Mind can also reflect on itself. In this respect, mind is like a flame, fed by the brain. Each flame has particular features. But in being a "flame", it is universal, – just "mind". Different people can observe different "samples" of "mind", that are "fed" by different brains. As different observers in different laboratories can study different samples of sodium, and discuss their results, coming to a common understanding of properties of sodium, something similar can happen when each person is looking at the "sample" of mind accessible to him, and communicating what he sees to others who are doing the same. So, in a way, a "science of mind" is possible. But here, the only instrument of perception is "mind itself". This science requires that "mind" be of a very high order of clarity and purity of perception. Most scientists, in their present states of mind, do not constitute sufficiently good "instruments" to allow the necessary observations to be carried out.

In a way, a "science of mind" would also have to be an "art of mind". For mind is so subtle and dynamic that in each manifestation, it has individual features of its DROPS that are like those of a work of art.

Understanding all this, let us now go on to see what it means to be aware of DROPS as the foundation of perception, experience, and even all existence. When you are listening to music, notice that the DROPS in the sounds correspond to DROPS in the feelings. Thus, in a drama, music may be played which stirs up feelings of fear or joy. Every feeling has its set of DROPS. Observe your feelings of fear and listen to them, as you would listen to a new piece of music. You will see that there is a pattern and rhythm in them, which is very interesting. The pattern of steady background and sudden "vibrating thrills" or "jolts" is expressed in the music that is used in the film to stimulate fear in the viewer. Or listen to Beethoven's "Ode to Joy" in the Choral Symphony. Watch how the feelings are always climbing in an ever rising hierarchy, moving up – up – up toward an all encompassing structure of indescribable and unimaginable beauty. Surely, Beethoven must have perceived all this and much more, when he wrote his music.

As you open up to the perception of all these DROPS in the feelings, you say that in turn, the feelings are "carrying" the DROPS of some yet deeper and more encompassing order of reality. This deeper level may also carry the DROPS of something that transcends it. Then, with a tremendous energy, one may have the impression of a vast totality that is the reality behind it all. Were not people like Einstein perhaps moved by such a perception? In this totality, subject and object seem trivial.

Now, let us look at nature in a similar way. Science has shown that the nerves in the brain have an immediate response in perception, which is followed a short time later by the response of memory, which shapes perception in terms of DROPS that are known. A young infant must be strongly aware of the DROPS of immediate perception, which in a way, might resemble what could be called "visual music". As we grow older, the response of memory gets stronger, until we hardly notice the DROPS of direct and immediate perception. Is it possible, however, to be sensitive, once again, to the structure of what we perceive, regarded as a kind of "music", rather than as "information", permitting the "construction" of the perceived "object" in awareness? Look at a tree and get a feeling for its structure, for the way in which light and space are essential to this structure. Be aware of how the word "tree" is always projecting a routine, mechanized structure into perception. See if it is possible not to think the word "tree". Rather, just be sensitive to the optically perceptible DROPS and let these work in the mind freely. Notice how the distinction of "self" and "tree" or "nature" begins to fade out. As happened with the music, one sees that the feelings are "carrying" the DROPS of nature. One can at least imagine how this can lead to a deeper reality that is the totality behind it all. However, because man's vision is even more deeply and broadly bound up with his whole mind than is his hearing, the effect may be much greater than that of music.

Very probably, this is the mode of perception of the real artist. But why should it be restricted to the artist? Do we not all need it in every phase of life? Can one be aware of another person through the totality of DROPS, rather than through images and stereotypes from past conditioning? Imagine what it would be if two people were aware of each other in this way. All barriers of "self" and "other" would come down. This would be such a revolution that we can hardly conceive of it.

Then we can be aware of scientific fact in this way. It is not merely "information" that may be useful for various purposes, or that may be the foundation of future achievements. It is, like music, a set of DROPS. One sees its beauty and ugliness, its harmonies and disharmonies, its truth and falsity. But there is no "self" who is "doing the looking". Rather, observation is just "going on".

When the division of subject and object ceases to be given top priority in awareness, then the DROPS of the mind are determined creatively, in a way that is appropriate to the whole fact, inward and outward. Indeed, as indicated earlier, these DROPS are always determined by the total content of awareness. But if this content includes the "false information" that is being observed by a separate "self" of sublime beauty and supreme value, then the brain will naturally produce a set of correspondingly false and arbitrary DROPS. To the extent that this illusion producing mechanism is not operating, the DROPS are appropriate to the total content of the moment. This includes not only direct and immediate perception, but also, all that we know, at that moment. The brain is aware of contradictions and disharmonies in this knowledge. So it is always spontaneously and naturally learning about them, thus coming to create new DROPS that are more nearly appropriate.

After all, as I indicated in my Relativity book, even science is primarily a mode of perception, rather than an accumulation of knowledge. What has been learned is often useful and relevant. But the essence of the life of the mind is the movement and process of learning, itself. When this stops, mind starts to deteriorate, no matter how much it may already know. Moreover it is meaningless for man to hope to know the totality of all existence, or even its essence. It is enough for him that he is of the same "energy-substance" as this totality. So he is of the essence of totality, and in particular, the DROPS of immediate perception are his most direct contact with the essential nature of this totality.

I hope that this begins to indicate what I mean by the role of DROPS in creation.

Up till now, I have been sending copies of these letters to Biederman. Since the typist is now rather busy, I am sending this to you directly. Perhaps, after you have read it and thought about it for a while, you could let Biederman have it, until he has time to read it.

With best regards

David Bohm

P.S. When I say that our most direct contact with the nature of reality is in direct perception, I refer not only to external sensual perception, but also, to "inner" perception of the state of mind, its feelings and thoughts, etc. For these too are apprehended in terms of DROPS. Indeed, one can go on and say that in the act of understanding, the mind creatively "grasps" or com-prehends a new totality of DROPS, in many cases referring more to the over-all structure of thought than to what is perceived sensually. In my view, the act of understanding is based on the immediate and direct response of the mind to the DROPS reflecting its own state of perception and thought. This is to be distinguished from the response of memory, which comes from the conditioning, and which is, generally speaking, not creative. Rather, it is at best adaptive and inventive. (Invention is the application of known means to new problems or the combination of known means in new ways – in a certain sense, a kind of synthesis.) Creation is of an entirely different order of mental process, immeasurably transcending mere invention.

This raises an interesting question. Is inventive art creative? Evidently, it can be creative. The aspect of invention seems to me to be the process of always trying out new combinations of things, new approaches, etc., to make something that is novel. As such, it is like the synthetic aspect of language and thought. Of course there is always the "analytic" approach needed to test what has been invented or synthesised. But something creative is not merely an invented novelty. Rather, it has the aspect of unity and totality of its DROPS, which can come only from deeper levels and higher orders of mental operations.

Biederman is able to go much more thoroughly into the creative character of inventive art than I can do. In any case, this letter is already so long that it would not be useful to extend it in this way.

Jan 28, 1967

Dear Jeff

In answer to your recent letter, I am glad to hear of your proposed research programme.

I enclose a copy of the paper in the Japanese Journal.[3] Part II is not written yet. I expect within a year or less to publish the mathematical basis, with some of which you are familiar.

Best regards

D Bohm

Feb 22, 1967

Dear Jeff

Thank you very much for your letter. I have read the proposals for your research program and think that they are very good. I have also sent the recommendation requested by you to the National Science Foundation.

What you say about Bohr in your letter is very pertinent. It is true that he saw that the quantum implies the wholeness of physical phenomena. We tried to show something similar in our Rev. Mod. Phys. articles when we suggested that the micro laws are dependent on macro environment. Neither point of view really gets the essence of what is meant by "indivisible wholeness", which I shall call henceforth by the name of "totality". Indeed, the trouble is that both the common language and mathematics are too crude and impoverished to allow the quality of totality to be discussed to any significant extent. Therefore, people are inclined to assume that "wholeness" is subjective and private, simply because they cannot communicate their genuine perceptions in the subject to other people. Thus, they fall into the confusion of assuming that only the parts and their consideration are real. But the consideration of parts is basically what is done in any good machine. So people are led to assume that only the mechanism is real, while all talk of wholeness and totality is at best a convenient fiction, introduced by the mind to deal practically with the vast complexity of what is in reality the universal mechanical process of external nature and human nature. And Bohr fell into basically the same trap, when he asserted that while reality was of the quality of wholeness, only the mechanical could be described in language. Therefore, he in effect condemned scientists to emphasize the computation

[3]Presumably Bohm (1965), also http://www5.bbk.ac.uk/lib/archive/bohm/BOHMB.149.pdf—CT.

of experimental results as the supreme value in science. For as long as one could in principle at least describe the mechanism, then one could get a glimpse, however inadequate, of how it worked in a coordinated way as a totality. This is indeed a pale and emasculated form of the genuine perception of wholeness or totality that we can get for example in listening properly to some of Beethoven's later quartets. Yet, it is at least something. By a curious irony of history, Bohr's wish to get a deeper intuition of wholeness in physics actually helped make it impossible for physicists even to get the intuitions that they had previously had, and left the latter with nothing to do but to seek wholeness in mathematical schemes of computation of the relationships between the fragmentary orders of different kinds of experimental results.

The key to really seeing wholeness in nature is both to enrich the common language on this subject, and to develop new forms of relevant mathematics. But today, physicists have an attitude which leads them to assume that the synthesis of existing ideas expressed in terms of existing languages will be adequate to the task. In this, they are very unlike Newton, who developed an entirely new mathematical language, the calculus, in order to express his new ideas on mechanics.

Now, the first step is to enrich the common language on the subject of order. I propose that a preliminary step in this direction to introduce the terms of similar differences and different similarities to describe order, as I explained in my Japanese article. So we can say a straight line is a curve of first order, determined by the first similarity in its differences. The circle is a curve of second order, the spiral of third order, etc. Thus we can go on to curves of ever greater "complexity" to curves of infinite order. Examples of these, I suggest, are the "chaotic" orbits of a particle in Brownian Motion.

Now, people generally say that these curves are "disordered". But it seems to me that the word "disorder" is a meaningless term, since it is impossible to have a curve with absolutely no order at all. Rather, the Brownian motion curve has a well defined order, which is infinite, in the sense that an infinite number of similar differences is needed to describe it. In addition, such a curve has statistical symmetry, in the sense that in the long run and the average, it spends nearly the same time in each unit volume of space accessible to it. All these features of the infinite order of Brownian motion are factual, communicable, and testable. They are in the no way subjective, private or personal. And they have nothing to do with "disorder", which would be an impossible situation, containing absolutely no order of any kind whatsoever. Indeed, the word "disorder" is an inappropriate name for the complex kind of order described above. This name is never useful and it is always a source of confusion. To remove this confusion, it is necessary to describe in each case the kind of order that is actually present. This can be done in terms of the language of similar differences and different similarities (as we use the language of units of length to communicate about measurements).

Now, in nature, we have a further kind of creative order. That is to say (e.g., in biology) each order has only approximate and relative symmetry. The breaks in symmetry of each order provide the basis of similar differences, that can constitute the symmetries of the next order. And so on in principle without limit. Thus, there is an unending series of hierarchies of order, leading to the evolution of new structures and new orders of structure. This order is potentially infinite. Thus, it resembles the random curve of Brownian motion. However, it is different, in that it does not lead to statistical symmetry, but rather, to creative evolution of new orders, structure, as described above.

In the evolution of order, there can be the growth of a state of harmony, within a structure of unified totality, within which all parts work together coherently. Or there can be a state of clash and conflict of the partial orders, leading to overall destruction and decay. (E.g., the unlimited growth of a cancer clashes with the order of the rest of the body, causing both the body and its cancer to die). Both harmony and conflict are factual real properties of the order, and are not merely private and subjective opinions.

Now, let us try to go further into developing a language for talking about creative order. To begin with, in physics, we are restricted today mainly to coordination (e.g., that is what are "coordinates" are doing – coordinating the movement of particles, to those of reference frames). As long as we stick to this language, we can never get out of mechanism, and into the field of totality and wholeness.

Consider instead the government of a country. To say that the prime minister coordinates his movements to those of each factory worker would be a very misleading way of talking. For in fact, there is a hierarchy of order of function. Unfortunately, this has gotten mixed up with a fictitious order of status. When the democratic revolution tried to overthrow the hierarchy of status, (without much success) it unfortunately confused the issue, by tending to imply that society could operate without a hierarchically organised order of function. The whole notion of hierarchy thus wrongly felt into disrepute. But this is clearly absurd.

Evidently, there is a two-fold stream of order moving hierarchically in every government. Thus the prime minister orders the general goals of his ministers, who order those of their department heads, and so on all the way down the line. This, I shall call downward ordination, or ordination of outgoing action. Then there is an upward flow of knowledge, going from one level to another. In each stage, there is a process of abstraction of essentials. On the basis of this abstraction, the downward ordinating action is guided and directed. So we have upward ordination, download ordination, and the links between them at each level.

I illustrate these below:

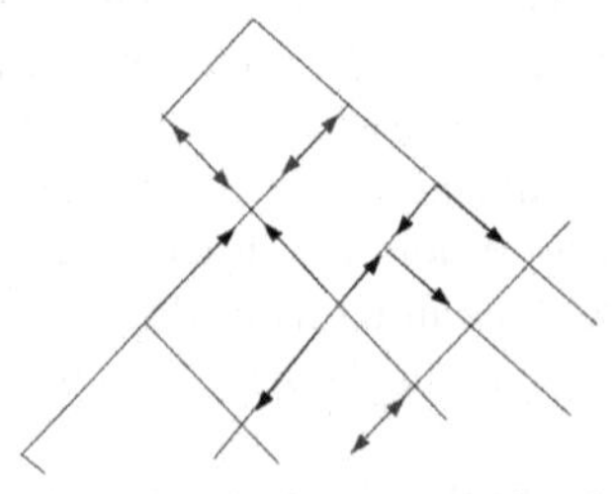

Now, if we look in any given level, the net result of all this is to produce what can be abstracted as ordered movement in that level alone. This I call co-ordination, or "sideways moving order". Basically, coordination is mainly the net result of up-ordination and down-ordination, but we tend to forget that we have abstracted it, so that it seems spuriously to have a kind of separate existence in itself.

Now, I would give to the whole process (up-ordination, down-ordination, their linkages, and coordination) the name "PANORDINATION". This is a new term, meaning "ordering of the whole" or "totality" (after the Greek word "Pan"). It is a quite new concept of order, not present today in mathematics, or in the common language.

To illustrate the new possibilities of this concept, note that the laws of physics thus far deal only with similar differences in the field of coordination (e.g. Newton's laws of motion). But now let us consider similar differences in the field of panordination. The simplest situation in coordinated movement is the free particle. Here, the differences in distance travelled in successive intervals of times are not only similar, but also, equal. Let us now look at panordination. The simplest situation would be to have not only similar differences in the structures of successive levels, but also, equal differences. This would produce a rather interesting looking hierarchy, that it would be nice to investigate. Of course, the real situation is more complex than this. But still, it would give us a nice paradigm case for study, especially to show how the coordination on each level is the outcome of the panordination, working through the upward and downward streams of ordination (E.g. as upward information and downward orders determine the "horizontal" coordination of a government department).

This view would imply that nature engages in abstraction in a way that is similar to man's, but different. I explained earlier that light rays abstract the whole structure of the environment, and even of the universe. But now, I propose that this abstracting movement is not contingent and accidental. Rather, it is universal and necessary – indeed the very essence of the order out of which all that there is is constituted.

This implies that nature has in it a series of orders of abstraction, plus the ability for outgoing action to be "guided" by these abstractions, that is akin to intelligent perception (similar but different). So no observer is needed. Man himself is a yet higher order aspect of this stream of abstraction, leading to yet higher orders of panordination. Indeed, while man's government is panordinated to a finite degree of order, nature's process and man's mind are panordinated to at least a potentially infinite degree of order.

The extreme paradigm case of panordination was to have a constant difference of order between the levels. In this case, any single "horizontal" cross-section of coordination contains the whole story. Thus, it can correctly be abstracted as mechanical and coordinated. The human observer can come into the stream of panordination at any level. This corresponds in Heisenberg's language to the arbitrariness of the location of the "cut" between observer and what he observes. But in a deeper sense, this "cut" is very misleading. For the very essence of the law is the "vertical" movement of panordination, so that a "cut" makes a description of this law of panordination impossible. Only coordination and mechanism are then left to us.

As soon as the differences of levels are no longer constant, panordination takes on a new content, which is essentially quantum theoretical rather than classical and mechanical. The usual case discussed in quantum theory is that as we consider successively higher levels of order, we approach the situation of constant differences. Eventually, then, at a suitably high order, it can be treated as classical. And here, the human observer can enter the stream of panordination without a significant effect on the lower levels.

However, in the lower levels, the laws are no longer those of mechanical coordination. Indeed, we can take two cross-sections. The "horizontal" discusses coordination – as Leibniz said, Space is the order of co-existence.[4] So the horizontal cross-section would refer to the space order. The vertical cross-section would refer to pan-orders, which transcend space and time.

Time is the order of successive existence of structures that are defined by their co-existence. So time is a mechanical concept of second order, space being of first order. In other words, until there are coordinated orders of coexistence, time has nothing to "sink its teeth into" in defining orders of successive existence.

My own feeling is that in simple cases, the pan order has to do with momentum space in quantum mechanics. This then determines velocity (space-time relationship) in the horizontal cross-section of co-order. But this determination is generally only statistical (e.g., as in the infinite order curve of Brownian motion).

To sum up, the concept of a hierarchy of discrete orders, panordinated into an undivided totality, has the following relations to quantum theory.

(1) Its discreteness is ultimately the origin of the discrete aspects of the quantum properties of matter.

(2) The creative origin of all "horizontal" mechanical co-orders in the "vertical" streams of pan order implies the indivisible wholeness of each structure, all the way up to the observer. Indeed, the observer is still part of the stream of pan orders.

(3) The wave-like properties are in the pan-order. Consider, for example, the structure of a government. Where is the government located in space? Does it "move through space" like a mechanical object? Actually, the space-location of a government resembles a wave. An action originating in one department spreads like a wave first to lower departments and then back up. In this movement, the government is not something that is located in the "government buildings", scattered throughout the country. For the movement of information and orders between the buildings is the very essence of what the government is. If we were conceptually to place a "cut" around certain government buildings, we would prevent ourselves from being able to understand the very essence of what is going on.

(4) The uncertainty principle is inherent in this view, in the sense that the more uniform the "vertical" order of panordination, the more the effects will "spread out" in lower government departments, so that the less precisely will we localize the activity on the levels of coordination. On the other hand, to try to localize the process of coordination is to make it impossible to determine the process of panordination as a simple one with constant differences.

(5) The statistical effects are in this point of view. For the panordinative process leads to coordinated processes of infinite order, resembling those of Brownian motion.

[4] Alexander (1956), pp. 69–70—CT.

Of course, this all has to be put in mathematical terms. I must emphasize that existing mathematics is too crude and limited to allow us to do this. We need a new mathematics. I have some general notions on the subject, which I shall send to you later. Meanwhile, let me say the following:

(1) The concept of "coordinate" is to be replaced by that of "panordinate," which refers to the vertical stream of order. Coordinates will be abstractions from panordinates.

(2) We must go beyond mere topology, or the theory of "places". We need to inquire into the order of space, i.e., the order of co-existence, and the order of time, (the order of successive existence).

(3) These orders are determined creatively, and not mechanically, as abstractions from the pan-order.

About our own Rev. Modern Phys. paper, I would say that the effect of the environment on the movement of the electron was a crude way of discussing pan-order. It contains, in effect, an ordering of the lower level quantum process by the macro-environment, but it fails to contain a clear expression of the process of abstraction of information about the micro-level at the macro-level.

I have recently given a talk on Creativity[5] to the Architects Association here. It will be stencilled soon, and then I'll send you a copy. Perhaps it will help clarify what I mean by the term.

I wonder if you could please also send this letter on to Biedermann, when you are through with it.

With best regards

David Bohm

P.S. I wish to add a few remarks on the "disturbance" due to an observation. Actually, since the lower orders are always being ordered by the higher orders, it is no disturbance at all, but instead, part of the normal order of movement. In this process, there is upward abstraction, symbolised by $(A \uparrow)$ and downward ordering action, symbolised by $(A \downarrow)$. Now $(A \downarrow)$ is dependent in some way on $(A \uparrow)$. But each $(A \downarrow)$ will produce a different $(A \uparrow)$ which in turn produces a different $(A \downarrow)$ and so on. Thus, there is a new order of successive "observations", which is simply not describable in the conventional quantum theory. It is this new order that will be the key test of the new point of view. We have to develop the mathematics far enough to see more or less what sort of order it should be.

This order is similar to (but different from) the order of intelligent perception. Here, the brain abstracts, then on the basis of this abstraction initiates an ordering action, directed outwardly. The results of this action are in turn abstracted, especially the difference between these results and inferences drawn from previous abstractions. This difference enables the brain to learn what it did not know before.

In a way, nature then has a kind of intelligent perception. Its evolutionary process therefore need not be a purely random one of "chance variations" followed by

[5] Bohm (1968) also Nichol (1998), pp. 1–18, or http://classes.dma.ucla.edu/Fall07/9-1/pdfs/week1/OnCreativity.pdf—CT.

"natural selection", and nothing else. Rather, nature can intelligently "learn" from its "mistakes" as we can.

If nature was able to lead to the human mind, with its intelligent perception, it is hard to believe that nature was, in the beginning, absolutely nothing more than a randomly functioning mechanism.

Of course, this does not imply that external nature is conscious. For consciousness is awareness of awareness. (Some people call it "awareness of self". But the term "self" is so confused that it means almost nothing.) Awareness of awareness implies a hierarchy of orders of abstraction, such that each order intelligently perceives and orders those below it. If this hierarchy is of a potentially infinite order, then there is consciousness.

It might be said that man is the highest known order of nature's awareness (i.e., intelligent perception). Or as Cezanne put it, man is the consciousness of nature.[6] This can be taken to mean that man is nature's consciousness, or that the content of one's consciousness is the whole of nature (including human nature). In my view, both meanings are essential parts of the whole truth.

To change the subject, I want to say a little more about the wave-particle properties of matter. In the pan-order, one might imagine a particular government department, coordinated to other departments in the same level. An action takes place there in a particular building. Because it is localised, we say it is like a "particle". The action drops to lower levels, where it spreads out like a wave, with complex inter-related movements resembling "interference". Then, in the abstracting movement, it comes up to another building, on the original level, where it is localized, thus resembling a "particle" once again. Thus, we have something like the particular interference experiment.

May 15, 1967

Dear Jeff

I have just returned from my trip to Israel, and am just now getting around to answering your letter of April 7.

I have read about Jung (and Pauli's) thoughts on synchronicity many years ago.[7] No doubt, there is something right about the idea. Nevertheless, it has so many confusing aspects that I was led to doubt that, on the whole, the idea is a useful one.

First of all, it seems to me that the notion of archetypes basically contradicts that of creativity. An archetype could perhaps be compared to a kind of hereditary conditioning. As such, it could be useful under some conditions, and a source of perceptual "blinding" under others, where the archetypes in question would not have

---

[6]This quote does not appear in Biederman's book on Cezanne, Biederman (1958), or in the published Bohm-Biederman correspondence, Pylkkänen (1999). However, as noted in the Introduction, the latter are only a small part of the Bohm-Biederman correspondence where Cezanne's views are often discussed, so it may appear there—CT.

[7]See Introduction pp. 4–5—CT.

an appropriate structure. But whether useful or harmful, archetypes would be, in essence, a denial of what is original and different in perception, and therefore a denial of creativity.

It is, in addition hard to know how much to believe about Jung's accounts of "meaningful coincidences", or else how much may be the result of perceptual distortion by wishful thinking on Jung's part.

Thirdly, I wonder whether meaning really comes from archetypal preconscious organization [of] your train of ideas. Perhaps conditional meanings arise in this way. But do new meanings arising in the act of understanding also arise in this way?

Finally I wonder whether von Franz[8] is right in saying, "I can watch disorder without doing anything about it". This raises the question "Is there such a thing as disorder, and if not, how can one watch it?" "If it is disorder, how can one establish order in it?" Don't forget that "disorder" means "no order at all". If there is a complex but as yet unknown order in it, it was never disordered. So I don't understand what von Franz is trying to say here.

I think the aspect of truth in Jung's ideas is that the causal order is insufficient, both for mind and for matter. But from this it does not follow that Jung's approach is a right one.

Even in classical physics the a-causal order appears in what are called "initial conditions". A tacit assumption of classical physics is that the a-causal order is completely contingent or arbitrary, while only the causal order is necessary and inevitable. On the other hand, it seems odd to suppose the arbitrariness of what is (i.e. the a-causal order existing at a given moment) while one supposes that the necessary order is merely the causal one, in which what is is followed by what will be. An alternative view of the world is that the order of what is is no more arbitrary than the causal order. However it has evidently to be determined by different principles. Thus, in a work of art, the order is determined by the principle of harmonious totality, and not by the causal order, describing how the picture was painted. The two orders do not contradict each other. But the causal order is basically determined by the order of the picture itself. Perhaps in the deepest sense, the order of external nature and the mind are determined in a somewhat similar way.

Anyone who said that a creative work of art was determined by archetypes would any fact be denying the reality of creation. The same holds true for the laws of mind and matter.

I think the solution is indicated by Biederman's approach to nature and to art. The order of nature is an infinitely rich one, with a tremendous number and variety of relative and limited symmetries. The breaks in these symmetries are the "elements" of the next order of symmetry. Thus, there is a dynamic hierarchy of symmetries of order.

Consider a tree for example. Look at it. You will see on each level a set of similar differences (in the leaves, the branches, etc.). The breaks in these similarities form the basis of the next level of order. And (in a forest) between different sets of trees

[8]Marie-Louise von Franz, a Swiss Jungian psychologist—CT.

we get more similar differences. Thus the whole forest is in a hierarchy of orders. And so on to the mountains, the sky, the clouds, etc.

Nature's order has in it the feature that each partial order is thus participating in an infinite hierarchy of orders. The principal is infinitely rich and yet harmonious.

On the other hand, most of man's structures are arranged in limited mechanical orders. For this reason they do not combine harmoniously with other man-made structures or with natural structures, to produce an infinite hierarchy of harmonious orders. Rather, they tend to clash, to produce an ugly and jarring impression of conflict.

Now, scientific thinking has thus far assumed that nature's stable forms are the "real reality" while all the rich orders that I have mentioned above are contingent and accidental, therefore of no deep significance. The causal laws are those that explain the evolution of nature's stable forms. These are assumed to represent what is really necessary and significant about the laws of nature. The same is assumed to hold for the as yet unknown laws of the mind. Such an assumption expresses itself naturally in mechanically ordered architecture, city planning, etc., which regards the solid mechanical forms and their causal function as the fundamental reality, and which ignores the creative aspect of nature's forms – i.e. the fact that they are in an infinitely rich order, such that, generally speaking, each aspect participates in an infinitely rich hierarchy of orders.

Now, I want to propose that this aspect of nature is even more significant than are the stable forms and their causal evolution. I propose that this character of nature is necessary and inevitable and indeed that eventually even the stable forms and their causal evolution can be understood through the infinitely rich order and its creative evolution.

I propose also that the basic order of the mind is similar to that of nature, in the respect described above. The mind also has stable forms, defined by thought. The order of succession of thought moves towards a logical order (analogous to, but different from the causal order of nature's forms). Modern culture assumes that thought with its logical order represents the fundamental reality of the mind. But I propose instead that there is a deeper creative order, similar to that of directly perceived nature. This is not the order of the archetypes of Jung, nor of the unconscious of Freud. Both these, if they exist at all, represent mechanically conditioned orders. Rather, it is an order such that each partial aspect of mental process tends to cohere with others in an infinitely rich hierarchy of orders.

How are mind and matter related? As is well known, the order of external nature functions in the brain through the sense organs and the nerves which abstract this function to ever higher levels. Thus nature's order, working in the brain, is an inseparable aspect of the total order of the mind. So when the mind is functioning properly, in a harmonious and total way, it will go on abstracting from the perceptual order, to ever higher levels. At each stage, the corresponding total structures in the mind will be infinitely rich hierarchies of order, harmonious and beautiful. The total harmony and beauty in this process is its meaning. Since nature's order is an inseparable part of the mind's order the "meaning" will refer as much to nature as to what is "in" the mind itself.

To understand this view, one must note that almost all previous activity in recorded history has been mechanical, conflicting, uncreative and essentially meaningless. Real creation has been very rare. In this regard, I suggest that you read my Creativity article carefully to see what I mean here.

It is possible in principle for man to be creative, not only in his immediate mental response to reality, but also in his outward actions (e.g. in science, art, or other fields). In such action, man creates material structures (e.g. in art) as structures, ideas (in science) each of whose aspects is in a vast and unlimited harmony, combining with others, in many levels of similar difference. In addition, these artistic structures are in harmony with nature's infinite structure of light, color, space, form. The scientific structures of ideas are in harmony with those aspects of nature that are perceived with the aid of instruments.

However I must emphasise that creativity has been very rare. Nevertheless, as I explain in the Creativity article it is in reality the necessary and proper norm for healthy human life. From this, it follows that healthy human life has also been rather rare.

I hope that this clarifies things a bit. Please let me hear what you think of these notions.

Regards

David Bohm

May 16, 1967

Dear Jeff

This is to supplement yesterday's letter.

Firstly, I want to emphasize the role of learning, in establishing the relationship of "mind" and "matter".

Now Jung's idea of synchronicity strikes me as a very mechanical solution of this problem. The "unus mundus"[9] imposes a certain order in matter and a corresponding order in the mind. If it were really all as simple as this, why would an active process of perception be needed? And how would error ever creep in? Does the "unus mundus" wish also to deceive us by setting up false correspondences? In Einstein's terminology, is God malicious?

In my view, it is necessary to go deeply into the role of learning (as I emphasized in the Creativity article). As far as we know, most knowledge requires an active process of learning based on sense perception. The kind of pre-conscious ordering of mind and external world that Jung speaks of is, at best, fairly rare, if indeed, it exists at all. Once we have learnt how to perceive something, then to perceive a similar thing, we may impose a previously learned order. It is possible that the human race

[9] i.e. "one world", the concept of an underlying unified reality from which everything emerges and to which everything returns. An idea popularized by Jung—CT.

learned certain forms of order long ago, which have since then been built into our hereditary mental structure. So there would be a pre conscious ordering of perception through something like an archetype. But the question of originality and creativity is precisely that of how we see something new, that has not been built into us, as a kind of archetype.

In my view, learning is a creative process. It "feels out" the perceived order, in an active way, by creating certain new orders and testing them against more directly perceived orders. Those orders that do not fit are discarded. That we learn from our mistakes and as we try out various orders, we note the order in which there is an improved fit. In other words, if we note that a certain difference of orders improves the fit, we try a similar difference in the next step. But as this order of fitting breaks down, the brain tries a different kind of difference. Thus, it is always moving toward an order that correctly abstracts the order in sense perception.

In this whole process, what is commonly called a "mistake" is actually an essential part of the process. No perception is "exactly right". Rather, by noticing what is wrong and what difference of order corrects this "mistake", we start to abstract the order of changing orders that is always moving in higher and higher levels toward approximate correspondence with orders in direct sense perception.

I feel that any theory of perception that does not incorporate the key role of "errors" in the development of correct perception is deficient in a basic way. As far as I can see, Jung's theory has no real place for the role of "mistakes". Does the "unus mundus" present us with mistaken correspondences between mind and matter? If so, what is the nature of these mistakes? Or even more important, what is the nature of the higher mental faculties that recognise these mistakes and correct them? If there are such faculties they imply an intelligence that transcends the "unus mundus". Therefore the "unus mundus" is not the totality. It is only one part, while the higher mental faculties are another part.

To change the topic. I do not understand the Chinese qualitative concept of number well enough to comment on it. There may be something in it. But even if there is, it does not follow from this that Jung's theory of archetypes renders an adequate account of the process of perception.

Best regards

David Bohm

May 18, 1967[10]

Dear Jeff

I would like to supplement my last two letters to you.

As I said before, I regard Jung's Theory of archetypes as too crude and mechanical to account for perception, especially its more creative aspects. But it might be useful

[10]First letter with this date—CT.

to show how it can be extended and transformed, so that it approaches what I regard as a more nearly adequate view. If you will focus, not on the resulting notions themselves, but on the hierarchy of similar differences that are revealed as they are developed, then you will perhaps see the directions in which I would like you to look.

Now, we first imagine the "unus mundus" creating a certain order in the external world and a corresponding order in the mind. As I indicated in earlier letters, this does not account for the role of "errors" or "mistakes" in this correspondence. Indeed, even to recognise such an error, there must be a higher order of intelligence that transcends the "unus mundus" and the orders that it creates. Thus, the "unus mundus" is seen to be itself nothing but an abstraction, created by this intelligence.

Can we get out of this division between the "unus mundus" and the perceiving intelligence, without assuming, with Bishop Berkeley, that the world is nothing but an aspect of the perceiver? We might at least try to do this by assuming that the "unus mundus" does not create perfect correspondence between mind and matter. But then, we assume also that in the mind, the unus mundus creates a hierarchy of correspondences, in which each level of correspondence of mind and matter is itself abstracted in a next order of correspondence. In this next order of correspondence, the unus mundus creates the order of truth and falsity, in such a way that some first order correspondences are felt to be true while others are felt to be false. But since this feeling is itself not always right, the unus mundus has also to create a third order correspondence, which orders the second order feelings of truth and falsity according to their truth and falsity. And so on ad infinitum.

Admittedly, this theory is very unsatisfactory. But it at least indicates the kind of problems there are in Jung's point of view. Let us now look at some of these problems.

Firstly, it seems wrong to attribute an infinite hierarchy of order to the mind while external nature has only a simple order on one level. How could the infinite hierarchy of mind evolve from a simple one level hierarchy of matter? So it seems necessary to assume that nature is also a hierarchy of infinite order, and that it contains correspondence between its parts and aspects that are in some ways similar to those of the mind. In other words, nature is itself already abstracting one part of itself in another, (e.g., as light rays through each point abstract the whole universe). And some of these abstractions provide relatively true correspondences of order, while others provide less true correspondences. So something akin to mind is already in the order of nature. Man's mind is an extension of this aspect of nature. One could say that nature is, in a certain sense, aware of itself, but that in being conscious, man is aware of this awareness. There is no sharp division of mind and matter. The intelligence in matter flows imperceptibly through light, heat, forces, etc., into the sensual nerves the brain, where it eventually reaches the quality of consciousness, which is similar to nature's abstractive intelligence, but also different. (Everything that exists is similar but different). Thus, there is no need to postulate an abstractive perceiving intelligence that is utterly distinct from the matter that is the object of its perceptions, and separated from the latter by an unbridgeable gulf.

Secondly, we must consider the fact that it requires a creative action for man to perceive the truth properly. It does not happen automatically and mechanically. It is not as if the unus mundus created the order of nature and the order of the mind in an infinite hierarchy that inevitably approached perfect correspondence in its totality. For then, how could we explain the fact that almost the whole human race in almost all of its history has been lost in confusion on almost all questions of deep importance, while only a few have only occasionally had correct and creative perception? Evidently we have to think of perception as a creative process, rather than as an automatically accomplished fact.

Here, we get into confusion over the notion of perfection. If man automatically had complete and perfect perception, he would be only a kind of machine. What is a essential to creative perception is not mainly the correspondence of perceived object to the order in the mind. This has a certain kind of importance, to be sure. But what is deeper and the very substance, energy, and order of perception is the movement in which man is learning to extend his perception in new ways, to new horizons, depths, etc. So what is the key point is not resultant adequacy of correspondence of order of mind and matter. This is necessary but it must come mainly as a by-product of a dynamic order in the mind, which contains a structure leading always to a movement toward truth. If man automatically had the truth, (e.g., from archetypes projected by the unus mundus), he could not have this principle of movement toward the truth. And if nature were finite in its order, man's mind could arrive at the complete truth, so that he would lose the possibility of realizing his deepest nature, which is to be in the order of moving toward truth. But because nature is infinite, this will not happen. And because the mind is infinite, it can always be in the state of moving toward truth, no matter how much it may already have learned in the past.

You may say that that all this is too vague, that you want a more accurate "pinning down" of the nature of creativity. But please remember that any definition that one gave for creativity would itself have to be created. The process of creation of this definition would inevitably transcend its mere result, i.e. the created definition. I wonder if our being in the academic profession does not cause us to give supreme value to results, because these can be pointed to, worked on, discussed, published, etc. Yet, this emphasis on results also leads to the mediocrity that is characteristic of most academic work.

What I'm trying to say is that you cannot do valid work on creativity without yourself being in the creative state about which you wish to talk. Many philosophers imagine that they can understand the essence of the work of a man like Einstein solely by analysing the results of this work. But in my view, unless they are in a state of creation similar in some ways to the state Einstein was in, they will miss the whole point and produce a superficial and mediocre piece of work, that is really not deeply related to what Einstein did at all.

In short, the main point is not to describe creation. Nor is it even to understand creation. Rather, the key point is to be creative. From this creative state, you can understand creation and describe it. But the purpose of such a description is mainly to help communicate to another what it means to be creative. It is not mainly to provide an interesting structure of ideas to talk about, as if it referred to something entirely external to the mind of the one to whom one is talking. In other words, when I talk of Einstein's creativity, what I really am referring to is creativity in me and in you. If this is not present, then to talk of Einstein's creativity is to make empty noises, that do not refer to anything at all. After all, what Einstein wrote is only the results of creativity, and not the creative process itself. Whatever I say about creativity (e.g., to refer it to the unus mundus) is also at best a result of some creative process. What is needed is to communicate, not merely the abstract structure of the results of creation, but also, the concrete reality of creation itself.

Now, I think that Biederman is suggesting that there is a non verbal way of looking at nature, which will directly show you the meaning of all this. Explanations are not enough. Eventually, one must come into direct perceptual contact with what is being referred to. Generally, we look at nature mechanically taking the stable created forms and their causal evolution to be the basic and fundamental substance, the "reality", while we regard the infinite hierarchy of similar differences leading to harmony and beauty as contingent, accidental, unimportant, etc. But it is possible to perceive directly that in our immediate contact with nature, it is this infinite hierarchy of ever-changing order that is the fundamental reality, while the stable forms and their causal evolution are abstracted from this hierarchy of orders of movement. To perceive this is to see what creation really is. One can then go on to perceive a similar but different creative process in the mind.

On the other hand, because of our professional training, we tend to proceed as if we could start purely from verbal definitions. Perhaps we vaguely think of some creative center in the mind, that would be creating the definitions. But we don't notice that an act of observation, of learning, is needed here. Actually, nobody knows what creation is. How are you going to find out? You must somehow perceive a creative process, at first non-verbally. And then, if you wish to communicate it, you will abstract it verbally, first at the descriptive level, and then at the inferential level. But where will you look for the creative process? You cannot find it, for example, by trying to observe the "unus mundus", because this is, at best, a verbal distinction. You can talk about an apple, describe it, make inferences about it. But if you are hungry, you will want a real apple to perceive, to handle, and to eat. If you are really "hungry" for creation, you won't be satisfied to talk about its being done by the unus mundus. Rather, you will want to pursue it directly, to come in contact with it, and assimilate it so that it is a part of your own very substance. Perhaps you can do this by seeing nature's creative process directly without words, and letting the perception "flow inward" to reveal the similar but different creative process of the mind.

If all of this is valid, you will have to think carefully of why you are doing research in this field of perception, in the first place. Do you wish to add your bit to the accumulation of man's "knowledge" of the creative process? If so, I suggest that your aim is confused. For whatever you say about the creative process will be empty words, unless it is the outcome of what is observed from the creative state, about the creative state. So one's first objective has to be to discover the creative state, to come upon it. I think that Biederman has discovered something of this, and that what he has to say about it is worth listening to with full attention.

Best regards

David Bohm

P.S. Perhaps I could now sum up my objections to Jung's ideas as follows: Jung suggests that you look for creativity manifesting itself in acausal synchronous combinations of events that are "meaningful". In my view, such events are at best, a very superficial manifestation of creativity, if indeed they even exist at all (which is not proven). I suggest that creation is always being demonstrated directly before your eyes in the immediate perception of nature, which has an a-causal synchronous order of an infinite hierarchy of similar differences and different similarities. Why not begin by learning what it means to be sensitive to this order, and to go from there to a sensitivity of the creative order of mind?

May 18, 1967[11]

Dear Jeff

This is just to add a bit to this morning's letter.

I wrote there suggesting a model in which the "unus mundus" creates a hierarchy of orders in the mind. The first order corresponds in some way to what is outside the mind. Let me add here the notion of a truth value as a domain of validity for each order of correspondence. In the second order, the "unus mundus" creates a certain "knowledge" of the domain of validity of the first order of correspondence. In the third order, there is knowledge of the domain of validity of the second order knowledge. And so on ad infinitum.

Now, this whole picture is still mechanical. For the "unus mundus" is just projecting a hierarchy of information into the mind, at the same time that it is projecting some hierarchic order into what is outside the mind. It is also projecting a hierarchy of knowledge about the truth, falsity, and domains of validity of its various projections.

In my view, the essential creative art of man is continually to see new limits to the domains of validity of his knowledge, and to create new orders in the mind that enable him to extend his perception in new ways. Even more, I have proposed that while the results of this extension are significant, what is most fundamental is the very movement of extension itself. It is this movement which is the creative process in the mind of man. The corresponding creative process in nature is the extension of evolution of natural process to new orders, and the continual creation of new orders of abstraction of one aspect of nature in another.

If we were to try to stick to the "unus mundus" hypothesis we could say that the "unus mundus" was always extending man's perceptions in new ways. But this would make it automatic and mechanical, hence not really creative. Moreover, it would not explain why most men for most of the time do not enjoy a creative extension of true perception, but instead engage in destructive extension of confusion of perception.

[11] Second letter with this date—CT.

Why does the “unus mundus” project ever increasing confusion into most people and extension of perception into a few?

I think that any theory of perception that leaves out error and confusion is so incomplete that it has missed the main point.

I think that if you follow the “unus mundus” hypotheses to its full conclusions, you see that it is not really an adequate one. Rather, eventually, you have to admit that creativity must first be perceived and experienced non-verbally, before you can meaningfully say much about it. In my letter of this morning, I tried to indicate further what this means.

Best regards

David Bohm

May 30, 1967

Dear Jeff

Thank you very much for your letter of May 25. On the whole I think it would be best if you discussed the artistic questions you raise there directly with Biederman, as a lot of it concerns his views on art. I will say a little about it, however.

Of course you are right to say that perception is unlimited in its possibilities of creative evolution. I am sure that Beiderman says the same. You are right to say that the mechanical recording of images is in itself a rather trivial goal (apart from its utilitarian significance). I would agree with much of what you say in your letter on this point. However, there are a few other points to keep in mind as well.

Firstly, whatever an artist sees in nature, if he is a mimetic artist, his goal must be to imitate what he sees. As his ability to do this improves, he eventually gets burdened down with a tremendous mass of of mechanical detail (moving the brush in certain ways, etc.) The invention of the camera made this unnecessary, and thus freed the artist to record what he sees by mechanical means. Naturally, better instruments have been invented (the movie camera, television etc.) and still better ones are called for. You are quite right to say this. But now, a creatively perceiving artist could use all these instruments to make a wonderful mimetic representation of what he sees. Unfortunately he rarely gets the opportunity to do so, because of the corrupt and confused nature of society. This latter insist mostly in the mechanical use of mimesis, either in news photography, in advertising, in propaganda or in stories and drama intended to “entertain”, i.e. to take people’s minds off the grim and ugly and dangerous nature of human reality, at least for a short time. So, very rarely does an artist get a chance to show us in color films and by other means what he really can see in nature. I have seen this attempted in a rather halting way, and even then, the result is very powerful and moving.

Can the artist by his use of a brush still make a contribution that the camera cannot? This seems questionable to me. Firstly, let us look at what such modern artists have actually done along these lines. I have never myself seen one who has portrayed some new aspect of reality to me. For example, my first reaction to Picasso was that this is very ugly indeed. Later, I tried to convince myself that it meant something, because it was taken seriously by so many people. But I can't say that I ever saw in Picasso anything that I did not already know before, and that I did not "read into" the picture from this knowledge.

Of course, I may be prejudiced, but in the last analysis, each person can only go by what he actually sees in a painting. When I think of Picasso's paintings, I see that he is in contradiction with his medium, and trying therefore to do what is actually impossible. As I see it, a painter works in a two dimensional structure of paint marks. With this, he tries to imitate the appearance of a three dimensional structure that he sees (or that could be seen). Everybody who looks at a picture understands this point. But when a painter (like Picasso) begins to break out of this tacit convention, then the interpretation of what is on the canvas is largely arbitrary. If I have complementarity in mind, I'll see that in the picture. If you have the idea of similarities and differences of woman A and B, that is what you will see. The main point is that you see only what is already in your mind. Picasso's picture becomes a vehicle for projecting your conditioning. Therefore, it is not a creative perception by the viewer. The same is true in essence about all of modern art.

On the other hand, a really creative work of art enables you to see what you have never seen before. Rembrandt's picture enables you to see aspects of human character that you never saw before, in just this way. You don't "read it" into the picture. Rather, it is just because the mimesis of three dimensional reality is so good that you can see in the picture an expression of what Rembrandt felt and perceived. Similarly, a man who makes a good film also expresses character by skillful use of modern means of mimesis.

Of course, Biederman is trying to do something different. He also enables you to see what you have never seen before. But this is not by imitating nature. Rather it is by creating in nature something that does not imitate what nature already created.

It is important to maintain a distinction between the two forms of art, as also there will be serious confusion. Each has its place.

When you engage in painting, the order of the brush marks on the canvas is very different from the order of what the person is to see. If you know that the brush marks aim to imitate the appearance of a three dimensional set of things then your brain knows tacitly how to translate the painted order, structure and form into the order, structure and form of objects that are in space and bathed in light. In other words, there is the possibility of a relatively unambiguous communication of what the artist has seen. Thus, the viewer can see more or less what the artist has seen, and does not have to introduce his own privately conditioned interpretation, which is different for each person, and which is irrelevant, in the sense that it is not a creative new perception, but rather, the imposition of something that one already knows.

To confuse mimetic creation and non-mimetic creation is therefore a very serious mistake. When one manipulates the contents of a painted image (as Picasso does) one alters the form and structure in an arbitrary way, so that the viewer has no real clue as to what led the artist to do just what he did and what he was trying to communicate. The viewer can only see whether what he sees is similar to some idea that is already known to him. It is impossible to communicate a new form, structure and order if one does not know how these are to be related to the form, structure and order, that one can see on the surface of the canvas.

This problem grows even more serious with some painters, who alter not merely the forms of objects, not merely their structures, but also the order of space itself. Some Surrealists do this to some extent, but people like Rothko specialise in making a depiction of spaces with highly ambiguous orders, (represented by changing shades of color). It is only your conditioning that can determine what you see in such pictures.

Not only is there the above described confusion, and ambiguity and arbitrariness in the mind of the viewer of a work of modern art. There is also a similar confusion in the mind of the painter. What is he actually trying to do? If he is trying to create something new, why does he still try to imitate the two dimensional appearance of nature's forms, structures, and orders? If you really were to create a new order, it could not be translated into two dimensions unless it was first present in three dimensional reality. And if people like Picasso try to transmit ideas like the similarity and difference of woman A and woman B, the proper vehicle for this is in words. To do it in a painting which inevitably evokes a three dimensional order, structure, and form of some kind is too ambiguous to be effective as a kind of communication.

Of course, you may argue that perhaps artists could make valid new discoveries in the field of mimesis by painting. In principle, this might perhaps be possible. But in fact, no real artist of great creativity (comparable to Rembrandt) has done this for over a century. Why not? Partly because mimesis had already reached the point when to proceed further by brushwork on canvas would have involved the artist in too much mechanical work. Another reason is that a man who wanted to express character as Rembrandt did would now do much better to make a film, than to try to paint a picture. The mechanical work is done by the camera and the artist can focus on what is to be seen and communicated. The camera is basically a much better mechanism for mimesis than is the artist's hand moving a brush (especially the cine camera).

Indeed, the really great artists of the last century have all been interested in questions that take them beyond mimesis. Of course, Biederman's attempt to continue their work in new ways is not the only possible direction. Biederman would surely admit this readily. But whatever the new direction may be, is it possible to mix mimesis and direct creation of something new? Biederman says it is not possible, and I think he is right. When you create something new within nature's structural process of form, space, color, light, etc, it is necessary that this creation shall not imitate something else. If it does, then you have no way of knowing what it is that has been created. Indeed, if you create something that imitates something else whose order, structure, and form are totally unknown and new, how can you find the thing that is being imitated, without first creating that too?

The scientist has a very different problem. Nature has already created the order, structure, and form that he studies. Thus, he has to create orders, structure and forms of ideas that reflect (or "imitate" in a certain sense) those that nature has already created. He does this when he explains the facts by means of a few assumptions. He tests his explanations by predicting new facts, that can be found in nature's structural process. He has, to some extent, to "make" these facts by creating new instruments. In this sense, his work is similar to that of a Structurist artist . But his creation here is intended mainly to test how well his ideas reflect reality. The Structurist artist is interested mainly in what has been created, and not mainly in testing whether his notions of nature's structural process are right.

I would ask you to tell me which modern artist has done work that is really creative and fairly free of confusion. Then I would ask you to see just what it is that has been created. What new perception did you get that you didn't "read into it"? Can you see anything new in Biederman's work that you didn't "read into it"? Did you see an order and structure that you had never seen before?

What has to be kept in mind is not only that perception evolves creatively, but also, that creative expression of this perception depends on harmonizing one's work with the laws (i.e., order and structure) of the medium in which one works. To try to create a vision of three dimensional reality that is contrary to nature's three dimensional order by manipulating paint marks on a two dimensional canvas leads only to confusion and not to creation of a new order. And to try to suggest verbal ideas (such as that my thoughts and memories of woman B are similar and different to those of woman A) by paint marks on canvas is also a source of confusion. For there is no clear correspondence between marks of paint and ideas about women. Any such correspondence that seems to exist is only the result of conditioning (e.g., what is done with images of women in advertising). It is not the natural result of a process in which the brain relates a two dimensional order to a three dimensional order.

So, in art, the distinction of mimetic and non-mimetic is a key one. It is not like the distinction of observer and what is observed, which is largely illusory. The key point is that mimesis involves a real fact – i.e., the (largely subtle and tacit) law of correspondence between the order of what is imitated and the order of what is "doing the imitating". The law does not permit arbitrary manipulation of the order of what is "doing the imitating" without confusion that is destructive. Modern artists generally do engage in such arbitrary manipulations.

To change the subject, Maxwell called me on the phone today, about my coming to Minnesota. I explained that I would like to come, but that there are difficulties. First, the department would have to make a strong effort to get me a visa. Secondly, at present, I can't come because of the crisis in Israel. My wife's whole family are there, and she is terribly worried about them. If and when the crisis is resolved, we can discuss more concretely when I should come. (Isn't your wife in Israel too? Do you have a similar problem?)

Best regards

David Bohm

P.S. I did read Whitehead, and got some of my ideas on structural process in this way. But actually, I feel that my correspondence with Biedermann was a lot more important, in this regard.

PPS. Perhaps after you have answered this letter, you can send it to Biedermann.

June 1, 1967

Dear Jeff,

I would like to add a bit to yesterday's letter.

You discuss the idea that our memories and thoughts respond in perception, to enrich what we see in various ways (e.g. Picasso sees woman B through her similarities and differences to woman A). It is, of course, a fact that this kind of process actually takes place, and is very widespread. But this process is one of the principal sources of destructive confusion in perception. It is therefore necessary to understand it fairly well, if one is not to get caught in the same kind of confusion.

Of course, I am not saying that thoughts and memories have no valid role at all in perception. They do have a right place, a right field of action. But when they overflow into another field, which I shall call that of direct and immediate perception then things begin to go very badly wrong.

Thoughts and memories contain structural influences, based on tacit generalizations of past perceptions and experiences. These influences are higher order abstractions. They may be right or wrong. But there is no way to test this unless direct and immediate perception operates at a deeper mental level in a simple way uncontaminated by these memories. Unfortunately, thoughts and memories tend to spring into action so rapidly and powerfully, that they overflow into the field of direct perception, where their structural influences are confused with directly perceived fact. Further thought then abstracts from this apparent "fact", and is "amplified" to be "fed back" once again as apparent direct perception. So a vicious cycle, a "feed-back" loop is set up. It is as if the output of a computer were fed back into its input, in the channels that were intended for the reception of "factual" data.

One of the principal reasons for maintaining this process is that thought contains a component that I shall call "self-indulgent pleasure". Of course, there is real pleasure, real enjoyment, which happens simply. Thus I enjoy a sunset or the appearance of a woman. Then the experience is over. Thought forms a "reflection" of the experience, which contains a reflection of the pleasure. This reflection is sensed as incomplete. So thought faces the challenge of continuing the "pleasure", enhancing it, securing it, etc. But this "pleasure" is only thought itself. So thought, experimenting to find how to continue the pleasure, discovers that this can be done by manipulating the contents of thought, in a self deluding way, so as to make it appear that the pleasure can continue. For example, when one has a sense of false euphoria, the brain maintains it by wrongly valuing all factual data to make it appear that "everything is going my way". Or when the woman is gone, the mind will develop images that seem to continue and enhance the pleasure, along with the often delusory over-estimation of its ability to do so, underestimation of difficulties and complications, etc. All this leads to pain. But in its state of self delusion, the brain attributes its pain to something else. Thus, the cycle goes on and builds up. I call this the "pleasure-pain principle", meaning that the brain deludes itself into believing that it will get pleasure and that it actually gets pain, which it attributes to something else.

As you will see on a little reflection, the "pleasure-pain principle" (PPP) is the major factor in all human actions, today and over past history. It is really the entire content of nationalism, ambition, fame, greed, lust for power or for sensual stimulation, aggressiveness, and countless other motives of human action. It is completely incompatible with creativity, love, and beauty.

Now, as a result of what Picasso's former mistress has written,[12] as well as a result of seeing his pictures, I feel that the structure of Picasso's pictures is determined mainly by the pleasure-pain principle. It very probably gave Picasso a brief thrill of pleasure to "express himself" by painting woman B over woman A. He may have increased the pleasure by self-deluding ideas, such as: "I am showing that I am all-powerful, that I can do as I please, by symbolically discarding A in favor of B". Or also he may have thought "I see that B has those features of A that gave me pleasure without those that gave me pain". Or else he may have had some correspondingly self deluding notions of yet another kind. Their precise nature does not matter. What does matter is whether or not his mind was operating in the structure of the PPP. If so, his work was necessarily destructive, in one way or another.

It seems to me that all modern art is determined basically by the PPP and necessarily has to be. For as I said in my previous letter, any mimetic work of art depends on the law of correspondence between its own order and the order that it imitates. Once you give up the correspondence between three dimensional reality and its two dimensional appearance, there is nothing in the intrinsic structure of the situation to determine any other correspondence. So, both for the artist and for the viewer, this correspondence will depend on arbitrary symbolic associations of one thing in the picture with another in the mind of the viewer or the artist. By accident, artist and viewer may have had similar conditioning, to establish similar symbolic correspondence. Then some sort of communication may take place. But more generally, there is no communication. Each person will establish a different and generally irrelevant correspondence, according to his special conditioning.

Even when artist and viewer are similarly conditioned about this correspondence, the result is destructive. For because the result is arbitrary from a logical as well as a perceptual point of view, it is in reality determined by factors outside these fields. And if you observe, you will see that these determining factors cannot be other than the operation of the PPP. Every "distortion" of modern art is determined by the fact that this particular "distortion" seems to give pleasure and satisfaction to the artist. It is not the pleasure and enjoyment that is the by-product of real perception of beauty. Rather, it is the delusory appearance of pleasure, maintained by the PPP, which manipulates the contents of perception and thought so as to give the momentary appearance of pleasure and satisfaction. The function of the modern artist, in manipulating the order, structure and form of his imitations of nature is a wonderfully exact external manifestation of how the PPP manipulates the internal order, structure and form of perception so as to try to maintain pleasure.

The key to a deeper understanding of the situation is to realize that there is a natural order of abstraction in perception. There is direct and immediate perception,

---

[12]Gilot (1964)—CT.

and there are higher order abstractions, based on words, thoughts, and memories. To confuse these is like allowing "feed back" from the output of an amplifier or a computer to its input. A genuine work of mimetic visual art is always conceived first with direct and immediate perception. It has to be seen, as far as possible, "as if for the first time" – i.e., free of visual conditioning. It is very hard for us to do this, because we are so heavily conditioned by our experiences with past art, memories of these experiences, inferences drawn from them, etc. These have their place. But it is poison for them to overflow into the field of direct and immediate perception, that is crucial for all art.

Of course, there is also the art of using words, and this depends on a certain amount of conditioning (e.g. learning the language) to establish the order of correspondence between words and their meanings. But ultimately, the content of what is being said has to be understood – i.e. perceived directly – beyond the level of words. It is crucial to realize that mimetic visual art and the art of using words operate by very different sets of orders of correspondence between the art work and the reality that the art work is "imitating". Modern art is, in a way, an effort to apply literary correspondence rules in the field of visual art. This cannot work, because the structure of the visual field does not really allow it.*

When you are finished with this letter, will you please send it to Biederman how long with the previous one.

Best regards

David Bohm

* It is possible to use "literary" associations with artistic images to communicate only what is in essence already known to the viewer, but not to communicate an order and structure not yet known to him.

PPS. I think you see that modern art is destructive, in the sense that it encourages the conditioning of perception. By breaking up the two dimensional order of paint marks in an arbitrary way, you force the viewer to supply a conditioned order from memory. Modern science tends to do the same. Recall how physicists are trained to start from a conditioned structure of ideas.

What is needed is that mankind should learn what it means to perceive directly without the contamination of conditioning. If this were possible, the Israeli-Arab dispute, for example, would collapse. For Jew and Arab see each other through a cloud of conditioned images. All human problems originate basically in conditioned perception. Art and science could have helped break out of this, by leading to a new order- structure perception. But modern art (and modern science) have instead helped trap men even more in the prison of conditional perception.

June 2, 1967

Dear Jeff,

I wish to supplement my previous two letters on art briefly.

As I understand it, Biederman's position is that the creative growth and evolution of perception is unlimited in its possibilities. There are also unlimited possibilities for this growth to work together with various kinds of mimesis, which can express this perception, as well as help it to grow (i.e., by means of new kinds of mimesis we are led to test our perceptions and see in new ways). However, he also thinks that painting as a means of mimesis has come to the end of its possibilities, not merely in the literal capacity to imitate, but even more, in its role of testing, aiding and stimulating creative evolution of perception. What has taken its place is camera art, especially the moving picture, with its extension to television and perhaps in yet other waya that will be developed in the future.

To understand this point of view, one must see clearly that by its very structure, painting is limited mainly to imitating what is literally perceived. It is true that by means of memory, training, propaganda, habit, etc., people may learn to associate painted (or photographed) images with various other things, represented by ideas (e.g., the photograph of a woman with the idea of sexual pleasure, the photograph of food with the idea of the pleasure of tasting, the photograph of a car with the idea of the pleasure of moving as one pleases, etc.). Or else, one can associate images with ideas of gods, mythological heroes, virtues, vices, etc. But none of this is intrinsic to the act of painting. What is intrinsic to painting is the correspondence between the two dimensional order of paint marks and the three dimensional appearance of objects, with forms, existing in space, bathed in light, ordered as near or far, bright and dark, etc., structured by a hierarchy of such orders. When this correspondence is modified or given up, the whole field becomes arbitrary and therefore it is determined mainly by the pleasure-pain principle (PPP). All these associations are contingent and external to the intrinsic order and structure of the painting and its correspondence with a three dimensional reality.

Unfortunately, artists did not generally realize this fact. They imagined that by re-ordering paint marks on the canvas, they were creating a new order of reality, or expressing a new perception of reality. Actually this is impossible. For the paint marks in themselves are not a new order of reality. There are in fact only paint marks on canvas. At best, they are mimetically leading the brain to perceive some order of reality. But in fact, all they can really do is either to suggest some kind of three dimensional reality, or to evoke conditioned responses to ideas (as is done in advertising). Most artists seem to agree that to go on simply with imitating three dimensional reality in paint on canvass is not worth while. Firstly, it involves the artist in tremendous mechanical work, that can be done better by the various forms of the camera and its extension to movies, television, etc. Secondly, this mechanical work now leaves little room for creative perception. Certainly, the movie camera and television in principle leave far more room for creative perception, and are therefore superior instruments for those who want to express perception mimetically. Therefore, most artists have ceased to try imitate nature literally, and instead, they try create a new order of reality in their paintings. But as I said before, this is in fact impossible. Paint marks on canvas are capable only of suggesting some kind of three dimensional order, either a reasonable one, or a confused one. When the painter tries to infuse perceptions in other fields (e.g., ideas) into the order of the paint marks, he actually succeeds only in evoking a confused three dimensional image. This fact is often hidden because of conditioned associations of these confused images to ideas. But you will notice but these are never new ideas. They are always familiar ideas. Paintings cannot suggest new ideas. This latter has to be done mainly through the use of words. You could never get the idea of complementarity across to a man who never heard it by Picasso's paintings. Rather, you must first explain it in words. Then, when the other fellow has already understood what you mean, he may react by projecting these ideas into Picasso's paintings.

This brings us to the role of direct and immediate perception. Unfortunately, mankind is now conditioned to project his words, memories and ideas, onto all that he sees, including works of art. Who can look at his friend or enemy without the tremendous weight of years of conditioning, with their memories of pleasure, pain, hurts, wishes, fulfillments etc., etc.? Mankind's whole existence depends literally on his ability to step out of this conditioning, and to see reality "as if for the first time". This is needed, not only in visual perception, but also in every other kind of perception including the act of understanding. Otherwise man will not see the fact at all, but will be lost in illusions, delusions and fantasies imposed on perception by his conditioning, according to the dictates of the PPP.

Of course, ideas, memories, etc. have their place. We need a certain kind of conditioning (e.g., to be able to talk, to work, to know our way around the world, etc). But when our perception is conditioned in its basic order and structure, then conditioning has overflowed into a wrong field. For perception requires that we see the fact as it is, whether it is pleasant or not, whether it is convenient or not, whether it is comfortable or not, etc. This fact is always in a new and different order from that which is remembered of the past. So the mind has to meet it directly, "as if for the first time." Our conditioned responses can then be useful as a means of taking proper action etc. For example, if you are in a strange country, a good map is useful. But first, you must look at the actual terrain, and then, you relate the map to it. It is wrong to take the map as basic reality, and to relate the terrain to the map. When we all allow memory to structure perception, something like that is what happens.

Beiderman feels that our conditioned response to art is seriously adding to the present crisis of human consciousness and perception. In particular, the modern artist is tacitly helping people to see in terms of the fragmentation of nature, the arbitrary breaking up of things and manipulation of them, according to what finally turns out to be the PPP. This is particularly tragic, since in principle, it is just the artist who could play a big role in helping people to obtain a new vision of nature, (and human nature) going beyond the immediate and literal appearance of things, to reach their basic order and structure of process. The same is true of science, which in principle could also help toward the same end, but which in fact favors fragmentation, mystification, and the arbitrary exploitation of nature according to motives arising in the PPP. The trends in modern art and in modern science are therefore extremely similar in certain crucial aspects.

Biederman feels that for the artist to play their proper role it is first necessary to see the distinction of mimetic and non mimetic art, so that this dangerous and destructive confusion between them can end. Painting is not a proper vehicle for non-mimetic creation. Rather, this latter has to be done in three dimensional reality, with its structure of form, space, colour, light, etc. Whether Biederman's particular approach to this is valid, as the only possible one, is another order of question, which can be discussed intelligently only when these deeper questions are clearly understood first.

Science is also moving away from the literal appearance of things to the study of this order and structure of process. But ultimately, science will have a basically mimetic content, in that it aims to produce ideas that reflect the structural process correctly.

Of course, some day there may arise the action of "Techne" which unites science and art. But this is so far in the future that it is almost entirely speculative. Nobody can actually do this today, under present social conditions. If he believes he can, he is deluding himself.

Best regards

David Bohm

Will you please send this to Biederman when you have finished with it?

References

Alexander, H. G. (Ed.). (1956). *The Leibniz Clarke correspondence*. Manchester: Manchester University Press.
Biederman, C. (1958). *The new cezanne*. Red Wing: Art History Publications.
Bohm, D. (1965). Space, time and the quantum theory understood in terms of discrete structural process. In Y. Tanikawa (Ed.), *Proceedings of the International Conference on Elementary Particles* (pp. 252–287). Kyoto: Kyoto University, Publication Office, Progress of Theoretical Physics.
Bohm, D. (1968). Creativity. *Leonardo, 1*(2), 137–149.
Gilot, F. (1964). *Life with Picasso*. New York: McGraw-Hill.
Jammer, M. (1966). *The conceptual development of quantum mechanics*. New York: McGraw-Hill.
Nichol, L. (1998). *On creativity*. Abingdon: Routledge.
Pylkkänen, P. (Ed.). (1999). *Bohm-Biederman correspondence*. Abingdon: Routledge.

Chapter 4
Folder C132. Perception Continued, Fact and Inference. July–September 1967

July 6, 1967

Dear Jeff

Thanks for your letter, which I'll answer in detail later.

My wife's family in Israel are all O.K. Nevertheless, we were very worried for a while, till we heard from them, because they live in Jerusalem. Even now, things don't look too good in the Middle East. Nobody knows when or whether there will be real peace.

How are you and your wife feeling about the situation now?

I enclose a copy of a reply to Jauch and Piron, and to Gudder,[1] whose articles I asked Dr Condon[2] to send to you. Because Dr. Condon is in a hurry to publish it along with the original articles, I have sent a copy directly to him assuming you will have no serious objections. If you have any objections, please let me know immediately, and I'll change the article.

Best regards

D.Bohm

[1]Theoretical physicists Josef Maria Jauch and Constantin Piron and, separately, S.P. Gudder had made criticisms of Bohm and Bubs' work. The reply was published in Bohm and Bub (1968)—CT.

[2]Edward Condon, a distinguished American physicist, editor of Reviews of Modern Physics at that time—CT.

C. Talbot (ed.), *David Bohm's Critique of Modern Physics*,
https://doi.org/10.1007/978-3-030-45537-8_4

[There follows a type-written document, which was published as the above Bohm and Bub (1968)—CT.]

JULY 17th, 1967

[Date added – CT.]

Dear Jeff

Thanks for your letter and article,[3] which I read with great interest. I shall comment in detail later. Meanwhile, let me say that Condon insisted on a short reply, so that subtleties of the type you mention cannot be put into it. All that can be done is to show that J. and P.[4] are in an absurd position. I'll leave it to you to analyze the whole story in detail.

I don't think that one would misunderstand Gudder's article in the way you suggest. I only said that Gudder had developed a broader class of models that are not compatible with hidden variables – not that his model excludes hidden variable theories. So there is no point in a serious modification in the letter. When I get the proof I may change a word or two, to clarify this issue.

With best regards

D Bohm

July 18, 1967

Dear Jeff

This is just a brief supplement to yesterday's letter. I'll send you the full answer to your letter a bit later.

You emphasize in your letter (and article) the notion that certain kinds of hidden variable theories are excluded by arguments, such as those of von Neumann. I would prefer to put it in another way. We can say instead that certain kinds of axiomatic structures (or "models") are not logically compatible with hidden variables. (Von Neumann's axiomatic structures are a case in point.) But each axiomatic structure is an assumption whose inferences have to be compared with fact. We have proposed another axiomatic structure, with hidden variables. This is a different set of assumptions, with different influences. In a limited range of experiments where times longer than τ, the relaxation time of the hidden variables are involved, the inferences from our axioms agree with those from von Neumann's. But more generally, we predict different results, and thus, it is possible to choose experimentally between the two sets of actions.

---

[3]See p. 102, n 6 below—CT.

[4]J. and P. refer to Jauch and Piron throughout the letters—CT.

In the article I sent you, I compared this to the choice between the axioms of Euclidean geometry and non Euclidean geometry.

I myself feel that no mathematical axioms can ever exclude hidden variable theories (any more than Euclidean axioms can exclude non Euclidean axioms). Rather, only the facts can exclude certain sets of axioms. Thus, the discovery that measurements satisfy the axioms of spherical geometry would exclude hyperbolic geometry. But the mere mathematical suggestion of spherical geometry cannot exclude hyperbolic geometry as the actual one that may be factually correct, over very long distances. Likewise, von Neumann's axioms cannot exclude hidden variable theories. They are merely logically incompatible with hidden variables, as Euclidean geometry is logically incompatible with spherical geometry.

So I do not think it is right to say that von Neumann's theorem factually excludes any kinds of hidden variables at all. One must remember that von Neumann's axioms contain assumptions that go beyond the facts (in this they are like all other axioms). We too make assumptions that go beyond the facts. But only the facts can ultimately show which set of axioms is a better reflection of reality. And there is no reason why any theory should be excluded, merely because it does not agree with von Neumann's axioms. Thus even the "naive" mechanistic interpretation of a particle criticized tacitly by London and Bauer[5] is not "excluded" by any axioms. It is given up only because it does not correspond to the fact of interference in the two slit experiment (We must not identify these facts with the contents of von Neumann's axioms).

This point is very important, if we are to understand how my 1951 papers on hidden variables fit into the picture. For here, I propose a "particle" model, in some ways similar to the model already "excluded" by analyses similar to that London and Bauer. But then I introduce a multi-dimensional wave, and include the apparatus as part of the whole system. In this way, I go beyond certain assumptions of mechanical separability that are tacit in classical physics. Thus, I come to a model that corresponds once again to the facts (at least in a certain domain of inferences). I think it introduces confusion to ask whether it is excluded by von Neumann's model or not. For this very question tacitly identifies von Neumann's model with fact or truth, and thus obscures the real issue – that we are choosing between two models, von Neumann's and my own. Indeed to fail to see this issue is the basic error of Jauch and Piron, who persistently identify their own models with "fact". I think you underestimate the importance of this point, when you bring in the question of what kinds of hidden variables are excluded by von Neumann's model. I would prefer to say that strictly speaking, von Neumann's model is logically incompatible with any kinds of hidden variables at all. But the question is: "Do we accept von Neumann's model, or some other one, that does contain hidden variables?" The respective advantages of these two approaches can then be discussed.

[5]London and Bauer (1983)—CT.

This is in essence the view proposed by Gudder, and it is the point of view that I wanted to express in the article.

Best regards

David Bohm

P.S. As a preliminary comment about your paper, I get the feeling that it combines two subjects that are rather disparate – a philosophical criticism of the notion of observation (as a structural process) and a detailed mathematical criticism of DLP.[6] Are the two subjects really all that closely related? Or is it not perhaps necessary to split the article into two papers, intended to follow directly on each other in the same issue?

PPS. I feel that we are also in danger of confusing the criticism of simple mechanistic explanations of quantum mechanics with the particular structures of axioms and algorithms proposed by Dirac, Jordan, von Neumann, etc. You are right to emphasize that the facts suggest very strongly that we need a structural-process point of view about the observing instrument. However, logically speaking, Bohr has proposed a set of axioms, in which everything is inherently vague in its definition. By this means, he avoids the detailed consideration of the mutual reflective functions of electron and observing instrument (though he considers them in a rough and general sort of way). His views are often said to be purely epistemological. But epistemology and ontology have a habit of mixing. Thus, Bohr's epistemology wouldn't be consistent, unless ontologically, nature were such as to produce an inherent vagueness in the reflective function. This vagueness is measured by Planck's constant, h, which is an objective quantity, not determined solely by epistemological considerations. In other words, Planck's constant, h, takes on a certain value. This value is an ontological feature of nature. Even if we accept Bohr's tacit assumption that the qualitative existence of Planck's constant follows epistemologically, surely its numerical value cannot be determined by such arguments, so that it must be an ontological feature of how things are. So I would insist that Bohr's axioms have an ontological side, no matter how much he and others might protest.

Now then, we have two alternative sets of axioms that are logically incompatible.
1. Bohr's involving inherent vagueness of reflective function, treated in the order of classical mechanics (x,p, etc.).
2. Structural process, involving clarity of reflective function, but not in the order of classical mechanics.

Unfortunately, structural process axioms have yet to be put in a clear mathematical form. So a test between 1 and 2 is not yet really possible.

But even now, the criticism of mechanism has very little direct relationship with Bohr's axioms or von Neumann's axioms. We can easily make the mistake of supposing that the aim of a deeper theory is to explain the axioms of Bohr or von

[6]Possibly an early draft of the two papers eventually published by Bub as: "Hidden Variables and the Copenhagen Interpretation–A Reconciliation", Bub (1968b), and "The DLP Quantum Theory of Measurement", Bub (1968a). See also C134, p. 219, n 12—CT.

Neumann. This would be a serious mistake. Its aim is to explain the facts, which are also explained by the axioms of Bohr and von Neumann, but in a very different way. Too much stress on the axioms may imprison us in unnecessary tacit assumptions, going beyond the facts, which prevent us from doing what has to be done.

In other words, it must be made clear that neither von Neumann's axioms nor Bohr's principle of complementarity are facts. They are both "models" from which inferences have been drawn that generally cohere with a certain range of facts. It is really very important to emphasize this point very strongly, as Rosenfeld also feels (like J. and P.) that complementarity merely expresses the facts. You are really underestimating the significance of this point.

All this is relevant to a deeper theory. What are the facts that it has to explain? Sometimes, it is said that it must explain classical mechanics. But classical mechanics is not a fact. It is a structure of axioms and inferences from these axioms. Not all these inferences are rigorously true (E.g., there may exist no orbit satisfying a second order differential equation). So our new structural process theory need not approach classical mechanics as such. It may approach a theory of rather different structure that explains more or less the same facts explained by classical mechanics.

The statement that in the large scale "classical mechanics is a fact" is the same kind of confusion as saying: "In a certain domain, von Neumann's axioms are mere statements of fact". It is not a sudden new development that people have been confusing axioms with facts. It is as old as the human race. It is crucial to clarify the difference between fact and axiomatic assumptions. It is not merely a "philosophical" question but a deep psychological and practical one as well.

About mechanism, one can be too strong and dogmatic about the need to transcend it. One does get a general impression that it should be transcended. But it is not possible to prove the need to do this. So there is no reason to exclude mechanical theories from consideration. Rather, they too have their place. By seeing in more and more ways how their inferences are inadequate, we may hope to be helped to see just what is the kind of non mechanistic theory that we need. But if a mechanical theory should work (and this cannot be excluded, a priori) then we would come to an opposite conclusion – i.e., that the facts under consideration do not force us to try to transcend mechanism.

In other words, it is best to avoid dogmatism. We have the facts, and we have assumptions, leading to inferences that imply an order in the facts. The inferences are observed to be either true or false. This observation is a higher order fact. Let us denote the two levels of fact by fact_1 and fact_2. Then we have

The fact_2 about fact_1 and the inferences_1 on fact_1.

This is a fact of second order. From here, we go to

The fact_3 about fact_2 and inferences_2 on fact_2.

Thus, the fact is built into an ever growing hierarchy of potentially unlimited order. It is this process that I have called "facting".

It is crucial to keep the order of fact and inference clear, or else we will be lost in confusion. Our inference is that mechanism should be transcended. But the next order of fact will be to observe whether in a better theory, mechanism is transcended or not. Thus we must keep away from the trap of dogmatism.

Notice that to see a fact requires an act of observation. Thus, it is not possible to infer a fact, or even to infer the correctness of inferences about a set of facts. Ultimately, there must be an act of direct perception, that is not conditioned and limited by the structure of inferences that is being evaluated and tested.

This act of direct perception may take place on the level of the senses or on higher levels, where we see the truth or falsity of certain inferences. In its totality, it is the act of understanding, which is continually putting the fact into a new order, as it tests the truth or falsity of inferences, and replaces false inferential orders and structures by newer ones that are in closer correspondence with lower order facts.

So we see that the act of understanding is always what is making the fact. Without understanding, it is meaningless to ask "What is the fact?"

I hope that you have begun to see the significance of direct and immediate perception on all levels. I shall go into this more in the next letter.

July 20, 1967

Dear Jeff,

Before sending you a detailed answer to your letters and discussion of your paper, I feel that it will be useful to go a bit into the question of the observer and the observed, which is really behind everything that you talk about.

Firstly, I want to emphasize that as I said in yesterday's letter, one set of theoretical assumptions can never exclude any other set. The two sets can either be logically compatible or logically incompatible. Only the facts can exclude one or the other of a logically incompatible pair of assumptions. I feel very strongly that you have not been paying enough attention to this point, especially in your article.

Now, I would say that Bohr's views are based on a certain set of axioms and assumptions. About these, we must first ask: "Are they logically consistent?" Bohr claims in no uncertain terms that his views are logically consistent. Whether his claims should be accepted is not yet clear to me. Nevertheless, let us, for the sake of argument, give him the benefit of the doubt. The next question is then: "Are these assumptions and axioms compatible with the relevant known facts?" Bohr claims that they are. This claim could also be questioned in certain ways. (In particular, I do not believe that Bohr's views permit a really adequate treatment of entropy.) However, let us again give him the benefit of the doubt, on this score. The third claim could then be that no other set of axioms and assumptions is possible that would be compatible with these facts, and yet have inferences that are potentially in contradiction with some of those that can be drawn from Bohr's axioms and assumptions. Whether Bohr actually claims this is not clear, because his language is ambiguous. Nevertheless, von Neumann did make an equivalent claim. And from the way Bohr writes, it looks as if he is least tacitly making such a claim.

It is here that Bohr is clearly wrong. The fact that his axioms are consistent and compatible with known facts cannot elevate these axioms to the level of facts. They are only ideas. These ideas cannot, by themselves, exclude any other ideas. I feel that you are trying to squeeze too much out of the work of men like Bohr and von

Neumann, when you try to use their results to exclude any kind of hidden variable theory whatever. Only the facts can exclude a theoretical idea. One theoretical idea can never exclude another. The belief that this is possible is what is leading to endless confusion in physics. Thus, classical mechanics is a set of theoretical ideas. We could never exclude another theory by saying that in a certain domain, where classical mechanics is supposed to be right, the new theory has a different structure from classical mechanics. Rather, we would have to see whether it leads to wrong inferences about the facts that have been regarded as basic confirmations of classical mechanics.

This can be done by showing that certain aspects of classical mechanics are some kind of approximate inference from the theory in question. But here, one tends to become confused. Which aspects of classical mechanics are to be recovered in this way? Evidently, it is not the whole structure of the theory that is to be recovered. Rather, there is a judgement of which inferences of classical mechanics are essential. But in this judgement, it is easy to make a mistake.

Thus, one now says that it is enough to recover from quantum theory certain average values of observables, which are inferred to satisfy the classical equations of motion. One does not require that one recover the property that the individual particle orbit is relatively well defined, and not spread over all the possibilities covered by the wave function. In my view, however, one should also require the latter. This is indeed the main motivation behind the introduction of hidden variables. And it is because DLP are confused about this requirement that their paper was written. On the one hand, they want to accept Bohr's point of view, which would make such a requirement both unnecessary and meaningless. On the other hand, they feel that something is wrong somewhere, and they would like to set it right. They don't want to see that their very feeling that there is a problem here that has to be solved is a sign that they don't really accept Bohr's point of view in its entirety.

What is Bohr's point of view? It is that the question is basically epistemological. Bohr's first assumption is that the division between observer and observed is of fundamental significance in every branch of knowledge and scientific research.

Whenever we talk about something or think about something, the observer is tacitly there, "doing the talking or thinking" from a certain standpoint. Thus, if I think about you, there is an image of you in my mind. But this image has a form which implies a perspective or standpoint of observation. (E.g., I could be "looking" from the right or from the left or straight on at you.) This perspective is seldom if ever mentioned explicitly. Yet, it is inevitable in every image. Similarly, every verbal description implies a perspective or standpoint, of the "observer" or the "describer".

Of course, we are talking only about an image of the observed object. Therefore, the implied standpoint is also only in the image. There is no real observer in thought. There is a (tacit) image of an observer, who is in the imaginary action of looking at the object of thought.

It is crucial to understand this point, or one will get hopelessly lost in confusion otherwise. For there is also an inference that there really is an "observer" in the mind, who is "looking" at the object of thought. But this latter observer is only an inference. One cannot locate him, either by anatomical analysis of the brain, or by looking inward introspectively. If one looks inward, one sees only thoughts and feelings. Among these is the thought of the object, with its tacit standpoint of an observer. (E.g., if I think of a picture of the Earth, there is a tacit "observer" standing somewhere in empty space.) There is also the inference "I am the observer, and I am really doing the looking." This inference is supported by certain inferred feelings, that resemble those that would seem to be appropriate, if there really were an entity who was doing the looking. But in fact, the inference of this entity leads to endless contradictions (as I shall explain in more detail in a later letter). Let us therefore drop this inference. The plain fact is that observation is going on. But there is no separate observer inside the mind, who would be "doing the looking".

When the brain is thinking, however, every thought contains a tacit observer (just as every picture does). Just as we don't imagine that the tacit observer of a picture is a separate real observer, so we need not imagine that the tacit observer of a thought is a separate real entity. In reality, there is only a thought, with a tacit point of observation, and possibly a false inference that there is a separate entity at this point, who is "doing the observing".

I emphasise all this because I am going to question Bohr's basic epistemological assumption about subject and object. So I am pointing out, from the very beginning, that it is founded on certain illusory views about the nature of thought and perception.

If however, we accept Bohr's assumption, then we are led to ask: "What does it mean to have an objective description (or image) of things?" What it means is that in some sense, the content of the description (or image) can be regarded as separate and distinct from the entity who is "doing the observing" and "doing the describing".

But if the observer were totally isolated from the object, nothing could be seen. Therefore, we postulate that the two are different and separate objects, in interaction.

What has happened is that we have gone to a second level of thought, in which the observed object and the observer are both being treated as objects, under observation by an observer of higher order. We can go on to observers of higher and higher order in this way, but we never get rid of an ultimate "highest order observer", who is merely tacit rather than explicit. So, from a fundamental point of view, this procedure never solves the problem, which is this: "Is the highest order tacit observer separate from what he observes, and if so, how does he interact with the latter?" There is really no way to answer this question. For since the highest order observer has to be tacit, we cannot possibly think about how he interacts, without confusing him with a lower order observer, who is really an "object", under the "scrutiny" of the highest order observer.

Because this basic question is inherently confused, one is led to argue that there is perhaps an inherent purely epistemological need to give up the notion that the process of observation can be treated ontologically in full detail. Of course, such problems will not arise, if we do not try to give a detailed and precise account of the interaction between observer and what is observed. In a vague and general sense, a lower order image of the observer can "stand for" the higher-order observer, to give approximately valid conclusions about the latter. But clearly, there is more to the highest order observer than can be correctly reflected in this image. So we are led to conclude that there is an inherent epistemological need for vagueness of all knowledge, inherent in the subject-object relationship.

One way to approach this problem is to assume that ontologically, we know the essence of the totality of natural law. The observed object is one part of this totality, while the observer is another. So we could perhaps investigate this relationship precisely, despite the above purely epistemological difficulties.

We may say that classical physics provides us with such a framework of concepts. It allows us to assume that the connection of the human observer to his instruments produces negligible effects. Thus, we are free to investigate the interaction of the observing instrument and the observed object.

But here, we come to the fact of the quantum of action. Because of this fact, we infer that not only does the response of the instrument reflect the state of the object, but that the response of the object also reflects the state of the instrument, in a way that cannot be known in detail. Within a certain degree of vagueness, determined by Planck's constant, these effects can be neglected, and the classical separation of instrument and object is recovered as an adequate approximation.

It seems then that nature itself is presenting us with the same confused questions about subject and object that we were led to earlier by purely epistemological arguments. Bohr therefore argues that we should treat the instrumental "subject" as tacitly present in the description of every object, rather than try to regard it as just another explicitly described object. Just as every picture has a tacit observer, so our physical theory has a tacit observer. When we look at a picture, all sorts of clues (e.g., converging lines) indicate the tacit observer. In physics, we get similar clues. Thus, in the position representation, the tacit observer is a position measuring instrument, and in the momentum representation, it is a momentum measuring instrument. Because the tacit observer is now seen to be inseparable from the object, it is wrong to go to a higher order of observation, which "looks" both at the object and at the observer, as if both were explicit objects.

Whether all this is consistent or not, I am not yet prepared to say. Bohr puts a lot of emphasis on the role of quantum mechanical algorithms here. The failure of various operators to commute is taken as a reflection of the incompatibility of different "standpoints" of observation. One could argue that the algorithms and their statistical interpretation provide a "metalanguage" within which the role of the observer is tacit. That is, instead of letting the observer be implied by the direct "language" of description of phenomena (taken to be classical physics) one supposes that the observer is tacitly indicated only in the "metalanguage" of operators and probabilities.

Be that as it may, it is my view that all this is irrelevant, because of the falsity of the basic assumption that the division of subject and object has a universal and fundamental epistemological significance (for all knowledge). In my view, the division of subject and object is a mistaken notion. Therefore, conclusions drawn from it do not really follow.

I would propose an entirely different point of view, which may be summed up as: The observer is the observed.

Let me explain. Consider, for example, a tree. Its leaves are moving, in response to the wind. I want to argue that the leaves are observing the wind. Similarly, the light responds to the leaves, and brings a moving structure corresponding to them to each point of space. So the light is observing the leaves. Likewise, the cells of the retina of the eye respond, so that they are observing the light. Cells further back respond to structural features (e.g., bright spot on dark background) of the responses of retinal cells. Similarly, cells yet further back respond to the structural features of the second order cells, and so on without limit, all the way into the brain, and up to consciousness. Each process that responds sensitively to the structure of another is observing the latter, thus fulfilling an abstractive and reflective function. This process takes place in nature, and is extended in man, in similar but different ways.

At no point is there a division or a break in this process. Thus, for a blind man, the stick is generally part of the observer. But for all of us, light rays act like millions of "sticks" probing the environment. Thus, the light is part of the observer. But then when we look at a tree, the leaves are part the observer. And so on without limit. As our attention moves outwardly, all that it encounters is part of the observer.

And if one looks inwardly, then as I have already indicated, there is also no division. Thus, genuine feelings can be regarded as "feelers", like millions of tiny leaves, responding sensitively to the "winds" and "currents" in the deeper layers of the mind. Thought is then a set of more definite "shapes" abstracted from these feelings (like the shapes one sees in clouds). The sensitive response of thought observes the feelings, and yet more sensitive responses observe the thoughts, going on to yield the totality of the act of perception and understanding.

At no point in this process is there a separate and distinct "self" who would be "doing the observing". As I indicated already, this is merely a false and delusory inference.

Anything whose rational order of movement is understood to some extent can serve as an observer of something else, because its contingent responses now reflect the necessary features of the order of what it is "observing". For example, we understand enough about the responses of silver atoms in photographic plates to realize that their responses can produce an ordered track that reveals various particles (electrons, protons, etc.) As we learn about these particles, we can use this understanding to allow the particles in turn to "observe" a yet deeper level of structure, by means of their responses (e.g., in a scattering process) and so on without limit. The more we understand, the more that our perceptions will be able to observe.

Vice versa, all errors in inferences produce further errors in observations. Thus, if the leaves of a tree were lacquered, they would not shake in the wind. And from this, it would be possible erroneously to infer that the air is still, when in reality, the wind is blowing. Likewise, wrong inferences higher up in order of abstraction can produce wrong perceptions. Thus, the inference that the "self" is the observer of everything implies that it must be protected at all costs, even if this requires manipulation of thought so as to produce delusion rather than true reflections of reality. Indeed, when the illusion of a "self " is operating, the associated inferences of pleasure and pain become the main observer. That is to say, all thoughts are evaluated by the pleasure-pain principle – so that ideas giving pleasure to the "self" are treated as true while ideas giving pain to the "self" are treated as false. Now, one usually supposes that the highest function of the "observer" is to see what is true or false, right or wrong. It appears that this function is being carried out by the inferred "observer". But in fact, it is being carried out by the observed sensations of pleasure and pain. Thus, even when thought goes wrong, the observed is still functioning as the observer. But because it is wrong, what is observed is delusion and fantasy, rather than fact and reality.

Each aspect of the world, inward or outward, that is observed can then respond to other aspects, and thus it functions also as an observer. (This is the reflective function.) So in reality, everything is potentially or actually observing everything else. The human being is part of this totality of observation, similar to the rest in certain ways, different in others.

In physics, this implies that the attempt to discuss the role of the observer is meaningless and futile, as well as a source of confusion. What is needed is to produce a theory of deep and comprehensive scope, with many related orders, constituting structures, processes, etc., and forming a very large totality. This theory is still however only a set of ideas. It is not by itself a fact; and never will be.

How can these ideas be related to reality, without bringing in an observer? The answer is that from the total theoretical structure, one has to abstract sub-orders, which are similar to certain perceived orders of phenomenon. For example, in physics, one may abstract a theoretical order similar to the order of droplets in a cloud chamber or grains of silver in a photographic plate. One then assumes that these theoretically abstracted orders correspond to the perceived orders.

To test the theory, one works out the inferred responses of the abstracted theoretical order to other theoretical orders of structures that are deeper. If the inferred responses correspond to the perceived responses, this is a confirmation of the theory. Otherwise, it is a falsification.

Thus, there is no need bring in an observer anywhere. But one needs to develop a vast, rich, unified totality of theoretical orders and structures. This is a slow, difficult process. It will not give quick "results". Yet, it is the only way really to avoid getting confused about the observer and the observed. Whoever wants to use sketchy theoretical constructions to get quick "results" is very likely to introduce all sorts of arbitrary dichotomies that get him into confusion on this basic question.

To return to the general question, I repeat that each aspect or part of the universe is both the observer of others and observed by them. But now, consider the totality. The totality of all the "observing" aspects is evidently the same as the totality of the "observed" aspects. Thus, we complete the explanation of the statement: "The observer is the observed."

From this point of view, when I am observing the world, the world is also observing me. If other people are present, this is obvious. But even when no one else is present, inanimate nature is observing me. For its responses reveal what I am, inwardly and outwardly. Thus, if my ideas are mistaken or confused, this fact about me is revealed by the responses of external nature, when I take action. It is really revealed for all to see. But because I can have more detailed knowledge of my thoughts than other people, it is potentially revealed in most detail to me. In this way, it can be said that I am observing myself. But more deeply, the whole world is always the ultimate observer, for everyone.

The above is the germ of learning and intelligence. Intelligence requires not merely a sensitive response to the structure of the world, but also an outgoing action related to this response. It also needs a sensitive response the structure of this outgoing action and a perception of how it corresponds or fails to correspond to the structure of the world.

Whenever there is mutual observation of one aspect of the world by another, there is a basis for intelligence. But this can be realized only when there is learning, i.e., the ability to form abstractions and to act according to these abstractions, with further abstraction of how the results of this action correspond to the actual structure of the whole situation.

Perhaps all nature has a kind of intelligence. But man has his own special kind. If nature had nothing akin to intelligence, how could intelligent beings ever develop from inanimate matter?

With best regards

David Bohm

P.S. It could be said that the illusion of a separation between observer and observed is due to a false comparison between the body and the mind. Physically, I look out at the world and see that there is an implied center of perspective from which observation is taking place. At the place implied by this center, I see a human body. I notice that as this body moves and turns, the center of perspective moves with it. Also, I see that these objects that are felt with the hands are those that are near the center of perspective. From all this, I infer that the body is what is carrying out the process of perception. As Piaget showed, the infant does not do this immediately, but learns to do it over a year or two. But in us, it is now habitual.

One of the earliest forms of thought is to imagine an object or a set of objects (in an imaginary space). Since this act of imagination is an internal imitation of what is seen in direct perception, the imagined set of objects will have an imagined perspective implying an imagined center of observation. And it is only natural that one imagines that at this center, there is another object, which is an imaginary image of the body. One projects this image of the body in the imaginary role or function of looking at the imaginary objects. Imaginary feelings (and) sensations are projected into this image, which imitate the feelings and sensations of the real body, when it is looking at real objects.

It is only natural then for the brain to abstract from all this the assumption that inside the body is a mind, and that inside the mind, is a "body of the mind" (called perhaps the "self" or the "soul") which is "looking" at the whole of the mind, as the physical body is looking physically at the whole of the physical universe. But I emphasise that while the physical body is real and is really the center of physical (sense) perception, there is no "body of the mind" at all. This is a purely imaginary inference, entirely without factual support, and full of so many contradictions that even to entertain the notion leads.to absurdities without limit (as I have tried to indicate earlier).

In reality the mind has no permanent center of observation. As I indicated in this letter, observation is taking place "everywhere", in the sense that whatever is in the contents of consciousness is a part of mental action that is "observing" other aspects of the contents of consciousness. Any part or aspect may momentarily take a pivotal or focal role. But it is an illusion to imagine that there is a permanent center of the mind, which is the necessary center of observation.

What confuses us on this score is always the tacit perspective or point of observation in thought. It is necessary to realise that this center is merely being imagined or constructed in thought, and has no essential connection with the real process of observation or scrutiny. If you look at a painting or a photograph, for example, you will see a tacit (or implied) center of observation. As you walk around the painting, you will notice that there is some confusion, because the implied center of observation does not move with the body. Some paintings have several possible centers, and the brain jumps from one to another as you move. But eventually, you reach a place where it is no longer possible to imagine that the body is at the tacit center of observation. Then if you watch carefully, you can see the imaginary point of observation, standing out in empty space, with no "observer" there, who would eventually be "doing the looking". Similarly, in thought, one can learn to see that the "observer" is just as imaginary as what is being "observed". It is all "going on" in a vast emptiness, without a center or a periphery.

P.P.S. I now want to return to the subject considered at the beginning of this letter.

Perhaps you will now see why I disagree with your idea that the main point is to show that hidden variables are not a mere return to classical mechanical concepts. To be sure, this is an important point. But what is still more important is to see the difference between inference and fact.

No idea is a fact: From each idea, we can make inferences which either correspond to the order of the fact or do not correspond. Within each structure of thought are theoretical orders that correspond to directly perceivable orders. These theoretical orders are never observed. Most of the confusion comes from identifying theoretical orders with observed orders. Theoretical orders are assumed to correspond to observed orders. Theoretical orders are then subject to the process of drawing logical inferences from them, based on the whole theoretical model, and these are compared once again with perceived (or observed) orders.

Whenever we describe any order, there is a tacit perspective or standpoint, from which our description takes place. This is like the tacit point of observation of the picture. In reality, it has no very deep significance, because the content of theories is always that aspect of the description that is invariant to a change of perspective.

However, once we identify a theoretical order with a really observed order, it is implied automatically that the tacit centre of "observation" in the theoretical description is a real observer. Thus, we introduce the spurious "problem" of how this "real" observer is related to the orders that are really observed. Once we admit that a theoretical order is never an observed order, the tacit standpoint of "observation" in the theory is seen to be a pure abstraction, of no deep significance at all.

So the essential difficulty is always the confusion between factual order and the inferential order of theory. Whether the theory is mechanistic or not, this confusion will bring in the false problem of how the "observer" is related to what is observed. So I must disagree with your assumption that to dissociate hidden variables from mechanism is the main point.

July 25, 1967

Dear Jeff

This is just a brief supplement to my letter on the "Observer and the Observed".

Firstly, I repeat that no theory (such as that of v. Neumann or J. and P.) should ever be accepted as completely equivalent to the empirical facts underlying quantum mechanics. Thus, even in classical mechanics, it is hard to tell which features are the essentially correct ones, that should be recovered in a broader theory. In quantum mechanics, this problem is even more severe. Should we, for example, regard every Hermitian as observable? (as v. Neumann does). This assumption goes enormously beyond the facts, and has been criticised accordingly (especially by Wigner in private discussions with me). Actually, only a few operators are known to be observable. A hidden variable theory might fail to agree with v. Neumann's axioms, by making only some Hermitian operators observable. Yet, it could agree with all known facts. So it is not necessarily reasonable to require of a hidden variable theory that v. Neumann's axioms can be recovered from it in some approximation.

I feel that all known forms of quantum mechanics are bad, in that they give a very poor treatment of processes in time. Usually, they treat each observation as essentially isolated, and do not attempt to put the time order of observations in any basic role. Yet, all observations tend to be ordered in this way. Consider, for example, a Heisenberg microscope, in which many light quanta were present. As the electron (P) crossed the field of vision with some momentum, p, it would scatter a series of quanta, which would be observed at Q_1, Q_2, Q_3, etc.

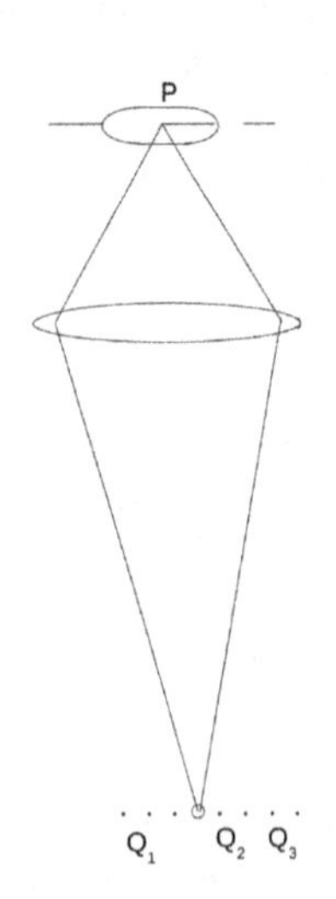

From these, one could deduce the momentum that the electron did have in its orbit. One would infer a process resembling Brownian motion. Why do we not regard the order of points in this "Brownian" orbit as the basic "observable"? In most cases, it is much closer to what we actually do observe than are the extremely abstract and schematic representations of "measurements" described by Heisenberg and Bohr.

If we accepted the above as the paradigm case of a measurement, we might make a new set of axioms, very different from those of v. Neumann, which still agreed with all known facts.

So I am dubious about the requirement of recovering v. Neumann's axioms from a broader theory. We did do this in our theory. But in my view, this need was psychological, i.e. to show that v. Neumann was either wrong or wrongly interpreted. Now that this has been done, we might be wise not to give too much significance v. Neumann's axioms. We did learn that these axioms can be recovered only if we take a view which is close to that of reflective function in structural process (as you describe in your article[7]). We could take it as a generally plausible inference that this is required in any theory that transcends quantum mechanics. But v. Neumann's axioms are too slender a reed on which to anchor a "proof" of such a requirement. Ultimately, "the proof of the pudding is in the eating". Will we actually develop a new theory of this kind? So our present considerations are at best heuristic. It would be wisest not to be dogmatic about the non mechanistic nature of this ultimate theory, before we actually have it.

In my view, the main point of hidden variable theory has been:
(1) To clarify the confused questions around v. Neumann's proof of their "impossibility".
(2) To indicate a few experimental conclusions of hidden variable theories that are different from those of the usual theory.

As a by-product, we have seen that axioms of v. Neumann's type seem to require that the micro-order of movement shall depend on the general environment – and thus imply a step away from mechanism. But since v. Neumann's axioms are only ideas, this is not a "proof" that mechanism is inadequate.

Incidentally, a crucial point of hidden variable theory was to make it clear that v. Neumann's axioms are only ideas and not "facts". J. and P. still think that axioms can be facts (as does Rosenfeld in a different way).

You are quite right to emphasize that in a structural process, there may be no property corresponding to the "conjunction" of two properties. Thus, consider an object which can rotate about various pivots, A, B, C, etc. Evidently, there is no pivot which is the "conjunction" of these pivots. Likewise, when you have an eigenfunction of an operator, this determines an "invariant" vector or "pivot" in the motion or transformation implied by the operator. One could suggest that in a process, the "observables" are the invariant features (like "pivots") and not some sort of object, entity, property, quality, etc., satisfying the usual logic of set theory, with its relationships and classes. The non-commutation of operators then refers only to the fact that the corresponding two kinds of "pivots" cannot both be stationary together because when one is stationary, the other "rotates" around it. This removes the mystery about "incompatibility of observables".

Our theory does involve notions of this sort tacitly. When a really good theory is developed, the whole thing should be a lot clearer.

On the whole, I do not like your idea of supposing that axioms like those of v. Neumann or J. and P. could "refute" certain classes of hidden variable theories. Only facts can refute theories, and no axioms are ever facts. Axioms can be compatible or incompatible with theories (in a logical sense).

With best regards

David Bohm

P.S. When you discuss Bohr's notions on the unambiguity of observation on the large scale level, you might find it useful to keep in mind the following:

[7]Probably Bub (1968b)—CT.

(1) Once we admit the separation of observer and observed, then as I explained in the previous letter, the observer is always only tacitly present in the observed. If you try then to represent the observer as a particular object in the total field of the observed, there is always some ambiguity in this procedure, as we cannot tell how much of the action that is observed originates in the object corresponding to the observer, and vice-versa.

(2) So there is a kind of epistemological requirement of ambiguity inherent in the notion of the separate existence of observer and observed.

(3) The assumption that action is continuous enables one to postulate conditions under which this ambiguity is negligibly small. Thus, in the classical domain, we could agree with Bohr that the significance of what is observed need not in principle be ambiguous.

(4) But the existence of the quantum of action denies this conclusion. So we are back in the field of epistemological ambiguity, (that is characteristic of the subject-object problem).

(5) It is really wrong in such a case to put the observer on the same level as the phenomena that are observed. Rather, the observer is tacitly on another level. So Bohr is right to treat the observer (the instrument) in a special way.

(6) Bohr says that what the observer observes is always described in classical language. (The spots on the plate, etc.) But then, there is a "metalanguage", based on the algorithms of quantum mechanics. The statements in the "language" (position and momentum) have to be translated into statements in the "metalanguage" (eigenvalues and eigenfunctions of operators). This "metalanguage" has, according to Bohr, just the right kind of ambiguity, needed to reflect the epistemological situation properly. But of course, it has this only when its terms are given the usual probability interpretation.

(7) One finds in fact that experimental results do not repeat, but are distributed statistically in the way implied by the "metalanguage". So both epistemologically and experimentally, one sees that the statistical correspondence of classical language and quantum metalanguage expresses the kind of ambiguity that seems to be inherent in the relationship of observer and observed.

Thus one can make a reasonable argument for the consistency of Bohr's views. But if one doesn't accept the separation of observer and observed as fundamental, new questions are opened up. These are:

(1) The observer is the observed. Each feature of the universe observes all the others and vice versa (through reflective function).

(2) Certain features are directly perceived by the senses. These are not separate "objects". Nor are they mere combinations of "micro-objects". Rather, they are higher order abstractions of the total order of orders of action.

(3) These abstracted orders may be those of classical mechanics (i.e., a series of positions, linearly ordered on a coordinate frame, and a corresponding series of changes of position, called velocities or momenta.)

(4) However, they may also be far more subtle, involving complex hierarchies of order and structure that transcend anything that can be described in the framework of classical orders of positions and momenta of particles. There is no reason why

these cannot be just as "unambiguous" as is anything in classical physics. (E.g., one may observe curves of very high order, and regard this order itself as a basic item of data, thus specifying something that is not describable in the language of classical physics).

So Bohr's basic assumption may break down. But I feel that one must produce a concrete example before people will take this possibility very seriously. To try to suggest it by analysis of hidden variables will not convince very many people.

JULY 30th, 1967
(?)

[Date with question mark added – CT.]

Dear Jeff,

I shall now try further to answer your two letters, at least in part.

In your first letter, you emphasize how similarity is what is basic for you. That is why you like Jung's notion of archetypes. It explains a possible creative origin of what is for you a basic perception of similarities.

You point out that everything is different, thus implying that difference is not all that significant, because in the quality of being different, everything is essentially similar. Vice versa, this implies that similarities are what are really significant – and this is because they are all different. So underneath, you recognise that what makes things significant is difference after all. And you come to a part of my thesis; i.e., Things differ in that they have different similarities.

But you have overlooked the crucial fact that all differences are not only similar. They are also different. It is the difference of the differences that is really significant. That is what makes the world so rich and complex in its structure.

In the enclosed manuscript,[8] you will notice a very important difference between constitutive differences and distinctive differences. The constitutive differences are those whose similarities define the basic orders, that determine the nature of things. The distinctive differences are then the differences between different orders. I fear that when the word "difference" is used, people usually are referring to distinctive differences. It would be absurd to regard these as fundamental. I want to emphasize that in structural process, the similarities of constitutive differences determine the basic orders, therefore what is fundamental is constitutive difference.

With regard to Jung, I feel that the notion of synchronistic archetypes is a way of avoiding the fact that differences are primary. It is similar to Plato's notion of eternal ideas or forms, to which matter approximates. Plato equates these forms with thoughts in the mind of God. Jung says that they are eternally in the "unus mundus", which injects them synchronistically into external nature and into the minds of men. But Plato and Jung are basically alike in the fundamental regard that they say reality is understood through its similarity either with the idea or with the archetype. So

[8] Waddington (1969), pp. 18–40, see Introduction, p. 3, n 10—CT.

they regard similarity as basic. And of course I want to regard difference as basic. So I am suggesting the need to transcend this whole position.

In your letter, you are overlooking the basic role of differences in many places. First, you seriously underestimate the importance of contradiction between ideas and observed facts. You say merely that some ideas are "better" than others. But there are cases where the order of experience inferred from an idea is in conflict or clash with the order of factual experience. Thus, my idea may be that I can step out of the window and fly by flapping my arms. Whoever tries to act according to the idea is practically certain to fall and hurt himself. So the order of idea is that I shall move upward and enjoy myself. The factual order will be that I will move downward and hurt myself.

Similarly, the Arab governments have said that they lost the war because of Anglo American intervention. You would perhaps say that this idea is alright, but that a "better" idea is that there is no Anglo American intervention at all, and that they lost the war because of incompetence in operating modern military equipment. Wouldn't you agree that these two ideas imply quite conflicting orders of action, as well as conflicting general inferences?

One must see that a crucial feature of intelligence is the ability to perceive the difference between orders inferred from an idea and orders observed directly in fact. As I explained in a previous letter, it is not enough to infer the difference between inference and fact. It must be directly observed. When observed in this way, it leads to a fact of higher order. I explained the scheme in the previous letter as:

"The fact_2 about inference_1 about fact_1"

"The fact_{n+1} " inference_n " fact_n"

Thus, we obtain an ever changing hierarchical structure of fact. The observation of this fact on all levels is the act of understanding, which is "making the fact" or "facting".

In this process of "facting" or "understanding" what is called for is neither to accept an inference nor to reject it. What is needed is simply to see the fact about the inference. As I indicated in the earlier letter, one fact is given by the answer to the question: "Is the content of the inference true or false?" Here, your notion that "some ideas are better than others" is relevant. Thus, Newtonian mechanics is a "good idea" in the sense that it gives true inferences in a broad domain. However, Einsteinian physics is "better" in that it gives true inferences in a yet broader domain. What we do is to propose ideas. The "higher order fact" is then given by delimiting the domain of truth and falsity of these ideas, thus establishing which are "better" than others.

However, there are deeper structural questions hidden in this procedure. What does it mean that one idea is "better" than another? It means that it has a broader domain of non-contradiction with the lower order facts. In other words, the orders inferred from this idea concord with the orders observed more directly in the lower order facts. But how is this discovered? Surely, it is not that the "unus mundus" injects into the mind a perception of the degree of accord or discord between various ideas and the facts. Such an assumption would lead to a very artificial theory, in which the "unus mundus" would sometimes inject into the mind a very confused perception of the accord or discord between ideas and fact, and sometimes a more clear perception.

Sometimes (e.g., in the case of the Arabs) it would inject into the mind a persistent tendency to resist perceiving that the idea is not in accord with the fact.

Clearly, we must see the fact about the structures of our ideas as well as about their contents. Do they contradict each other, and do they contradict the lower order facts? Are they clear or are they confused? Do they tend to change, in order to become more clear and to accord better with the facts, or do they resist changes of this kind, in order to protect and preserve feelings of pleasure that are associated with them?

How could the theory of archetypes deal with these questions? Why does the "unus mundus" inject confused ideas into the mind? I can understand that it might inject ideas of limited domains of validity, which are later followed by "better" ideas of broader demands of validity. But confused ideas have no domains of validity at all. Moreover, they often lead to a tendency to ever mounting confusion. It would seem that the "unus mundus" has a malicious streak in it, because it tries to deceive us and mix us up, leading us to destructive courses of action, that cause tremendous suffering.

Moreover, the archetypal theory seems to make perception purely mechanical. Not only does the "unus mundus" inject ideas into us. It injects into us a perception of the degree to which these ideas correspond to fact or not. Sometimes it injects us with clear ideas and sometimes with ideas leading to unending confusion. It even injects into us a perception (sometimes clear and sometimes confused) of the difference between ideas and facts. We have nothing to do but respond passively to the arbitrary and generally meaningless whims of the "unus mundus", and to jump like puppets, every time the latter injects us with an archetype, (whether clear or confused).

On the other hand, if it is we who see what is a contradiction and what is not, what is fact and what is inference, what is confused and what is clear, then why do we need the "unus mundus" to inject us with archetypes? For in my view, the ability to see the above things is already basically of the nature of creativity. Once we can see these, it is not too great a step to see also what it means to be creative.

Consider, for example, the ability to see the difference between fact and inference. This clearly depends on a mental process of higher order that transcends the limits of any particular inferential structure of ideas. For a mental process that works within a given order of ideas cannot see what it means that the fact goes outside that order. There is a primary and direct feeling or sensation or sensitivity, which lets us know that inference and fact are in different orders, so that there is contradiction between them. As the corner of the eye can sense a change without sensing what it is that is different, so the mind can sense the difference between inferred order and observed order, without seeing just what the difference is. This primary perception is aesthetic – i.e., it senses a certain lack of harmony between order inferred from idea and the observed order. So the possibility of a new step is opened up by an aesthetic sensitivity of this kind.

Later, the mind begins to abstract the differences in more detail, coming then to the similarities of the differences, until it is able to specify in just what way the two orders clash with each other. This kind of action is indeed quite a general process. Thus, when the eye looks at something new, it first abstracts the essential constitutive differences and their similarities, thus presenting the order of the thing to consciousness. If the order is really new, one may have no words for it. It will then take time to explore the use of language to discover an order of words that conveys or expresses the essence of the new order. But usually it is not new. In this case, the mind draws on memory, by evoking a remembered order of similar differences that is similar to the perceived order. But a vast and very rapid process of direct perception of the observed similar differences precedes this last step. We easily overlook the almost instantaneous process of direct perception of the constitutive order, and imagine that the basis of perception is similarity to the idea (or archetype).

We can thus be deluded because we pay so little attention to what is rapid and fleeting (differences) and so much to what is relatively fixed and permanent (similarities).

These remarks are relevant in connection with your illustration of perceptions of circular shapes. Thus you showed a lot of shapes, like

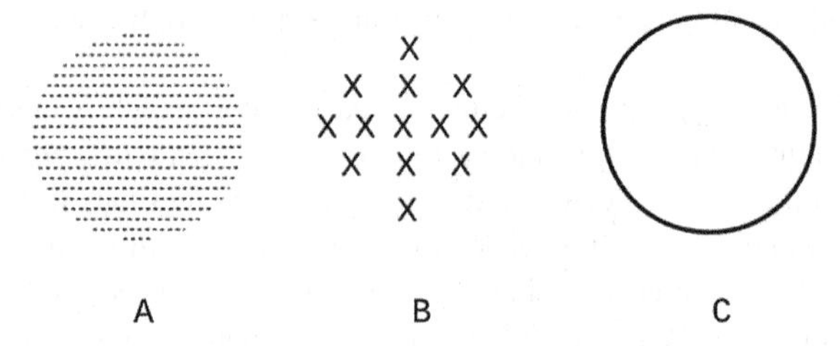

those above, and asked how one knew that they are all circles. You inferred that this is done by projecting the archetype of a circle into all these figures, and thus seeing their similarity to the archetype. I would suggest a very different explanation for this process.

In figure A, the eye abstracts the similar differences between the dots, to form the <u>order of the dotted lines</u>. Then it abstracts the similar differences of the dotted lines to form the <u>order of orders</u> of an array of parallel dotted lines. Then it abstracts the <u>breaks</u> (or ends or boundaries), of the dotted lines, which are indicated by differences between the area beyond the line and the line itself. The eye then sees the similarities of these differences, and thus abstracts the <u>boundary of the two dimensional order of orders</u>, presented as an ordered curve. The similar differences constituting this ordered curve are then related to similar differences in <u>remembered</u> curves, and it is noted that their order is similar to the order of similar differences in a circle. Then the curve is named a "circle". This whole process is so fast that only the last stage is likely to be noted. Thus, it appears that perception is <u>based</u> on projecting the archetype of a circle into what one sees. In reality, this is only a tiny part of the whole process, and not even its most basic or essential feature. Thus, to me, it seems that the archetypal theory of perception is an attempt to stand the pyramid of perception on its apex.

This process of perception of a whole set of orders and orders of orders of differences and similarities is, in a certain sense, "synchronistic". That is to say, although some time may be needed to "take in" the data, the <u>act of understanding</u>, which presents this totality of orders of orders is essentially something that does not involve a time sequence of causes and effects. We always feel that each act of understanding occurs "in a flash". My proposal is that this is a manifestation the true synchronicity of the process which establishes the perception of a new order in consciousness. I would say that we tend to fail to notice this, because of the tacit assumption that all necessary features of process are causally ordered in time, while all synchronous features are contingent (like initial conditions in the laws of mechanics). But I propose that both in nature and in the mind, some synchronous features are necessary, while some features of temporal order are contingent. In a way this unites space and time even more thoroughly than relativity does, for it attributes both necessity and contingency to both space order and time orders.

However, I disagree with Jung in his giving similarity a basic role, by saying that the "unus mundus" projects archetypes synchronistically into external nature and into the mind. Rather, I would say that both in nature and in the mind, the movement of process is such that the (synchronous) differences produced by it tend to have all sorts of similarities in them. Thus, the very law of process is that the new synchronous orders are always being created, both in nature and in the mind. We do not need to assume a special act of creation by the "unus mundus". For creation of new orders (and dissolution of old orders) is the very essence of the character of all process. The only question is how the creative process of the mind is related to that of external nature, so as to produce a reflection of the latter. And of course, all processes reflect the totality. The human mind creates a particular reflection. One of its special features is that the human being acts on the basis of this reflection, and the mind can then reflect whether the resulting fact is in the order of the reflection or not. If it is not, the creative processes of the mind naturally give rise to new reflections, until accord is reached. Exactly how this happens is a vast process, which science has barely begun to explore. But we understand enough of it already to see that it is not a mere mechanical copy of nature, injected by the "unus mundus" into the mind.

In the enclosed manuscript,[9] you will see some notions bearing on this point. Thus, I proposed that in quantum mechanical field theory, all movement is described as the "creation" of a new quantum state and the "annihilation" of an older one. But as explained in the manuscript, each quantum state corresponds to a certain order. Thus, movement is creation and dissolution of orders. In this movement, there is first of all a set of constitutive differences. The similarities of these constitutive differences lead to constitutive orders and to constitutive orders of orders (structural process). When these are analysed into sub-orders, one finds that the different sub orders are related, so that they reflect each other. Thus, the total creative process inevitably contains reflective functions, such that each aspect reflects all others. It is this which is at the basis of perception of what is new, and not the injection of archetypes.

As I indicated earlier, new orders are always being created, and this is indeed the basic law of movement and of the existence of the creative process. It is not that the world is basically mechanical, and that from time to time, the "unus mundus" projects or injects something new into it. (If it were that way, then one would in

[9] Waddington (1969), pp. 18–40. For details see Introduction, p. 3, n 10—CT.

fact need something like an archetype to explain what happens.) Rather, the world is basically a creative process, in which whatever is created is also reflected in all other aspects of the world. Thus, we need no further explanation of how the mind becomes aware of the creation of what is new. For the very law of the creative process is such as to be always providing the foundation of such awareness. The creative process is working reflectively everywhere, including the human brain. The process of abstraction is working everywhere also (E.g., the light rays reflect the universe by their function in each point of space). The human brain has a particular kind of abstraction, similar to that in nature in some ways, different in others. When the human brain abstracts the differences in the function of the creative process in the field of consciousness, it discovers the similarities in these differences, thus becoming aware of the constitutive orders of creative process. This perception is then tested by action in the way I described earlier and this leads in turn to new perceptions, in a process that is the essence of learning.

In this whole process, the ability to see contradiction is, as I have explained, a key feature. When contradiction is observed, the old order of ideas that is responsible for it must "get out of the way", because it is incompatible with the new order that has to be created. To allow this to take place is psychologically the hardest step of all. For it leaves the mind in an uncomfortable state of being empty and uncertain, as to what is the right order of action. The principle barrier to creativity is that the mind turns to "escape" this uncomfortable state, by "jumping" to another known idea. If however the mind can stay in this empty state, without falling asleep, one observes that it becomes extraordinarily sensitive and active in all sorts of subtle ways that are difficult to specify. In this action, exploration is taking place at a fantastic speed. What is being explored are all sorts of orders of orders that the mind is creating, not totally at random, but in the light of the whole fact as it is known thus far. Suddenly, the mind sees that a certain order of orders is right for this situation, and it is presented in a "flash" of understanding.

The total order of the mind in such a process is (like that of external nature) almost like a kind of music or art, rather than like a mechanical process of computation. But we have been trained for thousands of years not to notice this, and to pay attention almost entirely to the utilitarian mechanical order of action in time, which is given top priority in our perceptions. A few artists and musicians are still somewhat aware of the vaster non temporal order of reality (called beauty). But they are regarded as "odd bods", who are useful to entertain the rest of us, but whose perceptions are not to be taken too seriously. In my view, this is all a trap, that has caused untold misery for thousands of years, because it has condemned us to live in delusion and self deception, in all issues of real depth and subtlety.

The basic structure of the trap is in the confusion that arises between inference and fact. Thus, in propaganda, the basic trick is to slip inferences across in the role of facts. But this is always tending to happen spontaneously as well. For example, one may describe a fact, such as: "This boy is stealing". Someone else may say: "This boy is a thief", not noticing that he is making an inference of a "thieving structure", implying that the boy will always steal, because a thief is what he is. Therefore, he will cease to observe the fact about the boy's behavior which might well reveal, for example, that the boy steals to attract attention.

The error in the above is the confusion of an inference with a fact. Inferences are O.K., as long as one realizes that they are only inferences and not facts. Then one knows that it is necessary to observe the fact about the inference; (Is it true or false?) One thus forms the beginning of the hierarchy of "fact_{n+1} about inference_n about fact_n". But if one confuses "inference_n" with "fact_n", then the whole process screwed up, from there on. "Fact_n" is like the input of the nth stage of a computer, of which inference_n is the output. If we connect up "inference_n" in the place that is proper to "fact_n", this produces a "feed-back loop", which will disrupt the whole order of the entire structure of fact and inference.

The major confusion between inference and fact is in the "self". This latter is only an inference. Thus, as I indicated in the previous letter, one can never discover the "self" by anatomical analysis of the brain. Nor can it be seen introspectively by "looking inward". When this takes place, one observes only thoughts and feelings, and these are inferred to belong to an entity called the "self". This entity is inferred to be unseen, because he is the one who would be doing the looking, at the point tacitly attributed by thought to the observer.

When we have an idea leading to all sorts of contradictory inferences, we are led to consider the notion that the idea may be false. But the "self" has a vast set of contradictions that would, in any other case, have led to its being dropped as false and meaningless. For example, one of our favourite inferences is that the actions of the self are always right. This inference gives rise to pleasure. When someone presents evidence that the "self" is wrong, there is a violent reaction of pain, and a "self protective" effort to cover up and distort the evidence, to make it look as if the "self" were right, so that pleasure can be resumed. This reaction is contrary to one's real interests, which are to recognise one's mistakes and learn from them. Similar contradictions can be observed without limit in everyone, which develop from moment to moment. But each person is prevented from seeing his own contradictions by a "censorship mechanism" that makes him blithefully unaware of them, though everyone else can see them without any difficulty at all.

The essence of the "self" is the inference of pleasure. Now, there are real feelings of enjoyment and pain, which are informative. Thus, when all is going right, there is a sense of enjoyment, which is a by-product, and also an indicator that all is well. Pain should normally indicate that something is out of order. But there can also be an inference of pleasure (e.g., the "foretaste" of an expected pleasure) and a similar inference of pain (which we feel before we get into the dentist's chair). When the inference of pain or pleasure is attributed to the "self", it is confused with the fact of pain or pleasure. Thus, a "feed-back loop" is set up. The inferred pleasure is sensed as real but incomplete. It demands that something be done to "set it right" or to "complete it". Since it is all thought (i.e., inferences), this demand causes thought to move to increase pleasure, rather than to try to make inferences correspond with facts. So it all goes wrong. As it goes wrong, it leads to the inference of pain, which is called suffering. This creates a demand for yet more pleasure, etc., etc.

It is in this vast but delusory structure of inferred pleasure and suffering, mistaken for real enjoyment and pain, that mankind has been trapped for tens of thousands of years. Once it has been set into motion, this delusory order of operation tends to spread to other fields. For the whole structure of thought is one. If inferred pleasure is mistaken for real pleasure, the resulting distortion of thought may operate even in mathematics, to give the impression of pleasure coming from the delusory believe that one's ideas are right, when in fact they are wrong.

So it is crucial for mankind to see the difference between inference and fact. This may indeed be the most important thing that science has to teach us. Thus, in science, it is not proper procedure to "stick up" for an idea. Rather, we draw inferences from it, and see whether the facts confirm or falsify the inferences. Either way, we are learning. But in other fields (e.g., ethics, politics, etc.) people are afraid to operate in this way. They accept inferences as facts, and feel impelled to "defend" these inferences, as if they were necessary to the whole order of life.

Of course, everyone can see contradictions in other people (E.g., "he says one thing and does another"). But what is called for is the scientific approach of seeing contradictions in one's own ideas. One can only gain by seeing where he is wrong. Why do we defend our wrong inferences as if they were facts about something that is supremely precious?

Until the scientific attitude can spread into the whole of life, no real happiness is possible for people.

But this in turn calls for the artistic attitude. This is to see both beauty and ugliness, wherever they are.

We have a persistent tendency to want to see only beauty. But as I have shown, contradiction is sensed aesthetically as a kind of disharmony or ugliness. To refuse to see it because it is unpleasantly ugly is to doom oneself to delusion, and therefore to an ugly state of mind, which leads to suffering.

Real beauty can arise only in the mind of one who factually sees beauty and ugliness as they are. Then we have:

"Beauty_2 of Beauty_1 and Ugliness_1."

Beauty_2 is a higher level of beauty, open to one who sees the lower level of beauty and ugliness and how they are related.

All this is the artistic attitude. The scientific attitude is to see:

"Fact_2 about Fact_1 and Inference_1."

But it is necessary to approach life as a whole, not in a one-sided way. So each man needs both the artistic and the scientific attitudes. A man who recognises this has the religious attitude, which regards life as one, unbroken, and interconnected, in all its aspects. (As organised religion is not true religion, so organised art is not true art and organised science is not true science).

With best regards

David Bohm

P.S. If one were to ask what is the basic "theme" of religion, I would say it is:

"The Oneness (Undividedness) of the One and the Many."

To begin with, direct and immediate perception is One. Then, thought splits it into the Many aspects and parts. But these still belong to the One. First of all, there is the fact that:

"The Many reflect the One."

It is this that opens up the possibility of reflective function. Then there is the fact that:

"The One becomes the Many."

In physics, this shows itself as the wave particle duality. (The point event becomes the curve.)

Then we have:

The Many is the One.

That is to say, in the totality, the One includes the Many.

In ancient religions, the One was called by the name of God. Unfortunately, people began to regard God as an entity or a being possessed of qualities. But this in effect limited the One and tacitly distinguished it from Another One or Others. Thus, the One ceased to be a true unity and became One among Many. But the true One includes the Many. So every attempt to define the qualities of God in any way whatsoever (good, kind, merciful, like a father, etc.) made religion false and contradictory. The One is the Unknown. It is not the Unknown that we could in principle come to know. Rather, it is the Unknown that is, and in which everything known rests, and has its being as an abstraction.

Some people put the Unknown up in the sky or far away, or in some mysterious realm. Others say that it is immanent in all things. Both these views are confused, because they attempt to give the Unknown qualities and thus set limits on it. Once the Unknown is limited, then it becomes a known and partial sort of thing, set over against something else, and therefore divided as well. Thus, if it is immanent, it is not transcendent, and if it is transcendent, it is not immanent. Even if we say it is both, we are still caught in confusion. For what it is supposed to transcend is "things" and what it is supposed to be immanent in is "things". Therefore "things" are taken as logically prior to the Unknown. First, we imagine all sorts of "things", and then we suppose the Unknown to transcend them while it is also immanent within them. Thus, we "slip in" the tacit acceptance of the "thinghood of things". But this is just what has to be called into question. In the structural process point of view, for example, "things" are always abstractions from the ever changing flux of process (like the shapes that we see in clouds). In reality, there are no "things" in the universe (except as abstractions introduced by thought). What can it mean for the Unknown to transcend what does not exist or to be immanent within it? Evidently, this is absurd. The only sense in which the Unknown transcends "things" is that the Unknown is real and exists, while "things" are mere abstractions of thought. Indeed, "things" are abstractions from the Unknown.

Whatever we say the Unknown is, it isn't. What then can it mean to us? The answer is that we are the Unknown. We are not always aware of this however. So the real question is that of being aware of the Unknown that we are (and that everything else is). This Unknown cannot be something far away that transcends us (for it is the very essence of what we are). Nor can it be immanent within us. For the very word "us" represents only an abstraction from the Unknown that we are.

This may all seem a bit "metaphysical". And it would be metaphysical if it were taken only as a nice verbal formula or intellectual idea. But I mean it as a description of what is properly the norm for a healthy life, physical and mental. This norm implies the need for one's life and perception to be following the order indicated by the three great themes:

1. Seeing the fact about fact and inference.

2. Seeing the Beauty of Beauty and Ugliness.

3. Seeing the Undivided Oneness of the One and the Many (which is the Unknown that we are and that everything is).

These theme are common to the whole of humanity. At bottom, everyone can see their necessity and reasonableness, when they are properly discussed and understood. They deal with very deep aspects of the mind and of the whole being of each one of us, as well as with the whole of the universe. When we discuss these themes seriously, we inevitably come together on a common ground. But as soon as we begin to give first priority to our own personal peculiarities, our particular emotions and prejudices, our beliefs about religion, politics, nationalism, science, art, or any other subject, we start to be divided and to enter into conflict.

So what is needed is that the mind shall put first things first. To live in the order implied by these three themes is infinitely more important than is any particular thing, like nation, family, career, beliefs, etc. For unless one is living in this right order, the rest must go wrong. (Thus, to fail to distinguish fact from inference will turn family life into a destructive hell and misery.) And if these three themes are alright, then everything else is bound to come into a right order.

Before one can properly look into what it means to be creative, these three themes must be deeply understood. To try to discuss creatively without such an understanding is to make empty noises. But once you have some notion of these themes, you will directly see that they imply a creative order rather than a static order. Thus, to see the fact about inference and fact is just a step in an unending journey of discovery, which is always creating new orders of perceiving the fact and thinking about it. To see the beauty of beauty and ugliness is to see how to end ugliness. To see the Undivided Oneness of the One and the Many is to see unity of ever higher orders. For each perception of Oneness becomes a part of the Many and this latter has in turn to be perceived anew in a Oneness of higher order. So the whole of existence, physical and mental, is seen as creation.

Creation is not to be understood in terms of the order of time. At any moment, the whole creative process is. It is not the result of a chain of causes and effects (though it is compatible with the existence of such chains). Thus, the beauty of a tree or a flower is in the vast harmony of orders of orders with which the tree or flower operates within the brain, right now. The movements of the parts of the tree or flower may well follow the laws of mechanics, in quite a good degree of approximation. But the vast hierarchy of ever changing and developing orders of orders that is the beauty of the tree or flower is simply not capable of even being mentioned or described, in terms of the quantities appearing in the laws of mechanics. The same holds for the beauty of an idea, of a person, of a work of art or music, etc. It is simply an order that exists and whose essence cannot be referred to the order of time. Thus, it is a kind of quality of "synchronous order" (provided that we do not mean "exact synchronism" but only "general coexistence").

As I mentioned before, the human race is heavily conditioned to a belief that this synchronous order of beauty is of no real significance in nature. It is believed that the really significant order is the utilitarian order of cause and effect, carried out in time. Beauty is regarded as "subjective" and "personal", while the mechanical order is taken to be "objective" and "impersonal". What I want to emphasize is that reality (which is the Unknown) is in the order of beauty, rather than in the mechanical order of time. The mechanical order of time is an abstraction from the "synchronistic" order of beauty. And this is where science is going badly wrong today.

Jung's mistake was to identify the synchronistic order with an idea (called the archetype). It is too vast and dynamic, too much in a process of change of orders of order, even to be comprehended in an idea.

Ironically enough, the scientist's emphasis on utilitarian cause and effect is matched by the modern artist's emphasis on the apparently opposite belief that the artist can create anything that he wishes to, in quite an arbitrary way. The older artist recognised that he had to study the necessities and contingencies of his medium. When he understood these to some extent, he could create a new order in his medium, by taking advantage of the contingencies, without coming into conflict with the necessities. But the modern artist seems to feel that in his field, there are only contingencies and no necessities. So he can apparently do whatever enters into his mind.

There is a kind of "division of labour". Thus, the scientist is regarded as the man who deals with the dull utilitarian iron bound necessities of life, while the artist specializes in a domain where there is absolute freedom, and where all takes place according to his whim or fantasy.

But of course, this whole picture is a fantasy. The artist is still bound to the necessities of his medium (which involve the orders of space, form, color, and light). The universe that the scientist studies is still a universe whose basic order is that of timeless beauty, and not that of utilitarian time-ordered necessity. Nevertheless, the artist's belief that he is free of necessity helps confirm the notion that the scientist deals only with utilitarian time ordered necessity. Thus modern science and modern art work together to enslave the mind of man (in this way doing what the priest used to do in the past).

So I think it is crucial to see the timeless order of beauty as the basic expression of the Unknown, which is reality. Our conditioning against doing this is vast; having gone on in every phase of life for thousands of years. Unless mankind can break through it, there will be no end of delusion and suffering.

To this end, the understanding of what Jung called "synchronism" is crucial. However, the very word "synchronistic" is misleading, because it emphasizes time as being of the essence. Let us instead discuss the timelessness of the order of beauty. By this, we mean that the order is not essentially related to the order of time (and not that it is permanent or everlasting). The fact is then that the timeless order is our most direct contact with the unknown reality. The order of time is an abstraction from the timeless order, made in the process of thought.

What then does it mean to be aware of the Unknown, in its basically timeless order of beauty? Firstly, it means that the divisions and multiplicities introduced by thought are not to be given a fundamental role of top priority in our mental processes. To meet the undivided wholeness of the unknown, the mind needs an undivided kind of perception. But immediate and direct perception is always a totality, which is then analyzed and divided by thought. However, since thought too is real, this kind of immediate and direct perception is no longer adequate, since it is now set over against thought, and is therefore divided, rather than total. What is needed is a higher order of direct perception that is aware both of lower order direct perception and thought, and of how they are related.

To have this kind of higher order direct perception is crucial for the health of the mind. For ultimately, it is the only possible way of seeing that inferences are not being confused with facts (with the disastrous results that I described earlier). But even more, it means that there is no reason to seek the Undivided Oneness of the One and the Many. For in this higher order direct perception, the mind is this Undivided Oneness of the One and the Many. Therefore, it does not have to seek the Unknown, since it already is of the basic nature of the Unknown. Only when the mind thus is of the nature of the Undivided can it have a truly religious (unbroken, unfragmented) attitude to the whole of life. And with such an attitude, it no longer tends almost always to create divisions and conflicts. Thus it lives in harmony with itself and with the whole universe.

In such a state, the timeless order of beauty that is being observed is similar to the timeless order of beauty with which the mind itself is operating. Thus, the order of the mind is of the same nature as is the timeless order of reality as a whole. So the division between the observer and the observed has ceased.

In our ordinary state of perception, we feel that reality is more or less mechanical, while the observer (the "subject" or the "self") is an entity who appreciates beauty, sees truth, and has very delicate and wonderful feelings and emotions of every kind, whose value infinitely transcends the mechanical. Thus, there appears to be an absolute gulf between the nature of the observer and the nature of the observed.

But of course, in this state, the observer is actually the inference of a center, full of pleasure and suffering. This inference is the most mechanical thing in the whole field of perception. In reality, the apparently mechanical world and the apparently non mechanical observer are of the same illusory nature, projected by the same mechanism of thought and its inferences, mistaken for fact.

When the mind functions from the undivided state of perception, then the observed and the observing mind are of the same nature. Each is only an aspect of a yet broader and deeper reality, which is one and undivided, and which moves in the order of timeless beauty (while time is seen to be a functional abstraction from this order).

But as I indicated in a previous letter, one can say even more than this. For the observer is the observed. The process of perception is not only of the same order of timeless beauty as is what is observed. Even more, this process is being carried out by what is observed, inwardly and outwardly, in its totality. The total reality that is observed is also the observer. This is perhaps the deepest truth of all, and it is hard to appreciate simply, (in direct and immediate perception, rather than merely as an inference of thought).

Sept 8, 1967

Dear Jeff

Thank you for your letter of Aug 16. I think you are right in trying to work out these difficult questions for yourself, rather than to absorb what I say about them. Nevertheless, there are a few points which I would like to call your attention to, in your meditations on these questions. Perhaps it will help you to do so. At any rate I hope it will at least facilitate communication between us (which is, as you say, very difficult through correspondence). I shall look into the question of when it is convenient for me to come to Minneapolis, where we can at least discuss verbally. Please let Feigl[10] know that I am resolved to come at some time, but cannot yet say exactly when.

First, a few remarks on Fact and Inference. As I explained in an early letter (or did I?) , we can call Fact_1 the immediately observed fact and Inference_1 the immediate inferences about Fact_1 that spring to mind. Inference_1 imposes order and structure on Fact_1. Without it, fact_1 would be only a set of isolated bits of information. But inference_1 calls for a further act of observation to answer the question: "Is it true or not?" This leads to Fact_2 about Inference_1. From Fact_2 we go to Inference_2 about Fact_2, then to Fact_3 about Inference_2, etc., etc. Thus we see that the total body of fact has a hierarchical structure.

In this structure, each element can play two roles. It may at one level be treated as Inference_n while at the next level, it can become Fact_{n+1}. No "fact" is so elementary that it does not contain lower level inferences, which have been so well confirmed that they can be treated effectively as parts of the fact. For example, Australia is for me an inference, (as I have never seen it). But it is confirmed in such a vast hierarchical structure of observations and experiences that I can treat this inference as a fact, for most purposes. Nevertheless, to be clear about the order of operation of the mind, I must maintain the distinction between the inference of Australia and the observed fact of Australia. At this deep level, Australia is still for me an inference, but one that can in general be treated as effectively confirmed in every respect.

One sees then that the total body of fact is not identical with truth. For it always contains inferences that are yet to be tested observationally. Some of these may well be false. Nor is the fact identical with reality as a whole. For even the whole body of fact is but an abstraction from this reality, which is vast and immeasurable, being basically unknown in its totality.

---

[10]See C130, p. 21, n 9—CT.

The fact is what we have made or established. It is always being established and re-established. Thus, it is the outcome of a dynamic process, which is the act of learning. This learning is not merely the accumulation of knowledge. Rather, it is basically discovery, which is made possible by seeing the truth and falsity of inferences in the already established body of fact. This perception lays the groundwork for the drawing of new inferences, that become the "growing points" in the "living body of fact".

Where does this process go wrong? Of course there are the normal "errors" in our inferences, which are an inevitable part of the of the process of learning and establishing the fact. If the mind is operating in a healthy order, these are sooner or later noted and corrected. But there is another kind of "error" in the process, which represents not merely a "false content" in the process of thought. Rather, it is the result of a wrong order of operation of the mind, leading to a wrong structure of thought. This wrong structure does not tend to correct itself. Rather it tends to get worse and worse, leading to entanglement in ever growing confusion and delusion.

Basically the wrong order of operation is to confuse Inference_n with Fact_n. For example, the description of Fact_n may be "This boy is stealing". But the observer may say instead: "This boy is a thief", not noticing that this latter statement is an inference, which implies a "thieving structure-function", inherent in the boy's nature. Thus, we may infer of a motor car "It is powered by a gasoline engine". This inference can then be tested by further observation. But when one says: "This boy is a thief", one usually does not notice that this is an inference requiring further testing. Rather, one tends to suppose that his "thieving nature" is an observed fact (Fact_n), so that one knows factually (and not merely inferentially) that "he will always steal". As a result, one does not observed Fact_n carefully, and one fails to see for example, that the boy steals, not because of his thieving nature, but because he is trying to get some attention.

The above is typical of the wrong order of mental operation leading to a wrong structure of thought. What happens is either that one imagines one has actually observed the boy's thieving nature as Fact_n or else, that this is so well confirmed an inference that it is, like the existence of Australia, for all practical purposes equivalent to a fact. Whichever form this error takes, the results are destructive. For in effect, the function of thought is now caught in a "feed-back" loop. It is as if the output of a computer were fed back into the "factual input", rather than to a higher order computer or to a circuit controlling the activity of a machine. Such feedback would create self maintaining and self exciting cycles of meaningless activity, which could generate endless confusion. Similarly, when inference_n is taken for fact_n, then we form new false inferences_n based on the original inferences_n, rather than on fact_n. These in turn can be fed back as another part of fact_n. For example, one may infer that the boy's thieving nature can be "beaten out of him". This too would be taken as part of fact_n. When beatings fail to do this, one may infer that the boy has a "hopelessly criminal nature", which is also taken as yet another part of fact_n.

It is evidently crucial to maintain the distinction of inference_n and fact_n. This requires a never ending sensitivity, permitting a perceptive act of discrimination, from moment to moment. What is needed is thus to pay attention to the order of the fact, and not merely to its reflected content. By this, I mean the order of the process of learning or "facting", which is what establishes the factual content of this process, as this content changes from moment to moment.

We are generally taught to pay attention only to the factual content of our perceptions, and not to the higher order fact, about the order of operation of the process of "facting" itself. This is a serious and perhaps even fatal defect in our education. For we are taught "what to think" without being taught "how to think". It is urgently necessary for each one of us to start to remedy the defect, to the best of his ability. Only thus is there any hope of changing the generally meaningless and destructive course that most of our history has been taking, for thousands of years. For as long as we confuse fact and inference, the result is a tendency to systematic and ever-growing self-delusion. And whoever is guided by such delusions can never produce the results that he intends.

Here, one sees the root of the problem of evil. For the truth is that deep down, all men desire the good. But because they are deluded as to what is the good, they accomplished evil in almost all that they do. And this is inevitable, as long as men are thus deluded. But such delusions stem ultimately from the confusion of facts and inferences. Even the vast evil and ugliness brought about by the Nazis stemmed ultimately from such delusions, in which inferences about the sublimely beautiful and powerful nature of the German "Volk" were confused with facts (Incidentally, how would you reconcile the actions of the Nazis with your thesis that evil and ugliness do not really exist, in a deep sense?)

The notions that I have described above are summarized in the "theme": "The Factuality of the Total Hierarchy of Fact and Inference." In other words, while Fact and Inference are aspects of each item of information, the over-all structure gives rise to the ever growing and changing body of the "fact as a whole".

It may help clarify this structure to point out that thought is not merely an image or a reflection of reality. It is also a model. Such a model reflects not merely the structure of what is modelled, but also its function. So thought models the structure-function of reality as a whole. From this model, we draw structural-functional inferences, which test the model. Indeed, to understand a given field is to create a model of that field as a whole, from which coherent sets of structural-functional inferences can be drawn. A model may be visual, verbal, mathematical, or of some yet other form.

One can go further and say that immediate perception also contains a kind of model of the world. Thus, when I see a table, I "feel" its solidity and hardness before I even touch it. This is the result of a model, based on past experience going back to early childhood. It is crucial to note that this model (including mechanical, geometrical, and emotional features) directly interpenetrates the structure and function of all that we perceive.

One may convince himself that perception contains (among other things) a model of the world, by noting that the world remains unchanged in perception, as one turns the head and the body. This is the result of an orientation process, regulated by the inner ear. When this process goes wrong, one becomes "dizzy" and the world seems to "spin". What is happening is that the model of the world is no longer properly oriented.

It is crucial to note that generally speaking, both physically and mentally, we respond to this model of reality, and not directly to reality itself. Thus, when the model is wrong, our responses are wrong. By perceiving this, the brain is led to change the model in an appropriate way (provided that it is operating in a normal order). But since the model contains structural-functional inferences, projected directly into perception, this change is possible, only when the brain is properly and freely aware of the difference between fact_n and inference_n. If it takes inference_n as fact_n, its new model will be wrong. And the perceptual cues that would show it to be wrong are misinterpreted, because they are combined with other data, wrongly taken to be factual, rather than inferential (in the context in question).

In a way, this implies that there is no such thing as illusion. We see whatever we see. But we draw wrong inferences from it. All the magician's "illusions" are based on his skill in leading us to wrong inferences. When we see various patterns that lead us to wrong estimates of size and shape, this too is only a kind of wrong inference. So there is no illusion. There are only wrong inferences and delusions. (The latter being wrong inferences resulting from a wrong order of operation of the mind, mixing up inference_n and fact_n.)

It is important to note that even the physical, chemical, and emotional responses of the body and brain are determined largely by the model presented in perception. For example, the Nazis thought of the sublimely beautiful and irresistibly powerful nature of the Nordic German "Volk", and a model of all these qualities sprang into his consciousness. Thus, he apparently directly perceived what Hitler was talking about, not realising that words can give rise to models, full of feelings, color, and emotional effects. Then Hitler spoke of the dirty, cheating, cowardly, wicked Jews, with their crooked and long noses, who were corrupting the noble purity of the Aryan race. All this was "modelled" in perception. As a result, the adrenaline flowed, the glands functioned, the brain filled with blood and a feeling of "righteous indignation". At that moment, surrounded by a hundred thousand people all imitating each other's reactions, he felt an ecstasy of violence that swept away all reasoning power. This whole process was then recorded in the "memory banks" as a "programme". So whenever such a man saw a Jew, the "model" sprang into his perceptions so rapidly that he could never perceive its inferential character. Rather, it was fact to him. And remember, none of us can do other than act in accordance with the model of reality presented in his perception. Any man, poisoned mentally by the Nazi model, cannot do other than destroy himself and everything around him as he tries to do what appears to be right and good, but what is actually false and evil. More generally, as each man perceives so he acts, and so he is, inevitably and necessarily. A change of man's nature can follow only on a change of perception. If his perceptual models are right, his actions will be right, inevitably. If they are wrong, his actions will equally inevitably be wrong. Only a right order of mental operation can correct wrong perceptual models. So all depends ultimately on perceiving the order of mental operation, as well as its modelled and reflected content.

Here, one tends to become confused by thought. As direct perception contains structural-functional models of reality, so the brain is able to abstract these models, and thus to imagine or think about the structural-function of objects that are not present in direct and immediate perception. Indeed, the two year old child has learned to imagine a whole world of such objects, each ordered in its place in space, and one of these objects being himself. If you will keep such a process in mind, you will perhaps find an answer to your question as to the nature of a perception that is not direct. When you are thinking, you perceive the imaginary "objects" of thought. These "objects" are not being directly perceived. Of course, if you are sensitively observant, you will directly perceive something else – i.e., that you are thinking. So we have three things; direct perception, its model in thought, and a direct perception of a higher order, which perceives that thought is a model, and also perceives how this model is related to lower orders of direct perception. So we have an important further theme:

"The Direct Perception of Direct Perception and How it is Related to its Model in Thought"

This theme evidently gives rise to a hierarchy, since each order of direct perception can in turn be modelled in thought, and directly perceived at a yet higher order.

Evidently, just as the models of direct perception can fail to correspond to the order of reality, it is even easier for the models in thought to be arbitrary, and fortuitous, not having any essential relationship to the objects that they are supposed to model. Here, we must recall that the notion of fortuitous (essentially unrelated) orders is part of the basic concept of natural order as a whole. Thus, in the absence of traffic signals, the orders of cars in two intersecting roads is fortuitous. Hence, they will not be coordinated, so that collisions will be able to take place, which destroy the cars. A traffic signal serves to coordinate these orders. Similarly, arbitrary thoughts have only fortuitous relationships to what they are supposed to model. It takes an intelligent process of perception and learning to establish thought models that are properly coordinated to the objects that they are supposed to model. In other words, thought does not simply and mechanically "reflect" reality (as if it were a mirror). Rather, it is a special kind of reflection, which models the structure-function of real things. Such a model has to be created and established by the action of perception-learning.

Of course, thought models are full of feeling, emotion, color, etc., as well as capable of functioning and in other ways. So it is often difficult to distinguish thought models from models that are inherent in direct perception. Indeed, it is well-known that the young child often thinks that people can see his thoughts, as if they were objects in the room. It is therefore not surprising that adults as well as children often mistake the models of thought for directly perceived reality. For example, the convinced Nazi actually perceives the "Jewish" qualities that Hitler talks about, as if they were visible before his eyes, so that his wish to exterminate the Jew seems just as natural to him as would anybody's wish to kill a dangerous snake that was ready to bite him. One is not aware of how words can cause the corresponding "models" to be projected into what appears to be direct perception.

I hope you see that not only do we have perception of models in thought, but that also, such models can quite easily interfere with direct and immediate perception, even when one does not seem to be thinking. So what is called for is a sensitive discrimination between direct perception and its model either in thought or in projection into what at first sight appears to be direct perception. Our education totally neglects this key requirement, without which real mental health is impossible. As a result, confusion, delusion and neurosis are practically universal, and have been so for thousands of years.

It is essential to stress that thoughts can be false in two ways. First, they may fail to reflect the order of the real structure-function of things. Secondly, they may in themselves have a wrong order of structure-function, which confuses Inference_n with Fact_n. It is crucial to remember here that thoughts are not mere reflections of structure-function of reality as a whole; they are also in themselves structure-functions, which make up parts of reality as a whole (physical and mental). So it is necessary first that thoughts be in a right order. Then they will be able to reflect a right order, as well, provided that their order is coordinated to that of perception in the act of learning. But unless they are in a right order to begin with, they cannot reflect in a right order. Rather, they will be inherently deluded so that further perception will only lead to additional entanglement in confusion. This underlines the importance of perceiving the order of the thought process, as well as the order of what it reflects. So we need:

Direct Perception of the Order of Thought and Its Relationship to Direct Perceptions of a Lower Order.

It is evident that humanity is caught up in a vast structure of delusions about ambition, family, career, religion, nationalism, race, and a thousand other things. But these are all generated by what I shall call "The Central Delusion of the Human Race". This delusion has to do with man's notions about himself, especially how his mind functions and what he is, psychologically speaking. As we can see, it was only natural for man to develop such deluded notions about himself. Thus, the young infant develops a picture of an imaginary "space", full of modelled "objects". The model of the "object" is seen from a model of a "point of view" or "center of perspective". At this "centre of perspective" is placed a model of the body of the observer. This model is useful and necessary, as it helps to orient a person in his thinking about his own actions and experiences (E.g., it enables one to imagine how to go from one place to another).

Thus far, then, all is well. But trouble begins because the model of the body also contains a model of its feelings (pleasure and pain) and a model of its urges and intentions (felt as desire and will). So the model becomes very convincing, and is easily mistaken for a real entity or a "soul", which acts like an "essence of the mind". Terrible confusion may result when thought takes its challenge to be the preservation of the "welfare" of the model of the body, rather than that of the body itself. For the model of the body is merely a picturesque and lively way of presenting some of the inferences of thought or consciousness. If thought takes it as a challenge to have a nice pleasing model of the body, full of "good feelings" resulting from a model of "euphoria", then it will start to order its thought in a wrong way – i.e., to accept as true those thoughts that seem to "benefit" the model of the body and as false, those which seem to "hurt" it. Thus, one is entangled in the pleasure-pain principle, which is the road to delusion.

The pleasure-pain principle can work only by confusing $Fact_n$ with $Inference_n$. Thus, the thought: "I am a sublimely beautiful entity whose actions are always right" creates a model of pleasing and euphoric "glow of good feeling" in the model of the body. When someone shows that my actions are wrong, this model of a good feeling is interrupted. Thought takes its challenge to be the restoration of the model of "good feelings", rather than to model the true state of affairs. So it presents false models, in which my actions are still "right" in spite of evidence to the contrary (which is thus misinterpreted if not suppressed). The inference "I am in my very nature and structure a perfect being who is always right" is confused with a fact, and projected into the model of the body, as if it were actually an observed fact.

So one sees how the pleasure-pain principle leads necessarily to delusion, and to a wrong order of operation of the mind.

But of course, the model of pleasure is not real enjoyment. Everyone senses this, and can feel that "something is missing" in such "pleasure". How is it that we fail so persistently to see that this model is an unsatisfactory counterfeit, which leads, not to enjoyment, but to endless confusion and suffering?

The answer is to be found in a further feature of the "self". As I indicated earlier, the "self" is only an inference. But I now add that this inference is projected into consciousness as an active and functioning model, full of feelings, that are models of pleasure and pain, models of urgent needs and models of real intention. At first sight, it looks as if the model "feels" and "desires". Actually, of course, it can do nothing of the kind. It is the mind as a whole which attributes all this [to] the model. But even more, there is a yet deeper source of confusion. It is this:

The model of the body is "located" in a model of a "world" and is engaged in a model of a process of "looking" at this "world".

This is something very confusing and subtle, which requires extremely careful thought. Because the model of the body is engaged in a very convincing process of imitating the feelings, actions and experiences of the real body, it gives a deceptive imitation of the process of perception itself. Thus, the models projected by thought seem to be "perceived facts" and not models at all, because the model of the "self" seems to be "doing the looking".

In this whole deception, it is crucial that the model of the observer is projected as if it was separate from the model of the world that is observed. In reality, of course, both models are parts of a single unitary process of thought. But because what is observed (e.g., pleasure-pain) seems to be separate from the "observer", the former seems to be a real fact, "perceived" by the "observer". So the pleasure-pain in the model seems to be a real fact and not the counterfeit that it actually is. However this "fact" is interpreted as "incompleteness" of pleasure. So there is always a challenge to "complete" the pleasure and get rid of the pain. The mind perpetually escapes perception of the fact of contradiction in its "pleasures" by means of the delusion that these are "separate" from the "observer" who can "choose" to do something that will remove the contradiction. But this is impossible, since the contradiction is built into the whole process of thought, which projects the model, in the first place.

This delusion about a "self" who "observes" the mind is responsible for making all the other delusions plausible and credible (because it makes them seem to be "observed facts" rather than "models that are projections of inferences"). So the key to liberation from all this needless and useless suffering and confusion is to see that the "self" is only an inference projected as a model.

I am not sure whether all this is clear to you. But I wanted to emphasize that the understanding of the delusion of an "observer" would go very deep, and would indeed transform each of us individually as well as the society collectively. At present, social relationships are based largely on delusions, such as status, prestige, race, nation, etc. Such delusions seem credible only because the model of the body (called the "self") is engaged in a model of a process of perception, which makes the model of what is observed seem to be "Fact_n" rather than "Inference_n". This confusion spreads into all human activities, including physics (I shall discuss its implications for physics in another letter).

Now, to come to the question of beauty and ugliness. You ask: "Why is a city ugly while a forest is beautiful?" To you, the functioning of a city seems beautiful. But is it really so? Consider the traffic congestion, the slums, the mediocre dwellings of the middle class, the fouling of the air and water, the noise and confusion, the dirt, the loneliness of the family isolated in a flat or of single people living in isolated rooms, the lack of recreational facilities, the expensiveness of all entertainments, the poor public transport, etc., etc. How can all this be thought of as beautiful? Is it not the function of a city to make possible healthy, happy, fruitful life for its inhabitants? Do not modern cities lead to race riots, criminal violence, etc. on the part of the poor and drug taking as an "escape" by the middle class and wealthier youths (along with alcoholism and all forms of sexual amusement)? Are the cities themselves not paradigms of ugliness and mediocrity, as buildings, streets, etc., are laid out in a fortuitous manner, aimed only at profit or "saving public funds"?

I feel that there really is such a thing as ugliness, and that it leads to disharmony and destructive conflict in function. Isn't the Viet Nam war a very ugly thing indeed? Can you find beauty in it?

A city <u>could</u> be as beautiful as a forest. And <u>some</u> of the older cities <u>have been so</u>. But modern cities are generally ugly and disharmonious in their function. This is because their structure was determined in a fortuitous way, with no over-all view as to what was happening.

Of course, all ugliness is relative to certain levels and orders of process. More deeply, beauty is fundamental. That is what I meant by the theme:
"The Beauty of Beauty and Ugliness."

For example, one can see that no man can ever violate the real order of nature. No matter what he does, he is following the laws of nature. So when we say that science teaches us to <u>control</u> nature, this is a delusion. The real state of affairs is otherwise. When we understand the laws of nature correctly, our actions generally lead to their intended results. When we do not understand, we still inevitably follow nature's order. <u>But this order is such as to lead to results that were not intended</u>.

This is something very beautiful indeed. On the positive side, it is our crucial clue to nature's order. Whenever we don't get the intended result, this shows some sort of misunderstanding, which if inquired into, will bring us a deeper knowledge of nature's order. On the negative side, it means that nature's order is such as to make us suffer, as long as we persist in delusion. In all this, there is great beauty. But it does not deny the ugliness of disharmonious and conflicting actions, based on delusion. Our perception of this ugliness can be a significant clue to the fact that we are in delusion. But if we escape perceiving the fact of ugliness, we lose this clue, and condemn ourselves to go on with delusion and its inevitable suffering.

Finally, I would suggest that the key question is not the generation of illusion, but rather, the generation of delusion. As I have indicated, this arises when inference_n is confused with fact_n. The central delusion is the inference of a "self" who observes the "rest of the mind". This is projected as a functioning model, full of apparent feelings of pleasure and pain, apparent desire and will, etc. (actually, these are only models of real feelings). This model is engaged in a model of the function of "observation". So what is "seen" is apparently an "observed fact" rather than a model, expressing a set of inferences. This "fact" of the "self" is modelled on the "supremely precious and vulnerable center of existence". So thought takes its challenge to be the "protection" of this "center" and the securing and enhancement of the "pleasure" in it. Thus, delusions without end are introduced in the models, to accomplish this aim. Such delusions lead to endless suffering, followed by more delusions, whose aim is to relieve the suffering, but whose actual effect is to make it worse. This has been the fate of humanity for thousands of years. And unless it comes to an end soon, it will probably lead to the end of the human race.

I shall comment on the questions about physics that you raised in a later letter.

Best regards

David Bohm

P.S. Please send this to Beiderman when you are ready.
P.P.S. I hope you see that direct and immediate perception is not nearly as common as you implied in your letter. Consider, for example, how hard it is to see another person, without projecting what one knows about him as a model. Do we not see the Jew, the Negro, the Communist, etc., rather than the individual human being, in all the complexity of his qualities and behaviour? Do we not look at the wife, the husband, the colleague, etc., as a model based on past experience? Actually experiments show that people seldom notice others, except insofar as they are assimilated into certain standard features of the model. Likewise we seldom look at nature, except through our conditioning, expressed as a model. We say "This is more beautiful than that" not noticing that the model of beauty is thereby made to replace the infinitely subtle and complex fact of beauty. Whenever we compare two people with regard to their value, we get a similar substitution of the model for the fact (Inference_n for Fact_n). Whenever we compare ourselves to another (thus giving rise to envy and jealousy) we set the models of ourselves and the other in the place of the fact.

Very rarely do we have direct and immediate perception, uncontaminated by inferential models mistaken for parts of the directly perceived fact. Indeed, if we had such perception, the entire order of society would be seen immediately to be absurd, meaningless, dangerous, and destructive. Nationalism, racism, envy, jealousy, status-seeking, would be seen to be poisonous trivialities, and would cease to loom so large as to be major factors in determining the order of society. One would see that if we say "A is an American, B is a Russian, C is a Jew, D is an Arab, etc." we are putting verbal models in the place of real facts. Such a projection of verbal distinctions into models creates divisions between people that can never be bridged, until people see the fantastic triviality of this whole process, and how we are allowing these trivialities to destroy everything that is worthwhile.

It is like saying: "The boy is a thief", rather than saying "He is stealing". So one should say: "A is conditioned with American 'tapes', so that he believes he is an American, and acts according to this delusion. Actually, he is a human being, but doesn't realize that this means something, while being an American means nothing". Evidently, as long as A believes that he is an American, the "American model" of the "self" will be projected into A's consciousness. Similarly, the "Russian model" is projected into B's consciousness. Since each model of the "self" contains the inference of sublime perfection and supreme value, the other person's behaviour is taken to threaten the very essence of all that is good and beautiful. No wonder then that people are ready to destroy the whole of civilization in an atomic holocaust, in order to preserve the "American model", the "Russian model", "The Jewish model", "The Arab model" of the "self". For it seems that these models are the very source of whatever is good in life. Actually, of course, they are petty trivialities, put together by the manipulation of words. But very few people really see this, in direct and immediate perception. If more people saw it, it just couldn't go on. It is really too insane, as well as dangerous and destructive.

Sept 15th, 1967

[Date added – CT.]

Dear Jeff,

This is a continuation of my answer to your letter of Aug.16, to discuss the physics questions that you raise there.

I am not at all sure that everyone (including Jauch and Piron) would really argue that only facts can confirm or exclude sets of axioms (such as those of v. Neumann). Of course, everyone would be ready to give assent to these words. But deep down, many would continue, in a confused way, to assume sub-liminally and "unconsciously" that certain sets of axioms are either facts or practically the same as facts. If you read my answer to the letter of Jauch and Piron[11] (which I sent you some time ago) you will see a quotation from J. and P. to the effect that their "propositions" are in essence indistinguishable from facts. In my previous letter, I cited the existence of Australia as an inference which is so well confirmed that nobody would do other than take it as being effectively a proven fact. In my opinion, J. and P. regard their "propositional axioms" as being no more inferential than the existence of Australia. What tends to happen in such cases is that the imagination supplies a "model" in which it seems

[11] i.e. Bohm and Bub (1968) see p. 99, n 1—CT.

that vast quantities of data are "confirming" one's assumptions. The the fact is that J. and P.'s "propositional axioms" are confirmed in only a limited number of cases, to a limited degree of approximation, and in limited contexts of investigation (E.g., the randomness of successive observations has hardly been tested at all). But in the minds of J. and P., I feel sure that there is a vision of an overwhelming mass of data of every kind, which tests their axioms in every conceivable way, so that to question their axioms would be similar to questioning the existence of Australia.

Unless one can somehow bring this kind of confusion out into the open, it will be a waste of time to try to use axioms of v. Neumann or J. and P. to draw reasonable inferences. In my view, J. and P.'s paper is nonsense, based mainly on the pleasure-pain principle which accepts as true the pleasing assumption that the axioms of quantum mechanics are indubitable facts, and suppresses or discards all evidence to the contrary. It is no use trying to convince J and P of anything, as long as they think in this way. However, we must answer them, to help others to keep out of confusion, if they have a serious interest in doing so.

I do not think that J. and P. or von Neumann were trying to say that they had good reasons for proposing their axioms as reflections of the structure of certain real processes. If they had been thinking in this way, they would never have identified their axioms with facts. Rather, they would have said "I have such and such reasons for accepting these assumptions". Von Neumann insisted in many places that to contradict these assumptions would be to conflict with the facts, not with his reasons for accepting his assumptions. J. and P. talk in a similar way. It is not an accident that von Neumann's reasoning is so confused, that one cannot see what he has actually proved. It is, in my view, a consequence of a deep confusion in v. Neumann's attitude to physical theory. This kind of confusion is not uncommon.

I am sure that J. and P. do not think that any other axioms are logically compatible with the existing facts. Therefore, the first job was to show that other axioms are possible, which do fit the facts. Whether these axioms are "better" or "worse" than those of v. Neumann was not, in the beginning, the main point. Rather, the main point was to show that they are possible, while v. Neumann and J. and P. were arguing, in a confused way, that they are not possible. We can now come to the question of which axioms are "better" or "worse". But this is itself a confused question, because most of your criteria of "better" or "worse" are based on little more than the pleasure-pain principle. 19th century physicists would all have regarded v. Neumann's axioms as "terrible" and would have sought an ether-theoretical explanation of the whole thing. 20th century physicists have been conditioned in the opposite direction, to get pleasure out of purely mathematical models of the world. So they are ready to accept v. Neumann's axioms easily, to pay little or no attention to the large number of problems that these axioms simply avoid (e.g., how are "observables" really made accessible to perception, and how is a time-sequence of process to be described?) One of our problems is to be able to decide which axioms are "better" or "worse" in a way that is not largely subjective and fortuitously dependent on what happens to give pleasure to most present-day physicists (or to ourselves).

As to why v. Neumann proposed his assumptions, this is itself not a clear question. No doubt, he may have had some valid reasons for doing so. But the first question is: "Why did he forget (or become confused) about the fact that his assumptions were only assumptions?" The reasons for this were probably psychological, being grounded most likely in the pleasure-pain principle. Once this point has been thoroughly cleared up, then it is useful to ask what were v. Neumann's valid reasons for proposing these axioms, if any. Do you have some suggestions as to how to answer this question? I think it is an interesting but very difficult question.

In answering it, I wouldn't say that q. mchs excludes "simple" or "reasonable" classical models because, as you say, it is so difficult to see what these words really mean, in this context, other than "pleasing to present-day physicists". Rather, I would say that the classical models existing at the time of the onset of qu. mchs were excluded. For the very fact of stationary states and interference excluded the existing concept of the electron as a "little ball" moving according to Newton's laws in the force-field of the atomic nucleus. It did not exclude more subtle classical models. For example, if you make the equations of motion non-linear, it is easy to come to a set of stable orbits (called "limit cycles"), around which oscillations are taking place. A probabilistic theory of transitions between such discrete and stable "limit cycles" arises naturally, if one averages over an ensemble of states of oscillation around such "limit cycles". Non-linear equations would give rise to linearized approximations, resembling the super-position principle of quantum mechanics. Whether this theory could be made to work quantitatively is at present unknown. But qualitatively, it gives the right sort of results. However, nobody has seriously considered it, because it is mathematically difficult, and because current prejudices in favor of formal mathematical explanations make it seem hardly worth while to inquire into something so difficult and arduous, for purposes that appear to be so unimportant (i.e., to give a physical as well as a mathematical account of the process). On the other hand, in the 19th century, this question would have been given crucial importance. So the conditioning of physicists, according to the pleasure-pain principle, is deciding what is to be regarded as "simple" and "reasonable".

It is therefore not at all clear just what kinds of theories are "ruled out" by arguments of London and Bauer, v. Neumann, J. and P., etc. London and Bauer certainly do not rule out non-linear theories with discrete stable limit cycles that give a model of stationary states – or if they do, I am not aware of the arguments involved in leading to such a conclusion.

You say "This is physics, not logic." But I don't see the difference, in this context. If you can't rule out a theory logically, then the plain fact is that it just hasn't been "ruled out" and that is that. There could easily be an infinite set of very subtle "classical" theories, as yet unknown. How are these to be "ruled out"? Until we propose them, we can't rule them out. To rule them out before we propose them is to close our minds to them in a dogmatic way, so that research is channelled for inadequate reasons, away from the direction of exploring them.

Of course, it is useful to articulate in what way certain axioms are false or inadequate, as well as to give explicit reasons favoring their adoption. If you can do this with v. Neumann's axioms, this is fine. But aside from being false or inadequate to the facts, what else would be wrong with any particular set of axioms? Only that one has made a mistake in logic, or that he has made assumptions which were confused with facts. When one removes the latter from v. Neumann's "proof", all that is left is that these particular axioms are not compatible with certain kinds of explanations in terms of hidden variables (i.e., explanations in which quantum-mechanical averages would be obtained from statistical ensembles, like those explaining thermodynamics, in which micro and macro states were independent of each other). We have already done this in our papers. What more do you wish to do, other than to emphasize this point (which could of course be valuable, in a certain context)?

It might be interesting in this connection to ask how a theory of classical "limit cycles" escapes being "ruled out" by v. Neumann's axioms. The answer is that the quantum-mechanical "observables" are not basic concepts of the theory. Rather, they are abstractions, whose relevance may well depend on the general large scale environment (including the "observing instrument"). In other words, the limit cycles determining what are called "stationary states" in q. mchs. depend on the whole context in which the system under discussion finds itself. It would be useful if you could show how v. Neumann tacitly assumes that "hidden variables" never have this property, at all, so that his "proof" is irrelevant to this class of "hidden variable" explanations of qu. mchs.

In my opinion, there was a very peculiar state of mind prevalent among physicists when the quantum theory was being developed. It is difficult to describe it, from this distance in time. But it was based to some extent on the spread of positivist prejudices, resulting from misinterpretations of the theory of relativity. This latter theory had also predisposed physicists to accept purely mathematical explanations of nature. Notions of simplicity and symmetry were widespread. None of these were entirely wrong. But they were full of a great deal of confusion. The pleasure-pain principle is always operating, to make people ready to feel that they have solved a difficult problem, when in reality they have not done so. Classical physics always contained mechanistic features, which were ugly. It seemed that one could now quite easily "escape" all this, by exaggerating the extent of quantum-mechanical achievements. So one readily assumed that the great "anti-mechanistic revolution" had already been carried out, when in reality, only some semi-phenomenological theories of the new domain had been set up. Because of sneaking doubts about the whole thing, people were anxious to set up arguments "ruling out" the unwelcome possibility of new classical explanations. That is why people were so ready to confuse inference with fact.

It is now necessary to take a cold hard look at the whole story. What has really been proved, and what is wishful thinking? All of us would like to believe that quantum phenomena force science to transcend mechanism. In my opinion, such a conjecture is not unreasonable, in view of presently available facts. But it is necessary to be very clear, and not to get caught up in delusions once again, by treating this mere conjecture as a fact.

In my opinion, a worthwhile project would be more or less as follows.

(1) See if you can find some compelling reasons (other than the pleasure-pain principle) why the current axiomatic formulation of qu. mchs. (according to Dirac, Jordan and v. Neumann) was adopted. Of course, it fits certain facts (e.g., stationary states, interference, transition probabilities, etc.). But why was (and is) there such

confidence that no other scheme can possibly fit these facts? Why were certain questions of fact simply overlooked? (Consider, for example, the question of which Hermitian operators are actually observable, and how any operator is actually made accessible to perception, as well as that of whether we ever actually observe anything that is not a sequence of events in time, which is therefore totally outside the scope of a theory which can only discuss observations, all of whose aspects carried out at the same time). Why were physicists so ready to assume that the detailed order of a physical process in a transition is in principle unobservable, as well as meaningless?

In my view, the main reason behind all the above steps is that physicists wanted to feel that their scheme was in essence complete. It is disturbing in an emotional sense to have to commit one's whole life to doing work in terms of a scheme that is obviously provisional, partial, and subject to radical revision. The pleasure-pain principle leads one therefore to formulate delusory arguments, which serve to convince one that at least the basic scheme of things is clear, even if a lot of the details are still unclear or uncertain. This reaction is as old as the human race. It occurred with Aristotle's scheme. Then it occurred with Newton's scheme, then with Einstein's, and then with qu. mchs. If a new scheme going beyond qu. mchs. is developed, it will take place once again (as is happening today in biology with regard to genetic schemata involving DNA). So the key thing to call attention to is the falsity of the whole effort to get an emotional sense of security by seeking arguments that seem to guarantee the permanence of our general theoretical schemata. Rather we need arguments showing why our schemata are not likely to be permanent (if we are to avoid complacency).

What is called for is not only a revolution in the content of scientific theories. Even more, we need a psychological revolution in the way science is to be thought about. Scientific theories must always be seen to be inferential in nature, with a certain degree of factual confirmation and falsification, which varies from moment to moment. There is no room in science for the wish to feel secure emotionally about one's "fundamental ideas". Indeed, there is no room in any field whatsoever for such a sense of emotional security, which is evidently an invitation to accept delusory notions of every kind. Emotional security is appropriate to the infant or the small child. But it is poison for the supposedly mature individual. Indeed, it is essential to mental health and maturity that one be emotionally vulnerable. Only a complete vulnerability of one's state of mind will allow one to be open to the fact as it is, even when the latter is not as one would like it to be. But wherever "self", family, career, ambition, race, nation, etc., are involved, few are vulnerable to the fact, which might be very disturbing in an emotional sense. Nevertheless one should be disturbed, when something is wrong, rather than made comfortably somnolent by a pleasing sense of emotional security and invulnerability.

With best regards

David Bohm

P.S. If you are in doubt about how J. and P. are thinking, consider what happened when they learned from Bell[12] that v. Neumann's conclusions were based on the assumption of linearity, which goes beyond what is given by experimental fact. Instead of seriously considering the possibility that v. Neumann's conclusions might be wrong, they immediately assumed that some minor change in the system of axioms would save his conclusions.

Was there not here an a-priori conviction of the "rightness" of v. Neumann's conclusions, issuing from a certain state of mind, which is "committed" to these conclusions, rather than from "reasons" based in fact on theoretical considerations. In my view, they see a "mental model" of a vast body of facts, which directly confirm all of v. Neumann's conclusions, rather than his premisses. So when his premisses were shown to be wrong, this did not seriously disturb J. and P. For they "knew" that the conclusions were "right", and deduced from this that some other premises were needed, (which they were not slow in "finding"). Even if their new premisses were shown to be wrong in a way that convinced J. and P., the latter would still be "sure" that some further modifications of the premisses is all that is needed, because they are convinced of the factuality of the conclusion: "Hidden variables are impossible." Indeed, for J. and P. (as well as for v. Neumann), I think that the basic premisses were always: "Hidden variables are impossible" and that in reality, their choice of axioms is a conclusion based on this premiss. Only if they give up this premiss will they be open to real discussion on this point.

In saying this, I do not wish to blame or censure J. and P. In a way, we have all been trained to work similarly. Each of us has "pet" ideas which he apparently "supports" with arguments based on premisses. But in reality, this "support" is delusory. The premisses are usually chosen just because they do seem to "support" what appears to be the "conclusion" (but what is in reality the basic regulatory assumption). Part of the necessary "revolution" in science would be to bring this practice to an end, so that people would, for example, openly say: "I believe there are no hidden variables. From this belief I have been led to propose the following axioms, which are compatible with this belief, and which, as far as I know fit the facts."

If someone were to argue in this way, he could be making a useful contribution to science. For his statement could then reasonably be taken as a challenge by others, who could say "We believe there may be hidden variables." In answering this challenge, they would be led to propose an extended meaning to the term "hidden variables" (as we have done). The believers in no hidden variables would then have to meet this new challenge (e.g., by criticizing these extensions for lack of "simplicity"). Thus, the "dialogue" could go back and forth, and progress could be made, at least in clarifying the issue. But this cannot readily be done, while people like J. and P. refuse to admit that their ideas on hidden variables are no more than beliefs, and while they confuse them with "facts". And more generally, as long as people feel

[12]i.e. John Stewart Bell, well-known for his proof of Bell's theorem. See Freire (2019), pp. 156–162 on Bohm and Bell—CT.

the need to make existing general schemata appear to be permanent and necessary (thus allowing for a sense of emotional security), confusion on basic questions is inevitable.

PPS. I hope you see that v. Neumann and J.&P. have no really "good reasons" for proposing their axioms as "facts". Very probably, they imagined vast masses of data, confirming these axioms in every possible respect, and in this way their statements come to seem right to them. I can remember a similar experience when I first mastered the quantum theory. It seemed so powerful and overwhelming in its impact as to make it appear that all of its assumptions were indistinguishable from self evident facts.

Of course, you yourself may be able to discover "good reasons" for certain axioms. You won't find them in the writings of v. Neumann or J. and P. because these writings contain no such reasons.

But if you do this, you must also go into the "reasons" against these axioms (E.g., they don't discuss the "measurement problem" at all, they make the order of quantum processes incomprehensible, and they assume, in contradiction with evident fact, that all measurements consist of operations carried out at a particular moment of time, or in a very short single interval of time).

However in this regard, you must face the fact that most physicists do not regard these reasons against quantum axioms as having very much weight. Rather, they believe that facility and fluency in computing and predicting experimental results is the essence of understanding, at least in the field of physics. So the reasons "against" v. Neumann's axioms would not count very much with such people. When you come to decide which axioms were "better" and which were "worse" you would run into trouble here, based on different psychological points of view about what science is supposed to be. If you restricted yourself to purely physical arguments, this would be a source of confusion, interfering with real communication. For you would not be sharing a common set of premisses with the vast majority of physicists, about what the purpose of physical theories actually is. In my view, this is the real source of most of the confusion that hovers around the subject.

To understand why certain axiomatic structures were considered adequate at a given time, it is not enough to restrict yourself to fact and logic. You must 'project yourself' into the psychological state of mind of the people concerned. For this is the main factor determining our general metaphysical notions, especially when we are unaware of our metaphysics (as is indeed usually the case). And because this is very difficult, the problem that you have set yourself is a tough one. If you don't go into these deeper psychological and metaphysical questions, I fear that you won't really advance very much toward clarification of the issues involved.

Ultimately, you will probably have to state your views on the subject openly, and call attention to how they conflict with the prevailing views. Then what you say will at least be clear, and perhaps a few physicists will see that tacitly they have always agreed with your views and resisted the prevailing ones. These will be able to accept your arguments and seriously criticize them. The others will realize that argument is useless, because there is no agreement on basic premisses. This is the kind of clarification that one can reasonably aim for, in my opinion.

PPPS. A good exercise would be to "project" oneself into a state of mind that is probably similar to that of J. and P. It is important in doing this neither to blame them nor to accept their point of view. Can one just be "neutrally" aware of the problem? After all, when I think of certain ideas, I can also feel the impulses that are implicit and tacit in their structure. Usually, I reject such feelings when I think of the ideas of J. and P. But now, let me neither reject nor accept these feelings. Just let them "be", and reveal their full content to awareness, whatever that may be. In this way, my feelings will probably resemble those of J. and P. in crucial respects. For all thoughts contain similar structures of feelings. When I think of the Nazi ideology, I can feel the ugly feelings that this produces in me, the tending to delusion, etc., etc. – I can feel the impulse to violence, when I watch a riot on television. It is my own violence, my own delusion, that I feel. But as such, it is very likely basically similar to what happens in other people.

When I look at the ideas of J. and P. in this way, I get the following thoughts and feelings: "After all, any observation can be reduced to a series of propositions, having "Yes" or "No" character. When an electron goes through two slits, it either arrives at a certain spot on the plate, or it does not. Everybody knows this. It is absolutely certain. Only an idiot could try to question this fact and then, it is a fact that if one observes the electron going through one of the slits, this will interfere with the observation of the electron, as it arrives at the photographic plate. So the two kinds of proposition are evidently incompatible. Only a deluded physicist could doubt this. Two non commuting operators obviously express this incompatibility, in a way that is a perfect reflection of the observed facts. Surely not even Bohm and Bub would say otherwise. So how can anyone doubt that the structure of propositions is a fact, or at least a completely indubitable structure of inference? How could one even imagine an experiment that would not confirm this structure? After all, everybody knows that every experiment is built out of "Yes" or "No" propositions, and that it has been proved that some of these propositions are incompatible, and therefore equivalent to non-commuting operators. It takes only a little bit of elementary mathematics to show that hidden variables are not compatible with this propositional structure. Those who question it must have something wrong with their minds. It is entirely incomprehensible in any logical sense why they do so. Perhaps they are deluded in some way. They are not satisfied with the inescapable fact that physics consists of computing the results of experiments. They have romantic notions that physics should explain how things happen. These notions have been exploded by the factual development of modern physics. Clear headed individuals like us can see this quite easily. But confused and deluded physicists are still trying to do what is evidently impossible and meaningless. Why won't they accept the fact as it is? Perhaps it is because they are seeking in physics the kind of thing that one should properly obtain in other fields, such as art or religion or sex. Hard-headed individuals like us can see that computation of experimental results is the proper field of physics. Deluded and misguided romantics are always confusing the issue, on this point["].

As one thinks and feels in the way described above, it never enters one's head that all the statistical consequences of the propositional structure have not yet been factually confirmed in experiment. It seems impossible to imagine an experiment that would not confirm this structure. It seems meaningless to consider a set of hidden variables that would determine both the slit through which the particle passes and the interference pattern. For this conflicts with the propositional structure, which is an indubitable fact. Indeed, all possible consequences of this structure have already been confirmed, at least to a very high degree of approximation. There is nothing more doubtful in this structure than in the inference of the existence of Australia.

Finally, let me sum up what I feel to be the main point at issue in our correspondence. To you, it does not seem to be interesting merely to note that someone "has forgotten that his assumptions are only assumptions." You want to see what is wrong with his reasoning, if you do except his assumptions. But to me, the confusion of assumption (or inference) with observed fact is extremely interesting. As I indicated in the previous letter, it is the principal source of human misery, suffering and confusion, and has been so over the ages. It therefore appears to me to be very interesting to see how this is happening even in science, which is among the most rational of all man's activities.

In addition, it seems to me that unless we are clear about the confusion between inference and fact, which is the source of all delusion, our further deductions as to what is wrong with people's reasoning will be lost in confusion. For unless we are very clear about how people like J. and P. are confusing inference and fact, we will find ourselves "slipping in" some of their tacit premises, when we fail to notice just what we are doing. Here it is important to note that reasoning doesn't proceed along a straight line, like "Premiss through deduction to conclusions". Rather, the same premisses are often introduced into the argument at several stages. So you have to be on the lookout for confusion of inference and fact all the time.

To illustrate this, let me write P_F standing for factual premiss and P_I for inferential premiss. A typical chain of reasoning is (where S stands for intermediate steps in reasoning)

$$P_F + P_I \rightarrow S_1$$

$$S_1 + P_F + P_I \rightarrow S_2$$

$$S_2 + P_I \rightarrow S_3 = C = \text{final conclusion.}$$

Now, if we confuse P_F with P_I, it is not enough to correct this in the first step. You have to watch for this confusion in all the subsequent steps

What is even more important to emphasize is that P_F is being confused with P_I for psychological reasons (mainly pleasure and pain associated with P_I). So there is a tendency to "protect" P_I, by introducing a host of further premises ($P_I{}^1$,$P_I{}^2$,$P_I{}^3$ - - -), all of which are confused with ($P_F{}^1$,$P_F{}^2$,$P_F{}^3$ - - -). Unless you are very very alert, you will allow yourself to "slip into" confusing $P_I{}^N$ with $P_F{}^N$ (where N is fairly large). After all, the brain becomes weary after a while. Just to "be finished with the whole dull story", you accept $P_I{}^N$ as equivalent to $P_F{}^N$. And then, you are trapped once again in the confusion of J. and P.

So so it may well be that nothing is wrong with J. and P.'s reasoning, except a continual series of confusions of P_I^N and P_F^N. This is in fact what I think to be the case. But this kind of error is a paradigm on what tends to go wrong in all discussions of fundamentals in physics and in every other field. So when we look at it in this way, J. and P. have done us a service, by presenting us with such good paradigm of confusion in reasoning in basic problems.

In particular, whenever J. and P. talk of hidden variables, they are tacitly confusing P_I with P_F. For they have certain inferential notions about hidden variables, which they confuse with facts about hidden variables. Therefore, they probably do not even admit that what we talk about are really hidden variables. For they already "know" factually what are the essential qualities of all hidden variables. Therefore, to J. and P., what we call "hidden variables" are really "nonsensical statements in conflict with the evident fact." For this reason, when people like J. and P. talk of our papers, there is no real meeting of minds. The whole discussion tends to be lost in cross-purposes.

Sept 20, 1967

Dear Jeff

This is just a brief supplement to my two previous letters.

When I say that "if a theory of hidden variables is not ruled out logically, it just isn't ruled out", I did not mean to suggest that we have to consider all possible models as equally significant. Of course, one will use intuitive, heuristic and other "sub-liminal" cues, which will lead one to favor a certain direction of thought instead of another. But one also knows that these cues are very often wrong. So it is necessary to have a certain kind of humility about such notions, which allows them to be vulnerable to a broader perception of the fact, in the way that I described in a previous letter. It is advantageous if different people will explore different kinds of intuitive suggestions, and compare notes in a friendly and cooperative way. For a-priori nobody knows what is the "right" model. Arguments that "rule out" certain models on inadequate grounds may well be harmful, in this respect. They are particularly harmful when their tacit aim is to give one a feeling emotional security and safe invulnerability, with regard to one's fundamental ideas in physics. What is necessary, above all, in physics and in every other field of life, is to be able to live with the fact of uncertainty. Only thus can the mind be sensitive and open to the tiny promptings and cues which are its first response to the unknown.

Here, I don't mean to argue for indeterminism or for Heisenberg's "uncertainty principle". Indeed most physicists today are absolutely certain that the uncertainty principle is right in every respect. What I am emphasising is the fact of psychological uncertainty, which has no essential relationship to the question of whether physical phenomena are in principle predictable or not. Psychological uncertainty, when recognised properly, implies humility and vulnerability, two qualities not shown in any great measure by Bohr and Heisenberg, when they assure us of the absolute truth of physical uncertainty.

As I explained in my previous letters, the chief trouble with fundamental discussions in physics is lack of humility and vulnerability. It is this which leads to the search for security, which is "guaranteed" in a delusory way by the confusion of a large number of inferences with facts. It is no use blaming people for such reactions, since they are not even aware of what is actually taking place in their minds. However, if one simply rejects their arguments out of hand, as is justified by showing their contradictions, one is imitating the defensive attitude that is basically the source of all the trouble. For basically, all defense implies offense and aggression. Both physically and mentally, one can defend oneself only by taking violent action aimed at destroying the "danger", or else by setting up walls, to keep the danger out. But if one takes violent actions or tries to use arguments that "demolish" one's opponent, one resembles him in being proud and invulnerable, therefore basically impervious to the truth. Likewise, the wall that keeps the other fellow's ideas out also keeps out the truth. So we have to be open and vulnerable to von Neumann, to J. and P., to Danieri, Prosperi and Loinger and to Rosenfeld. If we do detect contradiction in their arguments, we have to try to "project ourselves" into their frame of mind, which would show us why they are so ready to accept contradictions that seems so ugly and meaningless to us.

In doing this, one will usually discover one of two things. Either the other fellow has confused inference with fact in order to give a delusory appearance of security to his basically metaphysical notions. Or else, one will find that his tacit assumptions about the basic purpose of science are different from one's own, so that he evaluates as unimportant certain difficulties that one is evaluating as crucial. For example, DLP do not regard a coherent logical physical model as an essential theory in physics.

To them this a dispensable luxury. They are not against it, but it does not disturb them if it is absent, provided that their ideas are in principle able to approximate the facts to any desired degree of approximation. A purely logical contradiction does not disturb them, because they see physics as a means of quantitative approximation to the observed phenomena. Or at least, so their attitude seems to me.

So if you want to meet their arguments, you will have to bring out their tacit assumptions to make them as plausible as possible, then contrast their assumptions with your own, and explain why you favor your own. You may never convince DLP in this way, but at least, the "neutral" reader will have a chance to see what all the fuss is really about, so that he can make up his mind for himself.

With best regards

David Bohm

P.S. With regard to basic metaphysical notions, I had an interesting experience at the Bellagio Conference in Theoretical Biology[13] for which the paper on Order sent to you was prepared.

I observed in the first four or five days that everyone was talking at cross-purposes, because his basic metaphysical assumptions were different. What was crucially significant to one man was of no importance to another, and vice versa. And this was inevitable, because each man's deep tacit assumptions led to different inferences as to what is essential, what is superficial, etc. So nobody really understood why others got so excited about certain things, and why nobody properly responded to his own contributions.

At a certain point, I plunged in and presented my own metaphysical notions.[14] In doing this, I explained that each person inevitably has some kind of metaphysics, which is just a set of general and basic assumptions about reality as a whole.

Even the most "practical" man has such a metaphysics, but it is tacit rather than explicit. Such metaphysics is the most dangerous, because does one doesn't know one has it. It seems to be fact rather than metaphysics. So it is important to be aware of one's own metaphysics, and to see its inferential and assumptive character. To have some kind of metaphysics is inevitable. The destructive thing is not to know that it is metaphysics.

Thus, the ancient creation myths were forms of metaphysics. Then came the Greeks with "All is becoming," "All is fire," "All is water," "All is air," etc., etc. The Pythagoreans said "Number is the essence of all things". Plato said that the essence of all things was in their idea. Aristotle compared all things to living organisms. Democritus put the basic and essential character of all things in the atoms. More modern physicists said "mechanics is the essence of all natural law." Present day physicists say that the essence is in mathematical formalism. Some people at the Conference wanted to explain all things as structures of automata or computers. Some biologists said that fortuitous mutation of genes selected according to survival is the essence of all evolution.

Whenever anyone says anything about "all," "everything," "essential," "basic," "fundamental," "always," "never," etc., etc. he is talking metaphysics. It is evidently some kind of assumption. For clearly he does not really know the totality of existence. Yet, science is impossible without some such assumptions. The important point is to see their assumptive character, not to confuse them with perceived facts. Failure to do this is the ultimate source of almost all confusion in discussions of the fundamentals of science.

I explained that when you are confronted with someone else's metaphysics, you generally feel seriously disturbed, because your own basic notions are thus tacitly if not explicitly challenged. But you should be disturbed. Otherwise, you will never

[13]The 2nd Symposium on Theoretical Biology, held August 3-12, 1967. The proceedings are published in Waddington (1969). See Introduction p. 3, n 10 for details. The paper for the first of Bohm's contributions "Some remarks on the Notion of Order" (pp. 18–40) had been sent to Jeffrey Bub, see p. 115, n 8 and p. 119, n 9—CT.

[14]Waddington (1969), "Further Remarks on Order", pp. 41–60. See Introduction p. 3, n 10—CT.

know that you have metaphysical notions, confused with facts. If you can face the disturbance and confront the other fellow's metaphysics against your own, then you begin to be aware of all kinds of metaphysical assumptions. Eventually you may drop your own assumptions, without necessarily accepting the other fellow's. Rather, your mind is now open to form new general assumptions, of which you are aware. These assumptions are always being questioned, and always changing. So your metaphysics is no longer rigid and fixed. Instead, it is a dynamic and living thing, which is useful rather than destructive.

The positivists saw the destructive effects of a rigid metaphysics. But their reaction was to condemn all metaphysics. In doing this, they did not notice that such an allness assumption about metaphysics is itself a form of metaphysics. Because they were unaware of this metaphysics, it became rigid and therefore destructive. What is needed is not to condemn metaphysics, but to be aware of it, thus allowing it to change and develop.

After I explained all this at Bellagio, I found that almost everybody wanted to get up and present his own metaphysics. In reality they had always wanted to do it, but had been afraid. For in their minds was the "model" of the scientist who is free of such silly things as metaphysics, and who has his "feet on the ground." When this model was dropped, everyone discussed his metaphysics, and thereafter, people began to see at least what the other fellow was trying to say. They didn't always agree. But at least, people's minds began to meet, so that a real challenge was possible, when they failed to agree.

In my opinion, our discussions of fundamentals of quantum mechanics should, as far as possible, have a similar aim.

Sept 28, 1967

Dear Jeff,

I have just been reading Feyerabend's chapter in Beyond the Edge of Certainty, edited by Colodny (Prentice-Hall).[15] He gives a very good illustration of much of what I mean by the notion of fact and inference. In particular, he points out how a great deal of confusion in science is the result of allowing inferences at various levels to be mistaken for directly observed facts. This mistake tends to support conservatism in scientific thinking. For it causes us to overlook the uncertainty of a great deal of what we "know", insofar as incompletely confirmed inferences are regarded as equivalent to "solid" and "indubitable" facts. This is one of the points that I have been making. The work of people like J. and P., D-L-P, Rosenfeld, etc., always leads to the conclusion that current ideas do not need to be questioned deeply, because they are either merely systematic statements of directly observed fact, or else inferences that are as well confirmed as the existence of Australia (and therefore equivalent for all practical purposes to facts).

[15] Feyerabend (1965)—CT.

He also mentions the empiricist notion that we should hold onto an idea until there is definite evidence against it. He contrasts this with the critical and pluralistic attitude; i.e., that we need many different theories, whose terms do not reduce to each other completely, either in content or in meaning. A new theory of this kind may call attention to different phenomena which refute the old theory, but only when the new theory is adopted. The role of the new theory is to connect phenomena (perhaps already known) with the older domain. Without the new theory, one would not realize that the "new" phenomena actually do refute the old theory. This is what happens in hidden variable theory. Thus, such a theory suggests that successive observations may fail to conform to the ensembles of quantum mechanics. Such a phenomenon would constitute a refutation of quantum mechanics.

Feyerabend brings out the essential point at issue here – i.e., in the confrontation of different theories of the same field, one either obtains a very strong refutation of the old theory or a very strong confirmation. In either case, knowledge is advanced significantly. The confusion of certain inferences of the old theory with facts leads to dogmatism and ossification, because it prevents such confrontation. This is the basic criticism of Bohr, Rosenfeld, DLP, von Neumann, J. P., etc.

With best regards

David Bohm

P.S. The situation with Bohr is not that he is an empiricist. On the contrary, he says that the structure of language is first. In other words, the structure of all possible unambiguous concepts (taken to be equivalent to that of classical physics concepts) is supposed to determine what is communicable and therefore what can be the subject matter of physics (and other sciences). Bohr seems to take it to be an evident fact that what is communicable is fixed within immutable limits. This inference is what he confuses with a fact. Once we accept this, the rest follows. As is usual, the confusion of inference with fact leads to dogmatism and conservatism. However, Bohr is concerned, not so much the empirical fact of physics, but rather, the fact about epistemology, which is a fact at a higher level of abstraction. Thus, Bohr is an anti-empiricist, in a certain sense, since he puts epistemological limits first, ahead of all actual sense perception. Nevertheless, he is still a dogmatist, who confuses inferences about epistemology with facts about epistemology.

References

Bohm, D., & Bub, J. (1968). On hidden variables - A reply to comments by Jauch and Piron and by Gudder. *Reviews of Modern Physics*, *40*(1), 235–236.

Bub, J. (1968a). The Daneri-Loinger-Prosperi quantum theory of measurement. *Nuovo Cimento B, LVI, 1*(2), 503–520.

Bub, J. (1968b). Hidden variables and the copenhagen interpretation-A reconciliation. *The British Journal for the Philosophy of Science*, *19*(3), 185–210.

Feyerabend, P. K. (1965). Problems of empiricism. In R. G. Colodny (Ed.), *Beyond the edge of certainty: Essays in contemporary science and philosophy* (pp. 145–260). Englewood Cliffs: Prentice-Hall.

Freire, O, Jr. (2019). *David Bohm: A life dedicated to understanding the quantum world.* Berlin: Springer.

London, F., & Bauer, E. (1983). The theory of observation in quantum mechanics. In J. A. Wheeler & W. H. Zurek (Eds.), *Quantum theory and measurement* (pp. 217–259). Princeton: Princeton University Press.

Waddington, C. H. (Ed.). (1969). *Towards a theoretical biology* (Vol. 2). Edinburgh: Edinburgh University Press.

Chapter 5
Folder C133. Fact and Inference Continued. Feyerabend. October–December 1967

OCT 10th, 1967

[Date added – CT.]

Dear Jeffrey

I received your recent letter and card, for which we thank you.

In answer to your question, I have no particular preferences about the duplication of my preprints. Let the university make the usual charges, and keep the money for me, if this is indeed the usual practice.

Your comments on fact and inference are very much to the point. The question of beauty and ugliness is involved in a very deep way. Probably we would need several days discussion to get really far into these issues. However, I shall try to sketch a bit of what is implied in the relationships of beauty, fact, truth, harmony, totality, peace, etc., etc.

Let us begin with the boy who is stealing. You say that there are two inferences:

(a) The boy <u>is</u> stealing.

(b) The boy <u>is</u> a thief.

How do I justify calling (a) a valid inference and (b) a source of delusion?

It is true that (a) and (b) are both inferences. But there is a tremendous difference in their basic qualities. (a) refers to the directly observable actions of the boy, and makes no assumptions about his deep more and or less permanent structure. In other words, (a) refers only to a contingent feature of the boy's actions. By observing these carefully enough, we can answer the question

(A) Is the boy stealing?

Note that in the step between Inference$_n$ and Fact$_{n+1}$, <u>Inference$_n$</u> is <u>first</u> changed into <u>Question$_n$</u>. (This step may be largely tacit and not verbally expressed.) <u>Question$_n$</u> calls for <u>Observation$_n$</u> and the answer thus supplied is Fact$_{n+1}$.

Let us now do this with Inference(b). We ask:

<u>Is</u> the boy a thief?

C. Talbot (ed.), *David Bohm's Critique of Modern Physics*,
https://doi.org/10.1007/978-3-030-45537-8_5

Note however that this question contains a deeper tacit assumption; i.e., that human nature is such a simple and mechanical sort of thing that this question can have a "yes" or "no" answer. Thus, we can ask "Is this an electric motor?" By studying the structure carefully we can answer "Yes, it is an electric motor", implying by this that when the switch is pressed, the motor must turn. But can we obtain a corresponding knowledge of the boy's structure, which would imply that when the boy sees something that he wants, this presses the mental "switch" so that he will steal, just as the motor will turn, when its switch is pressed? Evidently, this is a vast question that cannot be answered by a few observations of the boy's behavior. Nor can one even see how it could be answered by a detailed analysis of his brain and nervous system. Moreover, there is a vast quantity of knowledge available to everyone which suggests that the situation is more complex than this. For example, each man knows that he is similar to the boy in many basic ways. Yet, few men will agree willingly to having their own faults and weaknesses attributed to an "intrinsically wicked nature", implying that they must always be ugly and wicked, unless it is "beaten out of them." Rather, they ask for affection and sympathetic understanding of the deeper origins of their faults. But in thinking of the boy, they forget all this.

So what can we learn from this? Mainly that each question contains tacit assumptions, of which one has to be aware. The question "Is the boy stealing?" contains the tacit assumption that the boy is capable of a certain behavior called "stealing". We know enough of people to realize that everyone is capable of this behavior. So this is a valid question, because its basic tacit assumptions have been adequately confirmed. The question "Is the boy a thief?" assumes tacitly that the human mind is capable of having a thieving structure. This assumption is at best an inference from very limited and inadequate data, which moreover ignores vast quantities of contrary data that are available to everyone. So it is not a fact. Question (b) therefore confuses inference with fact, not in the form of an explicit statement, but rather in the form of tacit inferences, of which we are actually unaware. This is the really dangerous kind of inference. For without knowing it, we have taken certain inferences (e.g., about human nature) to be facts, before we have posed the questions and made the observations that could establish the real fact about them (i.e., truth or falsity). This very step is the beginning of delusion. And from here follows endless confusion. For as we obtain new data, we have to shift our consciously held inferences in order to accommodate this data, without altering our unconscious tacit influences. In this way, the very evidence that could show us the falsity of our tacit assumptions becomes the source of new false assumptions that make our minds more confused than ever.

For example, the inference "The boy's thieving proclivities can be beaten out of him" is generally found not to work. He goes on stealing. From this, one might be led to question the assumption that a human being can be a thief, who has a thieving structure. But one doesn't realize that this is an assumption. Rather, it seems to be an "evident fact". So instead one infers that the boy is an "incorrigible criminal".

If every question can contain deep and largely unconscious tacit assumptions which are easily mistaken for facts (i.e., highly confirmed inferences), what is to be done to be free of delusion and confusion? Evidently, there is no verbal formula that could adequately answer this question. Rather what is called for is a deep and extensive sensitivity to what is false, disharmonious, ugly, and full of conflict. Evidently, one gets a feeling that there is something wrong with the question "Is the boy a thief or not?" If one pays serious attention to this feeling, being sensitive to all its subtle nuances, one is led to a more detailed perception of what is wrong with the question. The same holds more generally. Thus, in science one may ask "Is the electron in a certain position or not?" This question presupposes a certain "particle-like" structure that the electron probably does not actually have. When we find that the question does not have a clear answer, we should be led to question the tacit assumptions behind the question. But instead, we may make new inferences to accommodate the facts (e.g., that the electron's structure function is incomprehensible, and that only computation of its phenomenally observed behavior is possible). These assumptions add to the confusion (as in the case of the boy).

So fundamentally, to see what is true or false (and therefore to perceive the fact) depends on sensitivity to beauty and ugliness, harmony and disharmony, etc. This evidently takes place at great depths in our mental processes. People like to say that it is "personal" and "subjective", therefore of no concern to the study of science, which is based on fact. Yet, without it, there can be no real fact, at all. Rather, a mind that is insensitive to beauty and ugliness must ultimately delude itself.

So you are right to say that we confuse fact and inference because the mind is already in an ugly state. But it is not enough just to say so and to leave it at that. Rather, we can go further. Firstly, we can be aware of the place where insensitivity to ugliness gives rise to delusion – i.e., it enables us to accept tacit assumptions or inferences as confirmed facts. To be aware of how this it is happening will be at least a step in ending the whole process of delusion and confusion, leading to evil. And secondly we can inquire more deeply into why we are insensitive to ugliness.

As a first step in this inquiry, let us notice that when I say "That boy is a thief", there is an implied comparison to me, meaning "I am not a thief but a good virtuous person". The thought of this comparison induces a fleeting sensation of pleasure, security, satisfaction, etc. If one doesn't look at it too closely, this pleasure seems to be "beautiful", to have in it a sense of "harmony", "well being", "euphoria", etc. Of course, if one does scrutinize it carefully, one sees that it is ugly, hence painful. That is why the sense of pleasure is so fleeting. For it can be "recognized" as beautiful only if one does not get a good look at it.

The real issue here is the confusion of glamor and beauty. Glamor is a superficial imitation of beauty, which "gets by" for a short time. Evidently, most of what is called "beauty" in modern society is actually glamor. And the same was true in ancient society. Thus, children were told stories about the "heroes" in which the murderous and destructive actions of these individuals were glamorized and made to appear to be noble and beautiful. Likewise, people tend to "glamorize" their own qualities, so that these appear to be beautiful, when they are actually ugly. For example, a person's violent and aggressive behavior is glamorized as "strength of character", while another person's timidity and fear is glamorized as "gentleness".

When the mind is full of false beauty, which is really glamor, then it has inevitably lost the sensitivity needed show up the deeper tacit assumptions that are in the questions, with which we establish the fact. For example a Nazi who was trying to be "factual" might say: "Are the Jews really of the same nature as Nordic Germans or not?" This very question contains the tacit assumption that people can be divided into well defined groups, which can meaningfully be compared with regard to the values in their deeper natures. This assumption makes possible the pleasant glow of euphoria and glamorized substitute for beauty, which comes when one thinks: "In a comparison between my group and other groups, my group comes out supremely superior." In the false light of this glow, the subtle feeling that "something is wrong" can no longer be noticed. So one is trapped. One has mistaken an inference for a fact. The resulting glow of "pleasure" creates a spurious impression of beauty, and therefore the tacit inference "Everything I think is right". Any hesitant step to perceiving the underlying ugliness interrupts this glamorized "beauty". The interruption of "beauty" is then interpreted as ugliness. So one confuses the perception of ugliness with the ugliness that is perceived. This confusion means that the very evidence of the falsity of the structure of one's thought becomes the means of preventing one from seeing this falsity. All evidence of this kind is reinterpreted and accommodated, in an ever growing structure of new false inferences, thus giving rise to never ending confusion.

Svetlana Stalin illustrates this nicely in her account of her father[1], who was always demanding "facts" about the people who were "scheming against him". This very question contained the tacit assumption that every man was either a friend of Stalin or an enemy of his. So all sorts of really harmless actions were inferred to be "plots" against him. Then when Stalin's wife questioned the actions of Beria, Stalin always asked for "facts" against him. Since Beria was the only possible source of such "facts", it was plain that none could ever be forthcoming. So Beria was always a "friend" and others against whom Beria gave "evidence" were always "enemies". In this way, vast evil and destruction were set in motion, from which tens of millions of people suffered and died.

It seems clear that when the mind is in a noisy state of mistaking glamor for beauty, then it cannot be sensitive to the subtle inferences in the questions, with the aid of which it is establishing the fact. For if there is already the roar of 100 decibels, the tiny "sounds" that indicate that "something is wrong" cannot be heard. So what is needed is a mind that is in a general state of harmony, which will be felt or sensed as "silence" or "emptiness", in the depths. The slightest disharmony on the surface will then "stand out" like a sore thumb. In the presence of such perception, the mind will explore and move until a new harmony is established. This exploration will contain as a by-product the perception of the falsity of certain inferences, and (where appropriate) the proposal of new inferences. It is essential that this be a by-product of a deeper non-verbal movement of harmony. For it is meaningless to try to establish harmony merely by manipulating the superficial content of ideas and feelings, (by adapting one's words and other symbolic structures).

In a way, we can compare the situation to the use of a galvanometer. In some circuits, this instrument is used "positively" to measure the electric current. But in

[1](Svetlana, 1967)—CT.

a Wheatstone bridge, it is used "negatively" to indicate "imbalance", which can be regarded as a kind of "disharmony". Here, its function is to enable the operator to change the circuit until it ceases to function. So its function is to bring about its non-function. Similarly, thought can function "positively" as a "measure" of the world in which we live. But its deeper function is "negative". That is to say, we think when there is a "problem" to be solved. A problem indicates that there is disharmony somewhere, either inwardly in the mind, or outwardly in the perceived world. This problem implies a contradiction somewhere, in our total structure of inferences and results of observations. This contradiction, if it is at all subtle, is first indicated by a feeling of disharmony, which "stands out" against the harmonious "silence" in the depths of the mind. Creative movement originating in these depths "alters the circuits" until the sense of contradiction ceases. So in its negative role, the function of thought is to indicate disharmony and thus help the deeper levels of the mind to bring about harmony, indicated by the ceasing of the function of thought. When this happens, our inferences will generally have a different structure. In the process, we also observe, inwardly and outwardly, to see the truth and falsity of our inferences.

You can perhaps see that the questions you raise go very deep. Indeed, if there were more opportunity, I would show that they lead to the questions of time, space, pleasure, the nature of thought, and the delusory character of the separation of the observer and the observed. However, this will have to be for the future.

Meanwhile, I am sending by surface mail some talks given by J. Krishnamurti, whom I once mentioned to you. Here, you will find all these questions discussed from a more psychological point of view. Perhaps this is what is called for, before going very much more deeply into all these issues.

Best regards

David Bohm

P.S. About von Neumann's saying hidden variables imply that qu. mchs. is "objectively false" I had an idea. The trouble is an ambiguity in these words. Do they mean only that qu. mchs. is an approximation and not an absolute truth? Or do they mean that qu. mchs. is already "objectively true", so that hidden variables are impossible? It was really v. Neumann's reponsibility to clarify the issue here. To do this, he should have refrained from discussing truth or falsity in qu. mchs. Rather he should have said: "The axioms of qu. mchs. are logically incompatible with hidden variables". Then it would have been clear that hidden variables are still possible, because the axioms of qu. mchs. have been verified factually only to limited approx., in limited domains, etc. By saying "objectively false", he confused the issue, because, for all we know, they are "objectively false" in more accurate measurements. We have in reality no way of knowing whether the axioms are "objectively true" or not. All we know is that inferences drawn from them have been confirmed in a certain domain up to a certain approximation (as is true also, for example, of Euclidean geometry). So the issue of "objective falsity" of any set of actions is a complete red herring. One uses these words in this context only because one has a tacit empiricist metaphysics, which identifies (and therefore confuses) truth with verification of factual inferences. No axioms are ever true or false. Only inferences drawn from them in the field of observation can be true or false.

This is a crucial point, which illustrates the structure

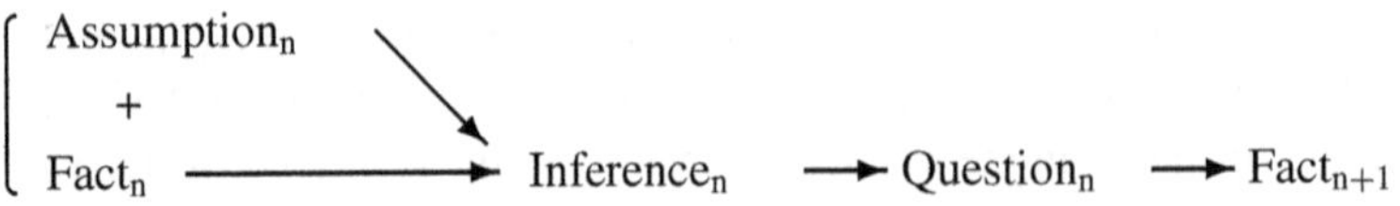

$$\left[\begin{array}{l}\text{Assumption}_n \\ \quad + \\ \text{Fact}_n\end{array}\right. \longrightarrow \text{Inference}_n \longrightarrow \text{Question}_n \longrightarrow \text{Fact}_{n+1}$$

The Answer to Question_n establishes the truth or falsity of Inference_n but not of Assumptions_n (which may be tacit as well as explicit).

Example:

Assumptions_n include the axioms of Euclidean geometry plus a lot of tacit "background" assumptions.

Fact_n is that we have been able to make measurements of various kinds, which form, as far as has been seen, a coherent structure, within these axioms and assumptions.

Inference_n is that the indefinite extension of such measures (over long distances) will yield similar results.

Question_n is "Will they do so?" This leads us to design instruments to answer it, as well as new theories showing how Euclidean geometry might fail.

Fact_{n+1} is either "They do" or "They don't".

Von Neumann would confuse the issue by saying that if the answer is "They don't", Euclidean geometry has to be "objectively false". Our response to this is that nobody can know finally and for sure whether Euclidean geometry is "objectively false" or not. Indeed, this is just why we are testing such inferences, with the aid of questions drawn from alternative systems of axioms. By using the phrase "objectively false", von Neumann tends to lure or trap us into the further inference that since Euclidean geometry is thus far "objectively true", this whole line of reasoning is unnecessary and fruitless. Of course defenders of v. Neumann can always say "He didn't really mean this". But this is like the question of "How to Win at Games Without Actually Cheating."[2]

It was really von Neumann's (and everybody's) responsibility to phrase his conclusions, so that a lazy or tired mind is not likely to dig itself a trap and fall into it. The fact that so many people were convinced by v. Neumann that "hidden variables are impossible" is a proof of the effectiveness of this "trap", made by the ambiguous phrase "objectively false".

Oct 11, 1967

Dear Jeff

This is a brief supplement to yesterday's letter.

[2]Title of a well-known book by Stephen Potter, originally published in 1947 (Potter, 1965)—CT.

First, let me give a more detailed description of the hierarchy of fact and inference. The first step is that one is always aware of previous inferences and factual observations. Then one has all sorts of assumptions. These too are inferences, but of a more general nature, drawn from a vast body of knowledge. Our most general assumptions constitute the metaphysics of the moment. I stress the "momentary" character of metaphysics, because it is really always changing in subtle ways. However, we tend to try to formulate it as if it never changed. This is, of course, a contradiction. The resulting confusion can be avoided by recognising and realizing (being aware) that we have a metaphysics and that it is inevitably always changing.

So we have the process:

Fact_n + Assumptive $\text{inferences}_n \rightarrow$ conclusive inferences_n

Conclusive inferences may include logical deductions from fact_n + assumptions_n, but usually more general processes are involved. Thus, there may be associative inferences (fact_n + assumption_n remind one of analogous situations). Then there are the inferences which generalize previous influences, as well as inferences which restrict the generality of previous inferences (thus generalizing the process of restriction). Besides, there are probably other kinds of inferences (e.g., hunches, intuitions, etc.)

The next step is:

Conclusive $\text{inferences}_n \rightarrow \text{Question}_n \rightarrow \text{Observation}_n \rightarrow \text{Fact}_{n+1}$

We have to be aware of the whole process. Thus, when we formulate Question_n, we may feel that "something is wrong", and this feeling may indicate that we are overgeneralizing our inferences about the factuality of some of our deeper assumptive inferences.

Our axioms are part of our general assumptive inferences. We do not directly test these assumptive inferences. Rather, we draw further inferences from our axioms, and test these latter inferences by observation. So our axioms can neither be "objectively true" nor "objectively false". They can be compared and criticized only through their ability to lead to lower order inferences that agree with what is observed.

Here is where von Neumann's ambiguity introduced confusion. When he said that qu. mchs. would have to be "objectively false" to allow hidden variables, he was playing on the ambiguity between the three phrases:

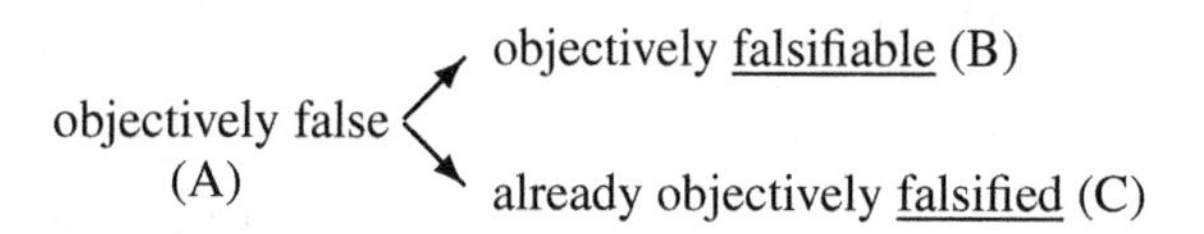

Phrase (A) may be interpreted as meaning either (B) or (C). Evidently, von Neumann's statement would be true, if one took it to mean "hidden variables imply that qu. mchs. is objectively falsifiable." This statement is obviously correct, but it is so obvious that nobody would even bother to say it, if that is all he meant to say. Therefore, when one reads von Neumann's statement, one is inevitably led to search for a "deeper meaning". This search is, of course, a trap, because his statement has no deeper meaning. But as soon as one begins the search, one considers meaning B. "If hidden variables exist, qu. mchs. would already have to be objectively falsified. But it has not been thus falsified, so that hidden variables therefore do not exist." This argument is of course a non-sequitur. Nevertheless, it is tacitly what any reader of v. Neumann would tend to do, unless he were extremely awake and alert. Because v. Neumann has so much status and authority, few physicists are ready to believe that von Neumann would say something obvious, in a form which suggests something deep. So even those who suspected that something is wrong were frightened out of really inquiring deeply into what is wrong.

Even in the relationship of statistical mchs. and thermodynamics, one can see that if statistical mechanical assumptions are true, then thermodynamics must be "objectively falsifiable" (e.g., by fluctuation phenomena). By using ambiguous phraseology (i.e., "objectively false") v. Neumann obscured the real significance of his theorem, which is this:

Quantum mechanical axioms imply that hidden variable explanations which attempt to treat quantum ensembles as simple averages of deeper deterministic ensembles will not work. (So that we cannot use ideas analogous to Gibbs ensembles, to explain quantum mechanics).

As you have pointed out, this means that the motion of individual members of hidden variable ensembles must depend in an inseparable way on the large scale environmental parameters.

I think that this is what has to be stressed in any further papers on the subject. The notion of structural process is a natural way of bringing in such a mutual dependence, through the idea of reflective function. Each level reflects the others. Mechanistic theories regard the macro-level as a reflection of the micro-level. But they regard the micro-level as "existing in itself" because it is the "fundamental reality". Its laws do not depend on macro-parameters. Only its contingent features (e.g., initial conditions) can reflect the macro-parameters. Our new idea is that laws of the macro-level reflect both necessary and contingent features of the micro-level, so the laws of the micro-level reflect both necessary and contingent features of the macro-level.

Best regards

David Bohm

Oct 16, 1967

Dear Jeff,

I have just given a talk on our paper in Birmingham, from which I have, I think, learned something about how to present our ideas more effectively.

I began by saying that the usual interpretation of qu. theory has as many shades and varieties as there are interpreters. But there are, broadly speaking, two poles, represented by Bohr and by von Neumann.

Von Neumann adopts the idea of mind-matter dualism as a basis. He defines subject and object, placing the former on the "classical" side of the "cut" and the latter on the "quantum" side. But by doing this, he effectively treats the "subject" as yet another object. Thus, he implicitly defines two totally different kinds of "objects", one of which is called the "observed" while the other is called the "observer". This is similar to Descartes definition of mind as "thinking substance" and matter as "extended substance". In both cases the problem arises as to how two such totally different domains (classical and quantum, or mind and matter) can be related. Descartes proposed that in his own mysterious way, God saw to the relationship of mind and matter. But von Neumann never proposed a similar explanation of how classical and quantum "objects" were to be related in his theory. The whole question is left "up in the air", in a very unsatisfactory way. (e.g., by appealing to a principle of psychophysical parallelism).

Bohr is unclear on this question. At times, he seems to accept the subject-object dualism, calling the observing instrument the "subject". But more deeply he has an entirely different notion, i.e., that the essential question is linguistic, having to do with unambiguous communication. Clearly, language is both subjective and objective. It is subjective with regard to its meaning, which varies subtly from person to person, and from moment to moment, even in a given individual. It can, however, be objective with regard to that part of its content that is unambiguously communicable. For example, when we use the word "chair", we can, generally speaking, always agree on which objects are chairs, and on what can be done with them.

In Bohr's view, the refinement of unambiguous language leads inevitably to the conceptual structure of classical physics – i.e., to the analysis of the world into particles and fields, which are completely described, in principle, by suitable sets of canonical pairs of variables (like position and momentum).

Classical physics is characterized by two related structures

1. Descriptive
2. Inferential

Clearly, we can use position and momentum variables to describe the states and changes of states of classical systems. In addition, we can draw inferences, in this framework, with regard to the basic laws of motion (e.g., Hamilton's equation).

When experiment disagrees with classical theory, the first step is to try to change inferential structure (e.g., change the Hamiltonian, introduce new fields and particles, etc.). But quantum mechanics went much further than this. It implied that the entire inferential structure breaks down altogether and even the descriptive structure underwent a decisive change. For quantum mechanics implies the denial of the classical notion that position and momentum variables can be described together, at least beyond the limits of precision set by the uncertainty principle.

Bohr interpreted this situation in terms of the principle of complementarity. That is, the arrangements of matter needed to measure complementary variables precisely are incompatible, because of the quantum of action. Therefore, there is an inherent ambiguity in these variables. When this ambiguity is relevant, there will be a corresponding ambiguity in the separation of subject and object (observer and observed). For it will no longer be possible to describe a physical interaction in enough detail to see whether it originates in the instrument or in the electron. This ambiguity is reflected in our language, which uses the same terms to describe the conditions of the experiment (momentum or position measurement) along with the results of the experiment, the two being interwoven in an inextricable way. So Bohr does not really assume mind-matter dualism (though at times he may appear to do so, because he doesn't make himself clear). Nor is he a positivist, since he puts language first, before the question of empirical observation. To Bohr, what can be observed depends on our language, which defines our terms and limits what we can recognize and communicate. In a way, this is like what Feyerabend said, in his article in Beyond the Edge of Certainty. But the key difference is that Bohr restricts the language of description of fact at the perceptual level to that of classical physics (limited by the principle of complementarity), while Feyerabend would allow all sorts of languages in this domain.

In Bohr's view, the algorithms of quantum theory play the role of a metalanguage, which makes statements about the language, that is used to describe the perceivable phenomena. Because there are probability statements, the metalanguage is ambiguously related to the language.

Now, in Bohr's view, the inferential statements are determined by the metalanguage, while the original (classical) language is being used only descriptively. Because of the ambiguous relationship of the metalanguage to the language, it follows that inferential statements have just the right degree of ambiguity to be compatible with the principle of complementarity. For example, the ambiguity of p (when q is fairly well defined) is reflected faithfully in the algorithm, which gives a corresponding probability distribution of p.

Now, my criticism of Bohr is that I feel he is wrong to project these linguistic questions into the quantum domain. In the psychological domain, one must consider the dual role of language, which is both subjective and objective. Some kind of metalanguage is needed to discuss this question. And eventually, it carries us to direct awareness, beyond what can be described in language, where subject and object are one. But in my view, there is no reason to regard the quantum algorithm as a metalanguage. This becomes necessary only when we assume that classical physics gives us the unique unambiguous language. I would rather explore the possibility of new languages (and conceptual structures) for describing the domain of directly observable large scale phenomena. These new languages would have room for a direct phenomenal interpretation of some of the terms in the algorithm of quantum mechanics.

One can further criticize Bohr's point of view in that it makes the details of the individual process undescribable. Is Bohr not here confusing an inference with a confirmed fact? Certainly if we accept Bohr's assumptions, the details will be indescribable, almost by definition. But the deeper question is that of whether Bohr's assumptions are inevitable transcriptions of already observed facts, or whether they are merely proposals that seemed plausible to Bohr and his followers. It is like the boy who is stealing. To some people, the question "Is he a thief?" is a relevant one, because the notion of human nature as divided between "honest" and "thieving" is a well established fact to them. They can point to the empirical possibility of dividing people between those who steal and those who do not as confirmation of this view. The critics of this view must question the entire set of structural assumptions about human nature. Similarly, to criticize Bohr adequately, one must question his whole set of assumptions about language. Are they really adequately confirmed or not? Our own papers help in my view, to make such criticism possible.

Here, I emphasized the need for alternative sets of metaphysics, as a strategy to help keep science from being frozen in a closed circle of concepts. As Feyerabend points out, what we are ready to observe depends on our general theoretical structures, which provide a basic language for description and communication of observed fact. For this very reason, we need alternative theories, leading to alternative languages, if we are even to see what it means to refute a general theoretical point of view. Theories are not merely being accumulated, to give predictions and explanations. In addition, they confront each other, to permit a dynamic process of fundamental change and development.

One can illustrate the strategic role of our ideas by reconsidering the von Neumann "proof" that there are no hidden variables. Of course, he starts from certain axioms, which are probably not unreasonable "models" of the kind of quantum mechanics that physicists have been using fairly "successfully" over a period of years. But he says that according to his "proof", quantum mechanics would have to be "objectively false" if hidden variable explanations of it are to be possible.

Here, we must ask "Isn't von Neumann confusing axioms, which are always inferential in nature, with what he would call 'objective facts'?" Axioms can never be true or false. This we can see to be a "higher order fact", by looking carefully at the real structure-function of axioms in the total process of theorizing. In other words, the "philosopher of science" is also trying to establish facts – That is, he must confirm or falsify his higher level "meta-inferences" about the "inferences" used by physicists in their theorizing. I suggest that any competent person will see, if he examines the questions, that axioms are neither true nor false. Rather, we draw inferences from them, which are either confirmed or falsified. Ultimately, as we go to "meta-inferences" and "meta-meta-inferences" we must always appeal to what a competent person will agree to be the perceived truth. (The same is true about the question of whether or not the boy can have a "thieving" nature). There are no "incorrigible facts" at a bottom level. But I do find, factually, that all those with whom I discuss the question seriously do actually agree that axioms can neither be true nor false.

Very probably, in von Neumann's day, this question was not too clearly understood. The old idea that a theory is either true or false still held sway. So von Neumann confused the inference that axioms can be true or false with a fact. We are now able to see that this is only an inference, and moreover, an inference that is not confirmed by observing the way physics can actually be done. We now see that empirical observation can falsify a set of axioms, without proving them to be "objectively false". So the real content of von Neumann's theorem can only be one of the two statements.

A. Hidden variables imply that quantum mechanics is potentially objectively falsifiable.

B. " " " " " " is already objectively falsified.

Everybody will accept the validity of (A) immediately. But then it seems so obvious that one does not understand why an outstanding physicist like von Neumann made such a fuss about this trivially obvious point.* So almost inevitably, one assumes that von Neumann must have had interpretation (B) in mind. But since quantum mechanics is thus far "objectively verified", it follows that hidden variables are really impossible. Thus, von Neumann's "proof" has led to confusion over the years.

Having cleared this point up, I now asked: "Is von Neumann's proof really only a triviality, or didn't he show something interesting, whose significance was hidden by the confusion between "potentially falsifiable" and "actually falsified"? What von Neumann showed was that if there are hidden variables underlying the quantum theory, these cannot be similar to the atomic "hidden variables" that underly classical statistical mechanics. For these latter satisfy the linearity condition

$$\langle aA + bB\rangle_{av} = a\langle A\rangle + b\langle B\rangle = a\int \rho(\lambda)A(\lambda)d\lambda + b\int \rho(\lambda)B(\lambda)d\lambda$$

Von Neuman showed that if such conditions were satisfied in quantum mechanics, then dispersionless ensembles are impossible. Or alternatively he showed that dispersionless ensembles imply the breakdown of the linearity condition. But the linearity condition is closely related to the assumption that whereas the macro-level is determined by the micro-laws, these latter laws are completely independent of the situation at the macro-level. (In other words, that observables like A and B depend only on the "hidden" micro-variables, λ, and not on statistical large scale parameters, like T and P, which appear only in the distribution function, $\rho(\lambda)$.) Therefore, we are led, by von Neumann's analysis, to consider the notion of a set of hidden variables that do not satisfy the linearity condition, and therefore do not satisfy the assumption that the micro-laws are independent of macro-conditions. This leads us to a new view of the structure of the world, which amounts to a new metaphysics. To some extent, Mach foreshadowed such a possibility by proposing that the inertial frame (even for the basic constituents of matter, whatever they may be) depends on the distribution of matter over the whole universe. But here, we are led to go much further, to propose that large-scale structures of many kinds may enter into the form of the micro-laws.

At this point, one sees the need for a more detailed model of such a new metaphysical point of view. Our proposals of hidden variables provide such a model. We see that, as von Neumann showed, our model implies the breakdown of the linearity assumption. In order to recover the linearity property, we propose a new process that randomizes the hidden variables in some unspecified period of time, τ. So by constructing a hidden variable "model", we actually see the deeper meaning

*E.g. Statistical mechanics implies that thermodynamics is "objectively falsifiable", for example, by fluctuation phenomena, but nobody would bother to construct a "proof" of this self evident point.

of von Neumann's analysis. The two steps (von Neumann's analysis and our model) complement each other. If von Neumann had properly expressed the results of his analysis (without bringing in the confusion of objectively false – falsifiable – and falsified) his analysis would have been a powerful impetus and guide for the inquiry into hidden variable theories.

Our model then brings out several further significant points.
(1) It brings in the possibility of new orders at the *level of the phenomenon*. For example, if the hidden variables are defined precisely, the order of the results of a series of measurements of "non-commuting observables" is completely defined. And more generally, a non random distribution of hidden variables implies a partial definition of orders of this kind.

Such new orders, if observed, would contradict Bohr's assumption that *all* unambiguously communicable orders have to be stated in the language of classical physics (e.g., positions and momenta of various objects). Here, what we can observe and communicate is a certain order of *results*, that is related unambiguously to the *extended algorithms* of quantum theory (which now includes hidden variables). So we refute the assertion that the algorithm is *only* a metalanguage, not directly and unambiguously related to the observable phenomena. In effect, we have brought in a new kind of language, both for describing the phenomena, and for drawing inferences about them.

One can compare the situation here to that prevailing in the beginnings of classical physics. By bringing in the calculus, Newton was able to talk about velocity, acceleration, etc., whereas before, there had been no way to do this. One could only vaguely describe a curve like this:

But now, one could be directed to observe and measure the various *derivatives* at different positions. Without this new mathematical language, this would have been impossible. Similarly, hidden variables enable us to define, talk about, and observe new properties, such as the *order* of a discrete set of results, as below:

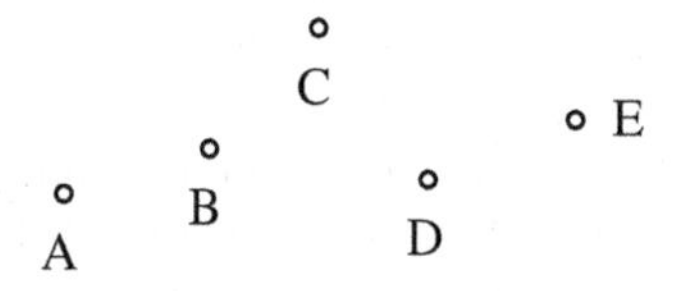

(2) By defining new "observables", hidden variables indicate what is needed to test the basic axiomatic structure of quantum mechanics as a whole. No experiment, unguided by new theories, can possibly provide such a test. For no matter what the results of such an experiment may be, one can always imagine a new Hamiltonian,

a new particle or group, etc., that might explain the results of a disagreement away, without altering the basic general principles (e.g., new epicycles in the Ptolemaic theory to fit any possible observational facts). A test consists in a confrontation of one theoretical framework with another, to see if the latter does not lead to a valid new descriptive language opening up a wide range of new kinds of observations, permitting new kinds of inferences, that could be further tested, etc., etc.

(3) Our model raises the question of the propagation of the effects of hidden variables in a relativistically covariant way. Our present model does not achieve this result, but it points to two possible modifications:

(a) Quantum-mechanical correlations may actually be propagated at the speed of light or less. This could be tested experimentally in principle. Also, one could try to make an extended model, embodying this feature.

(b) In any case, the question of the maximum speed of propagation is inseparably connected with the basic physical assumptions. Thus, in general relativity, it is given by $g_{\mu\nu} dX^{\mu} dX^{\nu} = 0$, so that it depends on the metrical tensor, possibly also the quantum fluctuations of the latter. If there are hidden variables, some new criterion for maximum propagation speed may arise, which involves the hidden variables. The exploration of such a theory is one of the possible lines of further inquiry into the subject.

(4) Our model embodies a dependence of micro-quantities on conditions at the macro-level. Thus, it leads, as I indicated before, to a basically new metaphysics.

Here, we can bring in the notion of a total undivided and indivisible structural process as a natural embodiment of this kind of metaphysics. Each aspect is an abstraction, which reflects the whole in its own way. It is this reflection that makes observation possible. Von Neumann's notion of a separate observer is thus a "red herring" that serves only to create confusion. Every level, including the human mind, contains a reflection of the whole, and therefore, a reflection of all the other reflections, as well. There is no need to introduce a separate "observer" because the "observer" interpenetrates the whole structure of all existence.

On the other hand, we also do not accept Bohr's notion of the indescribability of the processes by which we come to learn about things in the universe. For we have given up altogether the attempt to describe phenomena solely in terms of the language of classical physics (just as Newton no longer tried to use epicycles as his main basic description of the phenomena of astronomy).

One important point is to distinguish between the indivisibility of phenomena and their individuality. We account for indivisibility directly by assuming that phenomena are merely aspects, abstracted from a total structural process (e.g., doorways and walls are indivisible and inseparable). Individuality is a more subtle motion. Usually it implies:

(a) Particularity – i.e., distinct difference and separation from other similar phenomena (but of course, not from the structural process as a whole).

(b) There must be an order in the level of the phenomena themselves, both as different individual phenomena are related, and in the inner structure of each phenomena. This order is in essence what is observable in the phenomena. When this order is random, we obtain statistical inferences.

(c) Each phenomenon (as an individual) has a quality of wholeness – i.e., it is itself, on its own level, a limited sort of totality (as well as being inseparable from its deeper levels). This constitutes a kind of closure, that marks off one phenomenon from another. (E.g., a series of musical notes constituting melody A would be one phenomenon, while a series constituting melody B would, in certain contexts, be another phenomenon.) One is seeking a physical theory where the phenomena would have a similar aspect of wholeness, or relative completion. This means that while they can be divided, they will not in general produce the same kind of phenomenon on division. (E.g., the notes of a melody are a different kind of phenomenon from the melody.)

All these qualities have a very different meaning from that given to them by Bohr. This has to be emphasized very strongly. Bohr takes wholeness to be an undefinable linguistic term. That is, the language of classical physics, limited by the principle of complementarity, does not permit us to do other than to discuss the phenomenon in its not further specifiable wholeness.

With best regards

David Bohm

Oct 19, 1967

Dear Jeff

I have received your letter of Oct 9, which I found very interesting. I think it becomes clear from this letter where we are failing to communicate.

The key issue is one that I have already discussed in a letter to you about a week ago, i.e., the tacit inferences that are in the questions, which shape and determine the observations with which we establish the fact. Let me first summarize and extend my views on this subject, and then I will discuss your letter in the light of this summary and extension.

As I indicated, the question "Is this boy a thief or not?" already contains the tacit inference that human nature is a mechanical sort of thing (like an electric motor) that can either have a thieving structure or not have it (or at most, one that could be "altered" mechanically from one structure to another, for example, by beatings, punishments, and rewards). Only a mind that is insensitive because it is full of "noise" (i.e., the confusion of glamor with beauty) could accept this question, without feeling that there is something wrong with this whole way of putting it. Part of this insensitivity comes from wrong but generally accepted notions (i.e., inferences) about the nature of our inferences, which allows us to assume that the fact is always obtainable by answering whatever question the mind happens to pose, at any given moment.

I am now proposing another set of inferences about the nature of the process of "facting". As a competent person, you will have to observe whether my inferences are true or false (or whether the whole framework of my inferences is irrelevant, because based on wrong assumptions).

What we have been discussing this far is what I call "horizontal" facting. That is to say $\text{Fact}_1 \rightarrow \text{Inference}_1 \rightarrow \text{Question}_1 \rightarrow \text{Fact}_2 \rightarrow \text{Inference}_2$, etc. This is merely a time sequence, in which each set of inferences is regarded as independently structured, but having reference or relevance to the others (E.g., inference_2 is formed outside the structure of inference_1, but may imply the falsity of inference_1). But now, we must consider the vertical structure of facting, which is not a time sequence, but rather, the order of orders of the fact at any given moment. Thus, the question "Is the boy a thief?" contains, at that very moment, a certain set of tacit inferences about human nature.

Let us use letters (a,b,c & etc.) to indicate the vertical order of the fact, and numbers (1,2,3, etc.) to indicate the horizontal order. So we consider the question 'Is the boy a thief?" to be $Q^1{}_a$. We now consider inferences $I^1{}_b$ to be the inferences underlying this question. These inferences (e.g., that human beings can have a thieving mental structure) lead to a certain set of more subtle questions $(Q^1{}_b, Q^1{}_{b'}, Q^1{}_{b''}, \text{etc.}) \equiv Q^1{}_B$, which could ascertain the fact: "Is $I^1{}_a$ true or false?" These questions are answered by careful and sensitive observation of all that is at our disposal.

In answering these questions, we will be tacitly assuming yet further influences, $I^1{}_C \equiv (I^1{}_c, I^1{}_{c'}, I^1{}_{c''}, \text{etc.})$ of a lower and more primitive level. For example, we will assume that a given object (O) belongs to someone else and not to the boy. Where necessary, these influences can be tested, by answering further questions $Q^1{}_C \equiv (Q^1{}_c, Q^1{}_{c'}, Q^1{}_{c''}, \text{etc.})$ by means of yet further observations.

As we go on with this process, we eventually reach inferences, $I^1{}_p$, $I^1{}_q$, $I^1{}_r$, etc., which are in the domain of ordinary perception (E.g. that object before my eyes is a boy). As a rule, we do not question these, because we regard them as adequately confirmed. But sometimes, we are led to question them, when sensitive and intelligent perception indicates that there is something wrong, that may have its origin in this sphere.

In principle, this process goes deeper and deeper, into the non-verbal movement that is the foundation of awareness, where it "fades away" to merge and amalgamate with the vast and eternally changing unknown totality, in which all that exists, including ourselves, has its foundation, and its being. Ultimately, it is in these depths that truth and falsity are perceived. For what is false leads, as you say, to conflict, i.e., disharmony. As it begins to do so, the still deeper levels of movement begin to end whatever is responsible for this disharmony, and to create a new perception, that is harmonious, because it does not conflict with itself or with further observations of the fact, at various levels.

Some people might object that all this is too vague and perhaps even "mystical". Yet, there is one evident fact, which reveals an iron-bound necessity: i.e., that whoever is insensitive to beauty and ugliness cannot do other than delude himself on issues that are at all subtle. People who imagine that they can obtain a "solid" foundation for life, outside the need to be sensitive to beauty and ugliness, will inevitably get lost in the quagmire of confusion and delusion.

Nevertheless, it is important to probe more deeply into the question of why people are so anxious to find a "solid and secure foundation" for these deep processes. It is not adequate merely to condemn such efforts, without understanding how each human being, including oneself, has a structure of thought, which makes such an effort inevitable, until its deep roots are understood.

It is not hard to find the answer to this question. For a vast domain of past experience reveals all too clearly that human beings are prone to delude themselves, to believe what gives them pleasure, and to accept romantic and glamorous beliefs, ideals, and notions, even in the face of strong evidence that these are false or confused. Such thinking has led to disaster in a tremendous number of ways, as well as to the rigidification of pleasing metaphysical notions, in such a way as to freeze traditions and beliefs, and interfere with real learning and the genuine progress of science.

As a reaction to all this, people have tried to base everything on the directly perceived fact, at the level of "common sense", or something of a similar nature (e.g., general consensus, or common usage of language, etc., etc.). No more of these romantic and exciting metaphysical fantasies for them! Rather they base themselves solely on the "solid" ground of empirical observation.

Of course, the difficulty with this position that it too is based solely on "romantic" delusions, adopted because it happens to give a pleasing sense of security to those who can believe in it. Indeed, as we have been seeing in all our correspondence, the "solid" ground of empirical fact begins to dissolve and turn into quicksand, as soon as one begins to scrutinize it with some care. So the empiricist position can be held only if one is not too sensitive, i.e., if the mind suppresses or otherwise evades awareness of the difficulties in this position. Therefore, you are right to say that the empiricist position is "evil". Yet, you must be careful to note that it is merely a reaction to the opposite position, which is, at bottom, also "evil" and for precisely the same reason, i.e., it accepts pleasing delusions by suppressing sensitivity to disharmony and contradiction. Indeed, in a deep sense, all evil may be said to be just this – to suppress awareness of the whole fact, in order to impose a glamorously pleasing or exciting structure in the perceptions of the mind.

I feel therefore that you are wrong to identify empiricism as the sole root (or even the main root) of all evil. Rather, the truth is that "romantic speculation" and "hard-headed empiricism" are merely superficial variations on the same theme. If one simply condemns empiricism, one is tacitly starting back on the road of "romantic speculation", on which any idea that the theorist cares to formulate can be considered to be the truth. Of course, the fact is that one's theoretical ideas provide a descriptive language that shapes our observations. Here, Feyerabend is right to emphasize that the very terms of expression of the observed fact can be altered radically, by a new theory. Yet, this does not remove the need to formulate questions $Q^n{}_p$, leading to observations, $O^n{}_p$, etc., which can confirm or falsify the new theory. And here, the key issue is not in the formulation of facts and inferences. Rather, it is in the sensitivity of the mind, which can detect disharmony anywhere in the vast horizontal and vertical structure of the process of "facting". Both "romantic speculation" and "hard-headed empiricism" can be criticized as leading to insensitivity – i.e., to a kind of violent imposition of preconceived notions within the structure of perception. That the "romantic" does this is self-evident. But as Feyerabend showed so well, the empiricist does the same thing tacitly, by supposing that at a certain level, the general language for expressing the fact is incorrigible or unalterable.

So in answer to the first page of your letter, I would call your attention to the thousands of years of history, showing how people are ready to accept almost any idea as true, if it is pleasing, exciting, glamorous, convenient, apparently conducive to security, etc. It was in opposition to this tendency to delusion that people tried to insist on getting "incorrigible" facts. But as I have indicated, the notion of an "incorrigible" fact is just another such delusion, adopted because it gives a pleasing sense of security. So the real difficulty is that there was only a superficial change in the human tendency to accept pleasing delusions, a change which continued the same tendency at a new level, thus making the situation more confused than ever.

This is an example of what has gone wrong with human thinking, quite generally. That is, people see that something is wrong. But instead of getting to its roots, they make a superficial change, which continues the old evil at a new level, thus "confounding the confusion".

Consider, for example, the question of violence. Basically, violence is the imposition of arbitrary ideas or demands on a process, which are in conflict with the real nature of that process. The very structure of society is rooted in violence. Thus, people are trained to conform to certain standards of what they are (not merely as to how they behave). For example, there is the injunction "Be a good boy! Be brave, noble, gentle, obedient, considerate, etc., etc." Because, the order of the mind is infinitely subtle, it cannot conform to any such injunction. Thus, there is started a deep rooted conflict.

When people inevitably fail to conform to such standards, society sets up a compulsory order. That is, it tries violently to impose these standards. Such violence incites further violence, and counter-violence, etc., etc.

Eventually, people begin to see that violence is destructive. But instead of getting to its root, they form the ideal of "non-violence". They then try to impose this ideal on themselves and on other people (e.g., by persuasion or propaganda). But they do not perceive that such an imposition merely continues the reality of violence, at a new level, where its effects are even more confusing than they were before. And indeed, all who proclaimed themselves as "non-violent" (e.g., Gandhi, the American Civil Rights Movement, etc.) actually initiated a great deal of open and manifest violence (e.g., riots, etc.) This is no accident. For one who proclaims non-violence is by that very act being violent, in the root of his thought.

Indeed, all our thought tends to be violent in structure. For because of glamor and the pleasure-pain principle, it tends to suppress all thoughts and perceptions that are in conflict with it, and to stimulate others that appear to back it up. When these efforts at pleasure are frustrated, then violence is continued in an outward and manifest reaction, aimed at suppressing or destroying the frustrating factor, and encouraging factors that work in a contrary direction. So violence in action is merely a continuation of violence in thought (or as von Treitschke put it: "War is the continuation of politics by other means.")[3]

Therefore, to end violence, we need a deep and radical change in the entire violent structure of thought. Even when thought appears to be "peaceful" it is usually potentially violent. Either one is "worn out" from violence, so that the mind is momentarily "quiet" because it is deadened and insensitive. Or else, one is deluding himself into believing that all is beautiful and right. But let a disturbing factor enter our perceptions. We immediately experience a violent reaction. So at present, our "peace" is like the interlude between wars – not real harmony, but rather, a period of rest and recuperation, so that we can fight again.

And here, there is vast confusion. Many people glamorize violence as bravery, beautiful nobility, etc. Others say that it is not violent to react in self-defence, or when one is "in the right". Yet one cannot defend anything, even what is "right", except by some sort of violent attack on the other person. This in turn calls for a violent "defense" on his part, leading to a counter attack. Thus, defense leads inevitably to "escalation".

Then there are those who say "Don't defend. Just submit." This is just another form of violence. For in order to avoid insecurity and unpleasant disturbances, one violently imposes the other person's demands on himself. Actually, this cannot be done. The mind is too far beyond measure. Another part will resist, thus initiating deeper and subtler forms of violence.

I hope you see how deep and vast are the implications of the whole question of violence. Violence is rooted all pervasively in the over-all structure of our thought. Merely to change this structure superficially will be to continue violence at new levels that are more confused than ever. The ending of violence requires a total harmony, peace and silence, starting from the deepest levels of movement of the mind. And only this can lead to true perception on subtle issues.

The whole question is discussed in another way by Krishnamurti, some of whose writings you should soon receive. Meanwhile, I would like to suggest that when we respond to the ideas of another person, and find them wrong because they are arbitrary and violent impositions on the structure of the fact, it is necessary not to continue a similar violence at a new level, by defending oneself against the other person's ideas. For each defense is an attack, which violently imposes one's own ideas, in one way or another. Rather, one has to realize that as soon as one is aware of someone else's ideas, these ideas are his own. So when I attack your ideas, I am really attacking my own thought, which is "displaying" what I have "made" of your ideas. This is a meaningless and destructive kind of conflict. I need another response, of a very different kind. I have to understand all ideas, regardless of whether they are "mine" or "yours". Those that are false have to be dropped by the creative process that I have described in a previous letter. If I attack "your" ideas, in order to defend "my" ideas, I really continue both, in perpetual conflict. But if I understand all ideas, in a "neutral" way, then I may get free of what is false about both "my" ideas and "your" ideas.

[3] Actually the quote is from Von Clausewitz—CT.

Consider for example one's reaction to empiricism. Evidently, one may not like empiricism, because it contains arbitrary, violent elements. In order to defend oneself against these, one makes an attack on empiricism. But this is just what the empiricist is doing. He finds the "romantic metaphysician" imposing arbitrary notions, in a violent way. So he sets up a "defense" against metaphysics.

Thus, on P.2 of your letter, you say that you can understand my saying that a proposition, *P*, is not a fact but an inference, only within the framework of the radical empiricist's notion that there exist "incorrigible" facts. I wish, however, to ask you whether this assumption is not just an effort to "defend" yourself against radical empiricism. As such, it inevitably leads you to continue the basic structure of radical empiricism at a new level.

The essence of the radical empiricist position is that there is some deepest level of observational and linguistic terms that are "incorrigible". Probably he would say that the meaning of the word "fact" belongs to this deepest level. Similarly, you say that I must also be accepting the "factual background" as part of the incorrigible "soil" of observational results on which the "tree of knowledge" can grow. If I cannot tell when something is at least not a fact, I cannot talk about confusing inference with fact. But is not this just another example of radical empiricism, which implies that all fundamental terms can be defined explicitly (either positively or negatively) and thus observed?

I am saying, in contrast to the radical empiricist, that fundamental terms are never explicitly definable. Rather, their meanings are indicated by the order and structure of relationships, revealed in how one uses these terms. This holds true especially for words like "fact" and "inference". The first step is to establish a distinction between what is observed and what is inferred. This is merely a step in establishing the structure of our thought, not of its content. This distinction is not to be hard and fast, but rather, taken as two poles in a process (like moving Northward and moving Southward). We cannot escape that each element in our thought appears in its observational aspect and in its inferential aspect. Indeed, it is just because of this that inference can so easily be confused with a description of what has been observed.

I am not trying to say that what is observed is "incorrigible". Far from it, I say that inference can often show the need to "correct" previous observations. But what I emphasize is the structural feature that a given term is ambiguous, in the sense that unless further specified, it may be a description of observation or an inference. Not to be fully aware of this ambiguity is to become a slave of arbitrary factors, which will lead a person to treat the inferential applications of a given term as if they were observational applications. This gives rise to delusion.

Now the whole process by which observation and inference interweave maybe called "facting". In addition, common usage has it that an inference that is well confirmed by observation on many levels is to be called a "fact". (E.g., the existence of Australia is a "fact" of this kind). If you, in your defense against empiricism, try to get people to cease to take such "facts" seriously, you will be only ceding the field of battle to the empiricists. For then people will rightly worry that without "facts", they have no way to avoid becoming "romantic metaphysicians."

I emphasize that my thesis about the "fact" deals with the structure and not the content of thought. For example, consider the word "not". Can you define this word? Can you at least say what it is not? Evidently, you can't do it, without using the word "not" or some equivalent term (e.g., limitation, negation, etc., etc.). The word "not" has no content at all. It is a basic term that helps establish the structure of every language. And whatever "positive" meaning is given to this word, you can always add: "It is not just this, but something more besides." For this reason, no computer can ever really do reasoning. At best, it can imitate some aspects of reasoning, in a framework determined by how the word "not" is interpreted in terms of the operations of the computer (E.g., as an operation in a Boolean algebra).

Any notion that a fundamental word can be defined is in essence part of the radical empiricist position. Somehow, one supposes that one can directly perceive the unique meaning of words like "fact" or "inference". But if we give up the empiricist position in physics, we must, ipso facto, give it up in epistemology and in theories of language. The only thing we ask is that, given a structure in which words are used, can we discover in observation something that corresponds to whatever this structure may be discovered to mean to us?

When I say someone is confusing inference with fact, I mean that any competent person should be able, in his own thought, to see that the inferential aspect of a given term is being treated in an order that would be appropriate only for the factual aspect of that term. Thus, if someone asks: "Is the boy a thief", he is treating as already settled by observation the implicit question "Can human being's mind be categorised in such mechanical terms?"

Falsity or evil "slips in" through the insensitivity, which allows this kind of confusion to take place. This insensitivity is a manifestation of a violent general structure in the operation of the mind. This violent structure is mistaken for harmony, because it is glamorized as beautiful and pleasing. Such a mistake between glamor and beauty is based finally on confusion about the question of the observer and the observed, along with confusion about the meaning of psychological time. This is indeed the ultimate source of evil. Perhaps after you have read Krishnamurti's writings a bit, we can go further into these questions.

No single aspect of the whole process can really be identified as "the" source of all the trouble. Even the process of collecting "useful facts" at a given level of theory is in itself not wrong. Indeed, under certain conditions, it is really absolutely necessary. But what is wrong is that when sensitive perception could indicate the need for a new level of theory, the mind, drugged by its "glamorized" vision of the "successes" of the old theory, becomes insensitive, and goes on violently imposing the older ideas, suppressing the point in sub-liminal intimations of the need for change. So the real trouble is always violence. But violence is infinitely subtle, and can neither be identified as always being in a certain kind of process, nor as always being absent from such a process. Thus it would be quite easy for the "critical" views of Popper and Feyerabend to become a "mask", behind which one could go on with his own peculiar form of violence.

In this connection, you say that a certain amount of facting is necessary, i.e., a certain amount of destructive activity is necessary for creation. You rightly call this a "paradoxical" conclusion. But isn't the paradox due to the violence with which you defend your own views, by attacking those of the empiricist? If you did not wish totally to destroy (i.e., demolish) the position of the empiricist, you wouldn't feel impelled to say that "collecting useful information" is always destructive. After all, if I ask you how to get to a certain place, what violence am I doing to you, to me, or to anyone else's ideas? But if I suddenly did this in the middle of a deep discussion that was becoming embarrassing to my position, this would be violence, therefore destructive. The radical empiricist is more or less in the latter position, because whenever deep issues are raised, he insists on collecting more "useful information", when in fact, no more information is really needed, at this particular juncture. His "ploy" is a form of violence, similar to "How to Win at Games Without Actually Cheating."[4]

With best regards

David Bohm

Oct 20, 1967

Dear Jeff,

This is a supplement to yesterday's letter, answering yours of Oct 9.

Firstly, I want to emphasize the vast extent of the problem of violence. Violence is rooted in all thought, yours and mine, that of the empiricists, Rosenfeld[5], Johnson, Kosygin, etc., etc. Very probably, it originates in the survival of some of the old animal instincts, built into the structure of the thalamus. Sometimes, one experiences what may be called "unalloyed violence" or "pure hatred", which literally "fills the mind", causing the adrenaline to flow, the blood pressure to go up, etc. In such a state, all the higher functions of thought and awareness are overcome and overwhelmed. A person is in a kind of "ecstasy of violence" in which he really does not know what he is doing. Primitive people and young children experience this state frequently. In a moment, they are overcome by murderous rage and do great harm to others. A minute later, it is all over, almost like a summer thunderstorm, and the air is clear again. Violence is, in this case, a monetary phenomenon, that "just happens" without any tendency to continue over long periods of time.

As civilization developed, men began to realize that such violence is dangerous and destructive. It was only natural to think of controlling this violence. After all, man had begun to "exercise control" in some of the more mechanical aspects of his life. So he conceived the notion of "controlling" violence. To do this, he condemned violence as "evil" and set up the opposite, i.e., the ideal of the "good" man, who is not violent, but rather cooperative, kind, loving, and understanding, being thus "wise" instead of "foolish". Men were in this way enjoined to be "good" and not "evil". The

---

[4](Potter, 1965). See also p. 56, n 2—CT.

[5]A Freudian slip? Presumably this should be Roosevelt—CT.

"evil" was "what is" and the "good" was "what should be". So men made an effort to be "good".

The difficulty with this effort is that the mind is infinitely subtle and complex, so that such a mechanical division of "good" and "evil" has little or no meaning. Therefore, the effort to impose it led to confusion. Indeed, what happened is that men experienced violent impulses and then, in this state of violence, they remembered the injunctions to "be good". So, while moving in the order and structure of violence, (which is always imposing patterns, defending them, and destroying what gets in the way), it set up the imaginary ideal of "goodness" and tried to impose this, to defend it, and to destroy whatever was in the way. Of course the first thing in the way of such "goodness" is the whole set of spontaneous movements of the mind, which can never follow any pattern at all. So the mind began to be "good" by destroying its own creative impulses. Thus, it sets up a violent conflict within itself. In other words, the division of "good" and "evil" had the effect of continuing violence at a new level, where it was more confusing than ever. Indeed, the confusion became such that once started, violence could never end. For because men began to be frustrated in the creative depths of the mind, there arose a new violent demand for "expression". After long periods of suppression of his violent urges, he felt the need for orgies, binges, fights, "free for alls", wars, etc., etc., which he conceived to be "outlets" for his urges. These in turn led to new frustrations, which called for new violence at yet higher levels. Thus, unlike animal violence, (and primitive and childish human violence), civilized human adult violence tends to continue indefinitely, and to build up and spread out, to cover vast areas.

One of the factors that "confound the confusion" in this whole issue is the intimate relationship of violence and pleasure. As I have said before, there is real enjoyment or joy, which is not violent at all. It does not impose anything, nor does it defend itself, nor does it destroy what gets in its way. But there is a kind of imitation of enjoyment, which we may call "mechanical pleasure", and which is always "glamorized" to imitate a really beautiful state of feeling. This kind of pleasure tries to maintain itself by imposing various patterns felt to be pleasant and secure. When these are threatened, it initiates defenses, aimed at destroying what appears to be in the way of its continuity. Therefore, this kind of pleasure is a manifestation of violence.

Now, pleasure and pain are abstractions, like the North and South Poles of a magnet. Every sensation and every state of feeling contains both. Because violence is an intense state of feeling that is full of energy, one can by thinking about it abstract intense pleasure and intense pain. Thus, many people really enjoy a fight, while others find a fight very unpleasant indeed. It all depends on which end of the "pleasure-pain dipole" the mind happens to focus. And of course, to focus on either side is a distortion. Only a mind that can be aware "impartially" of the pleasure and pain in violence can see that both are irrelevant trivialities. Unfortunately, however, our tradition and background of thought lead us to treat pleasure and pain as two separate (or at least separable) "opposites". So either one focuses on the pleasure, and the brain wants more, or one focuses on the pain, and the brain wants less. Indeed, the man who enjoys fighting has totally suppressed or deadened awareness of the pain in it, while the man who hates fighting has suppressed awareness of the pleasure in it (perhaps because he wants the "higher" pleasure of thinking that he is a "good" man).

The essential point here is that the very movement of suppressing or deadening sensitivity to one side of the "dipole" or the other is itself a new order of violence, a process in which the mind tries to violate its own natural order of operation. Very often, one becomes so deadened to the pain in an urge or craving that one is not even aware of this process of violation. Likewise, one becomes deadened to the pleasure in violence, so that one ceases to realize that violation is going on. Consider for example, the "good" citizen, who is a "right-thinking" person. He sees how the children of the poor are wrongly treated, but he says "This cannot be helped. After all, their parents are shiftless. It is only a fact." What this man doesn't see is the deep pleasure in his feeling that his own children are, by comparison, secure and happy, along with the suppression of all the pain that is part of this thought (e.g., one sees that it really isn't right, that it is even dangerous for his own children if others become corrupt, etc., etc.). What is going on in his own mind is a vast process of violence, aimed at producing a pleasing sense of security of oneself (and of the inevitability of the situation of the others, implying no responsibility to do anything). Such a mind is literally deadened and made dull, so that it cannot be original and creative in any field at all.

From such inward violence, there proceeds outward violence. Thus, when others disturb one's pleasing sense of security, there arises a defensive movement, aimed at destroying whatever is conceived to be the source of the trouble. In this destructive movement is delusion. For the mind is always deadening itself to any idea that threatens its sense of security and opening itself up to any idea that does the opposite. So in its very thought, the mind is violating its own natural order, the more effectively to violate the order of whatever it is that outwardly seems to be in the way of the continuity of security and pleasure.

Almost all that is done in society proceeds from a violent structure of thought, aimed at imposing patterns, defending them, and destroying opposition. For example, all ambition is evidently violence of this kind. Much of what is called "love" is actually violence. A man who prides himself on being "good" is the most violent of all. For the "evil" movements of violence in himself are attributed to others. Thus, he is unaware of his violence, which can act without any check or hindrance at all. This point is significant even in scientific or philosophical controversies. One sees the "evil" views of the other person, which are "violating" the "right" order of things. So one condemns this "evil" and tries to "combat" it with one's own "right" and "true" ideas. In doing this, one does not notice that the "evil" in the other person is precisely his violence. That is, he is excited by his own idea, which gives him great pleasure. To defend this pleasure, his brain suppresses awareness of what is wrong with it, and tries to "demolish" the arguments of others, who wish to call the deficiencies of his idea to his attention. This whole behavior is violence, which has its dynamic source of energy in the sense of excitement, glamorized to appear "beautiful". But as soon as one calls this "evil", one is doing the same thing. One is excited, and this gives pleasure (as well as pain, of course). To defend the pleasure (and get rid of the pain) the brain suppresses awareness of what is wrong with one's own arguments, and tries to "demolish" the arguments one's "evil" opponent. He in turn feels that his opponent is "evil" and has to be "demolished". But evidently, this whole game is meaningless, and in fact, itself the very essence of evil.

Very much tied up with all this is the question of influencing and being influenced. Of course, a certain kind of influence is natural and inevitable. Thus, when I learn physics, the whole course of my life has been influenced in a great many ways. But there is another kind of influence that is actually a form of violence. I may try to impose my views by persuasion, trickery, pressure, or propaganda. Vice-versa, I may allow myself to be influenced by you. This too is violence. For I am now imposing your views on my mind, thus violating its natural order of operation. I may do this to please you, and thus to gain something from you, or to feel more secure. Perhaps I regard you as an authority, so that you know all about it. If I impose your views on my mind, then I too will "know". So to follow the authority of another is violence, when this following implies imposition or acceptance of his ideas without real understanding.

Seeing something of this immeasurably vast structure of violence, inward and outward, what are we to do? Evidently, to try to be non-violent is an absurdity. As I explained in the previous letter, this is merely the violent imposition the ideal of non-violence, a meaningless, destructive, confused and delusory process, if ever there was one.

What is called for is not merely superficial change in the forms of violence. Rather, the mind has to be aware, (directly and non-verbally) of its deep roots. Then perhaps, the violent structure-function of the mind will come to an end.

In this connection, an important source of confusion is the set of words: "I am violent". This implies that violence is a contingent feature of the "self", or in other words, that the "self" can sometimes be violent, while at other times, it can be peaceful and harmonious. But the real state of affairs is that wherever there is a "self" there is an inherently and inescapably violent structure of thought. For the "self" is always aiming at pleasure of the mechanical sort, which depends on imposing patterns of thought, defending them, and destroying (or deadening awareness of) whatever is in their way. So instead of "I am violent", one should better say: "I am violence". That is to say, violence is the very essence of structure-function of the "self". The ending of violence is the ending of the primary role of the structure-function of the "self".

To this end, we have to be careful about how we are using words. One can use words descriptively and inferentially. Also, there is the ostensive use of words, which "points to" an example of what they mean. But in addition, there is something entirely different. Words may be used to set up a mirror in the mind.

Now, a mirror doesn't tell you anything. It is neither descriptive, inferential, nor ostensive, in its function. Rather, with the aid of a mirror, you can see what you couldn't see before. For example, suppose you ask me to tell you about yourself. You are not satisfied with my descriptions, inferences and ostensive "pointing" to things that have qualities similar to yours. Imagine now that there are no mirrors actually available. So I give you directions as to how to make a mirror. When you have done this, you can see, not only your face, but something more; i.e. the relationship between your outward expressions and the inward moods, urges, wishes, drives, thoughts, etc. You may have thought: "This is a nice feeling." But now you see that it expresses itself as a silly or ugly appearance of the face. You wonder whether it is really so nice after all. Sensitive awareness discloses that there is a violent suppression or deadening of perception of the uglier aspects of this "nice feeling". Without this deadening, one sees the reality – i.e., a meaningless process of violating the natural order of operation of the mind.

All our outward actions can thus serve as mirrors. But sometimes, words may indicate how the mind can set up an inward mirror that is even more sensitive and comprehensive in what it reveals. Verbal communications having to do with the structure of thought rather than its content, can have this effect. Thus, what I say about the structure of violence will have a certain meaning to you. It is possible that through these meanings, you will cease to regard the particular content of violence as basic (e.g., I am violent because I was frustrated). Instead, I now see that the entire structure of my violence as it is at this moment is a mirror to the dynamic source of action that is really meant by the word "I". In other words, all my violent patterns of feeling and thought are mirroring the movements of the deep source of action which is meant by the words: "I – in my innermost essence." If awareness ceases to be seriously concerned about changing these patterns into "better" ones (because the futility of this is clear), then it can simply pay attention to "what is", without choice, neutrally, impartially, and factually. In doing this, one is aware of the relationship between these mirrored manifestations of violence and very subtle sub-liminal perceptual cues as to the activity of the central source of energy that is meant by the word "I". In this way, the mind learns the real source of violence, while it also learns about the real ugliness and destructiveness of the effects of this violence. Spontaneously and naturally, without choice, effort, or act of will, the violent structure of the central source of energy begins to come to an end. This latter becomes harmonious, peaceful, silent and empty, not merely on the surface, but down to very great depths.

In this peaceful harmonious state of mind, there is not stagnation, but rather, action of great intensity and subtlety, far beyond that of ordinary feeling and thought. Let us compare the depths of mind to a central fire or "sun", so "hot" that most of its "radiation" is X-rays and ultraviolet light, which are invisible. Therefore, it appears to be just "emptiness". As this energy moves outward, toward the "surface" of the mind, it can "cool down" to become "visible". Anything that is false or disharmonious on the "surface" works toward the depths, where its falsity is first "felt" and then revealed in more detail, in "thought", which comes "outward" toward the "surface". So the ability to be sensitive to beauty and ugliness underlies the ability to perceive truth and falsity.

In all this, it is crucial to understand the irrelevance and triviality of the content of our violent thoughts, feelings and actions. Rather, the whole of this content is now able to reveal the deeper structural process of the mind, which it is mirroring.

One interesting point. If you work in this way, day by day, moment by moment, being sensitive to violence as a mirror to the central source of dynamic action of the mind, you begin to get glimpses of that "pure" animal violence that we once experienced as children. However, we are now very afraid of it. So the mind tries to suppress it at one level, while it is continued at another. For as I have said, violence has a very enjoyable and pleasing side. The mind tries to retain this, thus continuing violence, while it tries to get rid of its dangerous and destructive side, by "controls" of the kind I have described earlier. But as has been indicated, these "controls" are delusory. Actually, they only proliferate violence into new fields.

The very effort to limit and control violence is what keeps it going indefinitely. The primitive man and the young child do not try to control violence, so that they are quickly "finished" with it. But is this not too dangerous?

This question implies a lack of understanding of the real nature of action. Generally, we assume that the essence of action is outward physical action, – i.e., to move, to speak, to write, etc. What we do not realize is that awareness is action. There is a vast and immeasurable action going on in the "central sun" of awareness. This action is creative. In it are formed the deeper aspects of our perceptions, feelings, thoughts, wishes, urges, etc., etc. The "overflow" of this action leads to the superficial activities of the mind, and to our outward physical actions.

So if we should experience the state of "pure" animal violence, this does not mean that this must act outwardly and physically. Rather, the mere "flowering" of this state in awareness is its basic action. Then, if awareness declares that this is appropriate, the action can proceed outward. Otherwise it is "switched off", and there is no doubt that if one is deeply aware of the full meaning of this "animal" violence, it will not proceed toward outward action.

Of course the child or primitive man could not realize that this is the case. So they can do nothing but allow animal violence to express itself outwardly, after which it "switches itself off". Civilized adults cannot do this, so that violence never "switches itself off", but continues indefinitely in ever changing forms. However, when an adult understands how the mind can use its own inward manifestation of violence as a mirror to the central source of mental energy, then it is no longer compelled to express violence outwardly. Rather, the "mirror" is already an adequate expression of this violence, which allows it to "flower" and "switch itself off". Thus, the mind is freed of the very root of violence.

Best regards

David Bohm

P.S. Could you please send this to Biederman when you are ready?

Oct 23, 1967

Dear Jeff

This is to continue the letters of a few days ago, concerning questions of fact, inference, beauty, glamor, violence, etc.

Now, you have objected to my using terms like fact, observation, inference, etc., because these have become very confused (essentially as a result of their being used as "weapons" in the struggle between the "romantic" metaphysicians and the "hard-headed" empiricists). At the same time, you tacitly adopt the empiricist position by saying that in using the term "fact", I should at least be able to say what it is not (i.e., how it differs from inference). Thus you imply that certain "fundamental" terms should be capable of clear and unambiguous definition, by being referred to other words, so that these definitions would constitute an "incorrigible" basis for our further discussions.

In response to this, I find the the word "love" springs to my mind. This word has been misused and confused all through the ages. And now, it is worse than ever. Every movie advertised is full of the word "love", which is equated with sex, violent excitement, mutual dependence and emotional exploitation, cheap sentiment, etc., etc. Likewise in books and in common usage. Following the spirit of your suggestion, I would ask "Why do we use a word that has become so confused? Can we at least say what love is not? Since we can't, wouldn't it be better to drop this word all together?"

The difficulty with this approach is that there is a vast domain of life to which people have always referred by using the word "love". If we stop using it, we simply become unable to refer to this domain. We could introduce another word, but this would only confuse the issue, because deep down, this word is imprinted into the structure of the mind. Besides, any other word would soon become involved in the same confusion, because the latter pervades all of our lives, and is not merely the result of certain words.

Basically, the confusion originates because of the struggle between the "romantic" metaphysical view of love and the "hard headed" empiricist view. The "romantic" is able to delude himself easily by accepting any glamorous or exciting notion that comes into his head, saying "This is such a beautiful feeling! Surely it must be real love!" Then along comes the "hard-headed" empiricist, who says "Away with such silly sentimentality! Let us get clear and well defined 'incorrigible' facts! Don't mistake superficial motions for a 'solid' and 'well-founded' mutually advantageous relationship, that is the only 'reliable' basis for real love". But of course, the "hard headed" realist is just as deluded as the "sentimental romantic". Love is actually too subtle and immeasurable a thing to be defined "solidly" and "incorrigibly". It can be perceived only with tremendous sensitivity. And because both the "romantic" and the "realist" are violent, neither can know love. Deep down, neither of them is very different. What is called for, therefore, is to end the meaningless struggle between "romantic" and "realist" that goes on within every human being. Neither point of view makes sense. We have actually to transcend the whole framework of violence, which imposes a point of view, and defends it, by attacking opposing points of view. Rather, the mind has to work from a basis of deep harmony and real peace (not the interlude between struggles and wars).

Now the words, "observation", "fact", "inference" refer also to a vast domain of real life. It is also no use to change them. Rather, as with the word "love", we have to learn to use these words properly.

How can we learn this? Firstly, we have to be free of the narrow minded utilitarian approach, which says that our whole activity is being directed towards some already defined end (such as clarity, freedom from confusion, etc.). Actually, the "romantic" shares this approach with the "realist". For tacitly, his end is glamorous excitement and pleasure. The "realist" also wants pleasure , but he emphasizes its "solid" security. But now, we perceive that any attempt to shape the mind according to preconceived notions is based on delusion, and is therefore doomed to inevitable frustration and disappointment. Rather, all that makes sense is to discover how the mind actually works. In the presence of the resulting awareness, the mind will spontaneously and naturally "find" its "way" to a new harmony, of a kind that is ever fresh and different. The mind is too infinitely complex and subtle to be described, predicted, or even "pointed to" in an "ostensive" fashion. But it is possible to use words to help the mind set up mirrors, in which it can "see" itself.

It is in this spirit that I am using words like "fact", "observation", "inference", etc. In a previous letter, I referred to the "pleasure-pain dipole", meaning by this that every sensation contains pleasure and pain, in the way that every magnet has a North and a South pole. Likewise, I want to refer to the "observation-inference dipole". By this, I mean that every statement and every perceptual abstraction, from which statements are drawn, must have an observational aspect and an inferential aspect. By analyzing the statement, we can make what was observational into something inferential, but then, we introduce a new observational aspect at another level (as by breaking a magnet, we introduce new North and South poles).

So I am suggesting that we regard terms like "fact", "observation", "inference", etc., as potential distinctions, which we intend to articulate as we go along. We do not and cannot begin by defining these distinctions "incorrigibly", with certain "utilitarian" ends in mind. Rather, we need a kind of artistic spirit here. When the artist draws a line or makes a colored spot on the canvas, he does not begin by defining "incorrigibly" what this action "means" or what it "does not mean". Rather its meaning is almost entirely potential. As he goes along, the meaning of the "whole picture" gradually takes shape, in a creative way. He himself didn't know fully what he "meant" until he finished. And even then, he sees new meanings all the time in what he did. If we regard opposing pairs of verbal terms as "dipoles", this gives us "room" to engage in such an "artistic" approach. For we realize that the meanings of the terms will emerge creatively as the "whole picture" begins to "take shape".

So now, we are approaching our subject in an artistic spirit rather than in a utilitarian spirit (which latter always moves towards predetermined ends, rather than allowing the ends to form together with the means, as the "whole picture" slowly takes shape). It is a bit like a child, who learns in order to play, while he plays in order to learn. So let us see if we cannot be a little less grim and utilitarian in spirit, not always trying to "do good" and "be in the right", and not always trying to "overcome the evil views of other people".

Now, as I suggested in earlier letters, because "observation-inference" is a dipole, it is easy to confuse one with the other. Thus, when you break a magnet, North and South Poles appear at what is, in the beginning, the same place. You have to be sensitive to understand that in the beginning the distinction of North and South is potential. Later, it will become manifest. But if you don't watch carefully, you may mix them up in the earlier stages.

What we often fail to notice is that an explicit question leading to an observation is based on a vast structure of tacit inferences. Either we may not know this, so that we regard the form of our question as inevitable. Or else, we may imagine that the deeper inferences are adequately confirmed, when in reality, they are not. Or else, we may misinterpret actual observational data, misconstruing it as a confirmation of these deeper inferences.

All of these mistakes can, at times, be "simple" and "honest". But usually, they are a result of insensitivity. This insensitivity is seldom due to fatigue or brain damage or some such "simple" cause. Almost always, it is the result of inward violence, which "defends" certain ideas, by "attacking" those that get in the way. The most common form of attack is by "deadening" awareness of them, "blanking them out", etc.

Of course, to end this kind of "evil", we must end the deep rooted violence in the mind, so that the latter is harmonious and peaceful. But this requires that we understand every link in the "chain" of violence, from its deep roots to its various outward manifestations. One of the key links is just the confusion between the observational (or factual) role of a term and its inferential role.

Here, I would like to call attention to the fact that violence always operates behind an apparently natural "mask" or "cover-up". For if it were seen to be the meaningless and empty movement that it actually is, the mind would immediately recoil from it altogether, as from poison. So there is a natural "cover-up", and an emotional "cover-up". Rationally, one "covers-up" by shifting the terms of our reasoning, so that what we are doing seems both logical and inevitable (E.g., we were only defending ourselves peacefully when the enemy violently attacked us). The emotional "cover-up" works by "glamorizing" our violent sensations, making them appear to be beautiful and enjoyable. Both "cover-ups" depend on "deadening", "dulling", "numbing", "anaesthetising" processes, going on in the mind. These enable one to be insensitive enough to confuse observational fact and inference, so as to "naturalize" our violence, while the ugliness and painfulness of this process is also blocked from perception, leaving only the "glamorized" imitation of beauty and enjoyment.

As I indicated earlier, the only way out of this is the deep and total ending of the structure of violence. This requires a "mirror" which reveals the "outward" manifestations of violence, enabling sensitivity to find the deep and subtle roots, that can never be described in words.

Here, I am reminded of a new invention to help amputees, with electrically powered artificial limbs. These are directed by nervous impulses in the stump of the limb, picked up electrically, amplified, and fed into the artificial limb in a suitable way. Such a person can learn to use these limbs almost as well as his own. What he does is to "try" to move his arm, and watch what the artificial limb actually does. By being sensitive both to the pattern of inward action and outward manifestation, he can learn how these are related. Nobody can describe verbally the inward pattern of nervous impulses that moves the arm in a certain way. He has to be aware of these, at a deep non-verbal level. The key point is then this: The outward movements of the arm are like a "mirror" of the inward pattern of dynamic nervous energy, that is the "source" of one's physical action.

Similarly, a certain structure of thought can lead the mind to regard the total manifest pattern of violence (in thought, feeling and action) as the outward "mirror" of the inward pattern of dynamic nervous energy, that is the deep source of all one's action. In this way, one discovers where violence really originates. And since one doesn't really want it, the mind just "turns off the switch" and brings it to an end. When the mind is thus peaceful and harmonious, it is not stagnant. Rather, there is a vast inward creative energy that manifests itself outwardly. At present, this energy is dissipated in friction and turbulence, which are due to violence. But when violence ends, there is tremendous creative energy.

Perhaps when you read Krishnamurti's writings on the subject, you will get another view of it.

Best regards

David Bohm

P.S. You can probably see that the utilitarian approach is basically violent. A certain aim is imposed, then defended, by attacking or destroying all opposition to this aim. Consider, for example, a modern city. Each person, each corporation, each group formulates its own utilitarian aim and imposes this aim, defends it, and attacks those who get in the way. The result is conflict all around. That is why I feel that a modern city is mostly ugly. Some of its functions do have a certain beauty. But basically, they all impose themselves violently on the individual who lives in the city. They leave little room for sensitive adaptation to the needs of the individual, even when they are not positively ugly.

This utilitarian point of view shows up in science, as the imposition of a certain view and its defense, by attacking other views. This in fact seems to be what von Neumann did. After working out what appeared to be a beautiful and comprehensive theory, von Neumann probably felt the need to make it "secure", i.e., to defend it by attacking views that would seem to threaten it (i.e. hidden variables). Being in a "violent" frame of mind, he was insensitive enough confuse the notions of false – falsifiable – falsified, as I explained in a previous letter. Thus, he could make quantum theory appear to be "invulnerable". People are always trying to make their achievements "secure" in similar ways, in many fields.

What we need is the artistic approach to science, rather than the utilitarian approach (i.e., the approach which sets up a certain aim, and defends the result that is thus achieved, by attacking or otherwise discouraging contrary approaches). But here, we need the genuinely artistic attitude. Most of modern art is in fact violent. Very often, indeed, the artist imposes his own emotional conflicts on the medium, and defends this imposition tacitly in various ways. Thus, he does art violently. Indeed some modern "artists" say openly that their creative work is to engage in destruction of various objects, such as furniture, thus "discharging" their violent impulses. (Even in a therapy, this is, of course false. For such "outlets" usually only serve to encourage a person in the habit of violence.)

Then there are those who impose rigid mathematical patterns on their work, and "defend" these by deadening their minds to the mechanicalness of what they are doing. Really, any artist who works from violence must be deadening his mind to the ugliness of what he is doing.

A really creative artist does not impose his arbitrary notions on the medium. On the other hand, he does not just allow his impulses to flow at random. For this too is violence. Creation is possible only when there is great sensitivity to the relationship between inward intentions and outward manifestations.

A fellow like Picasso seems to work by imposing his arbitrary ideas and feelings on the medium. He can do this because he tremendously overvalues all the arbitrary little things that come into his mind, mistaking them for something beautiful and creative. He talks of creation, by "destroying" the creations of nature. Is this merely an "outlet" for his deeply violent psyche?

Real creation is at the same time love for one's medium. Nobody with love will try to destroy what he loves. He destroys only what gets in the way of what he wants to impose and defend.

Oct 25, 1967

Dear Jeff

This is still in answer to your letter on fact, inference, etc.

In an earlier letter, I referred to a hierarchical "vertical" structure of fact and inference, as distinguished from its "horizontal" structure of time development. It is this "vertical" structure that allows fact and inference to be confused, when the mind is an a state of violence and consequent insensitivity. Thus, the inferences underlying a given question (E.g., "Is the boy a thief or not?") imply a tacit structure to the whole situation. Because of insensitivity and deadness (dullness, etc.), we may not even notice that the question contains such inferences. If one does notice it, one may imagine that these inferences were already confirmed in vast detail in the past. Here, recall that memory is tricky. It is always reshaping itself to emphasize what is pleasing and satisfactory, while it "blanks out" and suppresses what is not. It is not generally realized that this process is a basic aspect of violence, in which the past is being glamorized and made to appear beautiful. Such glamorization destroys one's ability to see what is true and false, because one can no longer sense disharmony, ugliness, etc. In particular, it can cause one to fail to see that there is little or no real confirmation of one's tacit inferences, and that there is often actually a lot of evidence against them, which is being "blanked out", suppressed, or otherwise overlooked. And even if one does try to go over this evidence, one's evaluations are distorted. For the brain tends to accept those evaluations that give pleasure, and to reject those that do not. So when one is in a violent state of mind (which is almost always) one can easily confuse what is actually a very dubious inference with a fact that has been thoroughly confirmed.

Of course, the confusion of the inference $I^1{}_a$ with a fact depends on the dullness and deadness of the brain as it deals with lower order inferences ($I^1{}_b$, $I^1{}_c$, - -) and their questions ($Q^1{}_b$, $Q^1{}_c$ - - -). This in turn is the result of glamorization of certain actually ugly sensations, so that these are confused with enjoyment, energy, liveliness and beauty, with the result that false answers can be accepted for these lower level questions. This process goes down deeper and deeper until it seems to reach our blood and our very pores. In other words, violence creates a similar state of confusion at all levels. And therefore the only way out is for the totality of violence to end, so that the mind is peaceful at very deep levels.

In your letter, you give an example of how the elaboration of a set of theories on a given level can be a "trap", because none of them is really right. Each contradiction leads to a minor modification of the theory, which in turn is contradicted in the next step, etc., etc. What is called for is a creative change of order or level of the whole structure of the theory. And you identify as "evil" the tendency to work in the old framework.

Of course, this sort of thing often does happen, and is indeed happening today in physics. Nevertheless your blanket condemnation of working in a given framework goes too far, and is therefore itself a kind of violence. Indeed, very often a new set of ideas has a great creative potential. It is reasonable to work creatively, developing what is in these ideas, until the potential begins to show signs of exhaustion. It is then necessary to begin to experiment with new frameworks. Unfortunately, because of the violent imposition of demands for "success", "results", "precision", etc., people keep on in the old framework long after it is exhausted. This is then the real source of "evil" – i.e., insensitivity to the signs that the framework is approaching exhaustion, as far as creative potential is concerned.

A similar situation arises in art. Mimesis had an enormous creative potential, over the centuries, which began to be exhausted in the 19th century. Most artists went on with it, anyway. A few experimented with something new, to change the framework, and open up new possibilities for creative work. These were first violently resisted. Then they were equally violently "accepted" by the vast majority of artists, who exploited these new developments as a means of "expression" of what was in the contents of the Ego. What these artists did was to "mix" the old security of mimesis with the new forms of art, thus leading to confusion.

Similarly, in science, a few scientists like Einstein opened up new creative possibilities, after those of Newtonian mechanics had been exhausted. These ideas were first violently resisted, then "accepted". But generally, those who accepted such new ideas "mixed" them with the old, and thus tried to establish a new "secure" framework in which they could satisfy the Ego with "successes" and "results". They did not see that (as in art) what is needed is now a continual creative development. This development must start with mastery of previous developments. It then goes on to see how these can be brought creatively to new levels, until they are exhausted. To leave a given framework before it is beginning to be exhausted is a false step. For then, you leave it for arbitrary reasons. Hence it is a violent imposition of your Ego on the physical situation. In seeing how the older structure is becoming exhausted, you get the "clues" for the natural and necessary changes toward a new order.

So creativity includes creative growth within a given framework, along with radical mutation of the framework from time to time. Each growth implies mutation in more limited aspects of the total structure. So it is all a question of the scope, extent, degree, and order of mutation, which must be appropriate at each stage of the process.

The conservative and the radical are both violent, hence uncreative. The conservative violently imposes continuation of the old order. The radical violently imposes a struggle against the old order. So deep down, both are similar in what counts most, i.e. the quality of violence. The truly creative man does not struggle either to continue the old order or to change it. Rather, from a state of peace and harmony at the deeper levels of the mind, he understands the old order, helps to bring it to full creative flower, and to die naturally when its time for ending has come. Meanwhile, he is always sowing the seeds of the new orders, and allowing them to come to maturity, in their own natural times. There is no violence in this process anywhere.

One more point – the modern city. As I explained in a previous letter, it is basically violent. It does not sensitively adapt to people's needs, but forces people to adapt to its structure. Those who are poor evidently suffer, as they are forced to live in decaying, noisy slums where violence is everywhere. But even those who have a bit of money have to conform to the pattern of the city. Even those of its functions that may seem "beautifully" ordered on a superficial level are actually generally violent impositions on the pattern of life of the individual. Consider the road system, for example. Once you are caught in it, you are like a straw in a roaring torrent. Watching the moving lights of the cars from high up may give an impression of "beauty". But to drive on the "freeways" day after day is to be violently assaulted in one's nervous system. So basically the modern city is violent, hence ugly.

Some of the worst ugliness of a city is in the division between the "ugly" and "beautiful" parts. Thus, the middle class lives in quiet, "beautiful", "peaceful" sections, while the poorer people and the Negroes live in noisy, decaying, ugly and "violent" sections. The middle class individual says: "My family is growing up in peaceful, harmonious, comfortable surroundings". As he says this, his mind "blanks out" the nearby slums, or says that only "shiftless" people are living in them. Thus, his mind becomes dull and insensitive in vast areas of its functioning. He has done tremendous inward violence to his mind, and from this, outward violence tends to follow, both his own and that of his children. The latter, feeling the "dullness" of the "peaceful", "harmonious" surroundings begin to take drugs, to become hippies, and to engage in other forms of violence. So the man was deluding himself when he thought of the "beautiful" surroundings in which his children were growing up.

Best regards

David Bohm

Nov 6, 1967

Dear Jeff

This is a brief letter, to register receipt of your two letters, and to invite you to the conference[6] (symposium) described in the attached sheet? Can you come?

[6]Presumably the "Informal Colloqium" held at Cambridge in July 1968. The papers given at the colloqium were put together in a book: (Bastin, 1971). See C136 p. 250, n 1—CT.

I am afraid I can't come to America this Spring, as we are going to Israel again for a month. Could I have a rain-check for a year, on this visit?

I won't comment on your very interesting letters in detail till later. I would only want to say something about the man who (like the hippies) finds pure beauty and joy in what you call the "micro"- aspects of the "music of life". To me, this seems to be a disguised form of violence. After all, violence is refusal to look at the whole of the fact, insofar as this is accessible (or establishable) in the mind of man. Glamor becomes a substitute for beauty whenever the mind "anaesthetizes" or otherwise escapes awareness of any aspect of reality. It is not really the object of awareness to produce a state of joy. Rather, this is properly a by-product of complete, total and harmonious movement. Real joy contains a perception of disharmony and sorrow, insofar as these exist. If these are not perceived, they will work anyway, to produce delusion, chaos, destruction, violence.

Let me put it another way. Chemical studies show that LSD cuts down the oxygen content of the blood (alcohol may do this also, to a lesser extent). This weakens the clarity of perception of contradiction and conflict. At present, we are conditioned to inhibition of certain perceptions, because these could lead to awareness of painful conflicts. With LSD (and alcohol) these inhibitions are themselves inhibited. So certain capacities, previously blocked, are now released. Thus, the hippie can feel: "It's here, It's here!" However true awareness requires all our faculties, including those usually called "critical". These are knocked out by LSD. Hence, such a man is not really aware.

The key point is that to be "cured" of violence, man must be aware of the factors that are deadening, dulling him, making him insensitive. LSD can knock these out altogether. But if the inhibitions are "asleep", then they cannot be perceived. So no true liberation can be produced by LSD. On the contrary, man becomes a slave of the drug, because he needs it to produce his desired experiences. This drug is a violent interference with the natural order of operation of the mind; – i.e., it violates this order by putting the critical intellectual faculties to sleep. These latter are just as integral to the mind as are emotions and beautiful perceptions.

In any case, any imposed pattern of perfection is violence. If I find the "macro" aspects of life confusing, and refuse to look at them, I am imposing the pattern, defending it, and attacking those functions of the brain that inform me about the macro aspects (e.g., by becoming "dead" to these functions). So the man you describe in your letter was actually practicing an extreme form of violence on his own mind.

Real harmony is incompatible with any fixed pattern, defined as "perfection". Thus, if I say "Joy is only in the micro-aspects of perception", this fixes and limits the patterns of the mind. We need to break each order of symmetry or harmony, to make possible the next level of richness of harmony. This break produces danger i.e., that it opens the door to violence, delusion, chaos, etc. But if, through fear or wish to be always secure in our joy, we refuse to open-this-door, we condemn ourselves to just the kind of violence, delusion, and chaos, that we want to escape. For real creativity calls for eternally breaking the old order and ordering these breaks, to form the next order of harmony. And the moment we cease to be creative, we do this through violence, so that we are destructive. Whoever is not creative is violent. Mediocrity is violence. Covering up mediocrity by taking a drug is a still deeper form of violence.

What is called for is the deep perception of the root of violence. Then you won't need a drug. Nor will you need an "insane" concentration on the "micro" aspects of life. Real joy will then occur spontaneously, naturally, creatively. It will include all violence, and yet be utterly beyond this violence, in its deepest being. In a state of real joy, a man can feel all the sorrow of the whole world, and yet realize that there is a deeper level still, where this sorrow has no place, no meaning. No "ugly" fact is left out of this perception. But one sees that nevertheless, this "ugliness" is only a superficial "wave" on top of the "music of life". Yet, it is important to perceive it. For otherwise, we do not and cannot function in a natural, coherent way. Rational, coherent orderly function is part of real joy. In addition, it is necessary for maintaining the kind of state of health, (physical and mental) in which real joy can take place. When social function is irrational and violent, the state of real joy can perceive this, without losing the harmony of the depths. From this deeper harmony can arise the action needed to correct the superficial disharmony, leading to a state of creation.

I am happy to hear of your expected child, and hope that all goes well with your wife (and the child).

Best regards

David Bohm

Nov 7, 1967

Dear Jeff

This is a brief supplement to yesterday's letter.

First, one of the previous series of letters on violence contained a request to send this letter on to Biederman. If you have not already done so, will you please find this letter, and send it on to Biederman.

About drugs, such as LSD. I have been interested in them since about 1958. I have talked with psychiatrists who use them (in treatments of patients) and I have read several books written by people who experimented with LSD. At first I was attracted by the idea of "liberating creativity" with LSD. But the more I looked into it, the less enthusiastic I became.

A typical account of LSD experiences was by a woman, being treated by a psychiatrist. (The theory was that it would help reveal the unconscious.) Before her first treatment, the woman happened to be reading Life magazine, all about evolution. So after taking the drug, she literally experienced the process of evolution. In boring detail, she described, step-by-step, how she was an amoeba, how it felt to be a worm, a dinosaur, etc., etc. Evidently this was all an elaboration of the article she had been reading. In subsequent treatments, one also saw, from her reports, that she was elaborating various kinds of conditioning, mostly fortuitous in nature. Other authors describe experiences which constitute similar elaborations. E.g., the psychiatrist told a woman to try to remember the sexual behaviour of her mother and father, and not surprisingly, she experienced herself as a baby watching the behaviour, (though both she and the psychiatrist agreed that this was probably fabricated by the mind).

It began to be clear to me that LSD inhibits the critical faculties of the mind. In this way, it allows every form of fantasy and delusion to be projected, as if it were being directly perceived. This is a spurious kind of "creativity". For its basic source is evidently in the "memory tapes" of conditioning, which act as "programmes" that determine the images and feelings that are perceived in consciousness. Far from being creativity, this is a form of mediocrity, therefore, of violence. Real creativity comes from the unknown depths, not from the computer "programmes" stored up in memory.

Indeed, many takers of LSD report violent and disturbing hallucinations, which are fantastically terrifying. It is all a matter of fortuitous events, determining whether the pleasant or painful aspects of one's conditioning will be "replayed" at a certain moment.

Even those who seem to see the "music of life" at deeper "micro" levels are suspect. After all, they have heard about this sort of thing from other "hippies" or from reading about "mystical" states of ecstasy, or from perusing books on Zen Buddhism, or in other ways. Perhaps they may have accidentally "seen" the "micro-music" once, and this was "tape-recorded" as a "programme". In one way or another, they come to experience "the tape-recording" of the "micro-music" as if it were a real perception. After all, if a woman can literally "experience" the process of evolution or the sexual behaviour of her mother and father, it is equally possible to "experience" a good imitation of "mystical ecstasy". Anything is possible, once the critical faculties of the intellect are inhibited.

As I indicated in yesterday's letter, what is called for is that all faculties of the mind be at their highest level of activity, needed for true awareness. The critical faculties of the mind may well now be inhibiting the free movement of images, feelings, and "models" needed for creation. Nevertheless, the right answer to this is not to inhibit the critical faculties by a violent, therefore mechanical, interference with the chemistry of the nervous system, by means of a drug. To inhibit an inhibition is generally a wrong approach. What is needed is by clear and sensitive perception, to "feel out" the real root of the inhibition and bring it to an end. Then the mind will be harmonious and creative, from its depths. Even if the LSD taken really sees the "micro-music" (which is, as I have pointed out, very doubtful) this is a wrong thing to do. For if one inhibits the inhibiting factor, one will never perceive it. Rather, one will be compelled to end the inhibition mechanically by introducing further chemical inhibitions, that destroy the clarity of the intellect.

In real creativity, there is no attempt either to hold a fixed pattern or to change it. Rather, the mind is participating in the total process of reality, including external nature and the processes of the human being. It is learning what these processes are – what is their natural order. From this learning, is acting creatively to form new higher orders, that are in harmony with the whole natural order. This creation has neither beginning nor end. To live is to be creative. Each level of harmony begins to break spontaneously. The sensitivity of the mind sees here the possibility of new orders of creation. On the other hand, when the mind is dull, insensitive, dead, it either tries to hold the old order or to impose arbitrary new orders (which are really "unconscious" features of the old order, stored in the "programmes" of the memory "tapes"). All this is, of course, violence.

So to fail to be creative is, in the same step, to be violent. Mankind has been almost entirely mediocre and violent for thousands of years. Only a few have been somewhat creative. But their creative work was applied in a utilitarian way toward mediocre and violent ends. When someone is creative, what is called for is that other people respond equally creatively, first by learning from him, then by extending what he did to new fields, and then by going on to new levels of creation of their own, (extending to yet higher levels and newer orders). Whoever does not do this is doing violence to the creative work of others, and the inevitable result of such a violent state of mind is destruction.

Nevertheless, one must be clearly aware of the trap of violently reacting against the violence of others. The fact is that whoever is violent is thereby dull, insensitive, deadened, numbed, and anesthetized. Thereby, he does not know what he is doing, and cannot do other than he does. One must understand the absolute inevitability of the behavior of a man who is trapped in his own violence. When he does not understand, he inevitably deludes himself. And when he does understand, he equally inevitably stops deluding himself. Whether he understands or not depends on largely fortuitous characteristics of his conditioning, and how it responds to what is said (whether true or false). It may happen that A, who is deluded and deadened by violence, hears something from B that "penetrates" and ends the state of violence. But C, hearing the same thing, may violently resist what B says, because of his peculiar conditioning. Yet D (or even B) may later say something else that happens to "penetrate" C's conditioning. So the question of breaking through violence is one requiring exploration in a provisional and tentative way. To lay down hard and fast rules on the subject is itself a form of violence.

To put it in other terms, choice is violence, when applied to the order of the depths of the mind. Choice makes sense in superficial and mechanical questions, where the evaluation of the factors that determine the choice has a meaningful basis. But the mind is too vast and immeasurable to be thus evaluated. Therefore, every such choice (e.g., to be a "better" person) is an attempt to impose an order violently on the mind. It is doomed to fail and to add to confusion, because it has no basis in the real order of movement of the mind. Rather, what is called for is a sensitive and uncommitted "choiceless" awareness, that sees the factual order of the mind and in the same step takes the right action, in an order that is far beyond the "micro-level" of thought.

We may sum up by extending the discussion of fact and inference a bit. With Korzybski[7], let us consider the four broad levels of operation of the mind:

A. Event level
B. Object level
C. Descriptive level
D. Inferential level

[7] See Introduction, p. 10, n 59 and C130, p. 48, n 28—CT.

Actually, these are interwoven and interpenetrate. Yet, in some rough sense, we may order them as above for analytic purposes.

The event level is what you have called the "micro level" of awareness. It contains the "music", the vibrations of sensation, the depths of beauty and love, far beyond anything that can be put in words.

On top of the event level is a sort of "wave" of variation of quality and intensity, from which is abstracted the object level. The "wave" gives information about relatively stable forms, orders, structures, etc. It also refers to their separation, difference, similarity, relationship, etc., etc.

In the descriptive level, what we see of objects (and to a slight extent, of events) is abstracted further, into the structure of language and associated thoughts (which latter form a yet slower and more irregular "wave" at a yet more "macro" level).

In the inferential level, a yet higher level of distinction occurs, in which the orders and relationships, structures, of the described objects is abstracted and generalized.

Of course, the process moves both ways. The inferential level helps shape and structure the descriptive level. In other words, what is an inference in one context becomes a descriptive term in another context (i.e., we have a "description-inference dipole"). The descriptive structure influences our perception of objects. The structure of objects influences our perceptions at the event level.

Moreover there are higher order inferences about inferences. These constitute abstract thought.

Now, delusion is a wrong order of abstraction, resulting from violence, which dulls and deadens the mind, so that it lacks sensitivity for seeing the wrongness of this order.

The simplest delusion is to mistake the inferential level for the descriptive level (E.g., "this boy is a thief" is taken as a description of fact, observed directly at the object level).

The next order of delusion is to mistake the descriptive level for the object level. When this takes place with regard to external objects, it is called a "hallucination".

The mechanism of hallucination is probably based on the fact that all perception contains "models" of the world. When the mind is dulled and deadened by drugs or by the intensity of violence that leads to insanity, these "models" can no longer be distinguished from the direct perceptions of objects that are, in normal perception, the observational, factual basis for eliciting each particular model and correcting or altering it when it is wrong.

But of course the "self" is an object that depends on the projection of descriptive and inferential levels of thought into the "models" of perception. Everyone has a "model" of his "self" and the "selves" of other people. But it is very hard to see that there is no real object, corresponding to this "model". Only a mind deeply free of violence can see this. On the other hand, this delusion helps sustain the violence that deadens the mind, and thus further sustains the delusion. It is a sort of vicious circle.

The deepest order of delusion is to project thought onto the event level, without knowing that this is happening. In this way, the brain can create glamorized substitutes for beauty and harmony, while it substitutes violence for true passion, and self-indulgent pleasure for love. When a man is under the influence of drugs or insane, he can go further, and mistake the "noise" of his conditioning for the "micro-music" that is the essence of real joy.

One more point. You are right to emphasize the value of Piaget's notions of assimilation and accommodation[8] . Yet, you must also notice that what we assimilate is an abstraction from the whole of reality. We do not assimilate all of it, in its full concrete detail. So "accommodation in order to assimilate" creates an abstraction. And from this abstraction, we act, either creatively or violently and destructively. When our abstractions are in a right order, then from this assimilation flows a creative action. Otherwise there is a destructive action.

What Piaget fails to mention is creation (and what gets in the way of it, which is violence and destruction). This is a serious omission, as it leaves out the whole point of the assimilation process, its whole reason for being, as well as its crowning point and culmination.

Best regards

David Bohm

P.S. Piaget implies that assimilation is basically for a utilitarian aim, either of the individual or of society. To put utilitarian values at the foundation of perception is a form of violence. Piaget shares this kind of violence with most scientists in the psychological, biological, and physical spheres of work. Either they imply that certain recognized "use values" are the basic determinant factors or they imply that "survival" is the basic value, or they cynically say that the basic value is to keep ahead in the "rat race" of society.

Of course, survival is, in a certain way, a necessary by-product of the basic order of life. But in the long run, this makes sense only in a context of overall harmony and creativity.

Nov 15, 1967

Dear Jeff

Received your letter of Nov 12 on my review of your article on DLP etc.[9] As I am now rather busy, I can make only a few preliminary comments in this letter. More will follow later.

I will leave most of the minor technical details to your own discretion. If there is something I didn't understand, this may perhaps indicate to you the need for a slight rephrasing, or for adding a sentence or a paragraph here and there. For example, (P.9 near top) a brief statement, telling that this is a consequence of quantum mechanical laws, would suffice. P.10 near top requires a brief summary of some of the conclusions of our article on this point. Don't require the mainly philosophical reader to pore

[8]See C130, p. 38, n 22—CT.

[9]See C132, p. 102, n 6—CT.

through technical details to get hold of your points essential to the philosophy. Same about P.10, middle.

About Bohr, Schumacher is claiming that he is seriously misunderstood. I am now discussing these issues with him, and will let you know what I manage to understand from these discussions, if and when I do understand something. Schumacher agrees that Bohr was wrong to assume that the only unambiguous language is that of classical physics (suitably refined). But he claims that Bohr's linguistic thesis is not all that different from what I am trying to say about structural process. This I haven't seen, as yet.

All this emphasizes the need for more care in discussing what Bohr says. Is it really possible to extract a few strands from Bohr without serious distortion? I would very much object to it if a "Bohrian" were to do this with structural process. Doesn't one need to understand the essence of Bohr's position, in order intelligently to compare it with a structural-process position? Is empiricism the essence of Bohr's position? Schumacher claims, very seriously, that in many ways, Bohr is doing what Feyerabend advocates in his article in Beyond the Edge of Certainty.[10]

Does Bohr accept "the myth of the given"? Schumacher denies that he does. Schumacher claims that Bohr also dismisses epistemology as well as ontology. To be precise, he wants to assert the lack of a sharp division between the two.

All agree however that Bohr is wrong to restrict language to a classical form (however it is refined or developed). It is not certain that he believes every observation has a certain "objective control" (which would be "incorrigible", the "factual stuff" out of which knowledge is built, etc.).

You ask why Bohr could not describe the sequence of spin measurements? The answer is that to Bohr this description would have no phenomenal significance. According to the laws of qu. mchs, the sequence would be random (within q.m. probabilities). Therefore, it could tell you no more about the individual electrons than the sequence, heads, tails, etc., tells you about the individual coin.

Because we propose a theory in which this sequence is not random, but related to hidden variables, the sequence could, in our theory, have phenomenal significance. This is characteristic of the role of mathematical concepts in indicating what might be "observable". Thus, before the invention of the calculus, nobody would have thought of acceleration of a body as an "observable phenomenon". Likewise, given a theory of hidden variables, an ordered sequence of spin measurements becomes an "observable phenomenon" revealing an aspect of the state of the hidden variables (as acceleration reveals an aspect of the state of motion of a particle).

It is not merely the "radical empiricist" philosophy that is involved. There is more; i.e., the mistaking of an inference for a fact. The inference is that the probabilistic laws associated with qu. mchal algorithms are universally valid.

If the article is to be stimulating, it must be clear, at least, a lot more clear than it is now. The reader should not require an "immense effort" to see what the point at issue is. (However much effort he will need to "make up his mind".)

[10] I.e. (Feyerabend, 1965). See also C132, pp. 148-9—CT.

Your clarification about the meaning of the top paragraph of P.3 seems O.K. to me.

It may be useful for you to explain all this in a letter to me, rather than in conversation. It will perhaps help you to write a better article.

Regards,

David Bohm

P.S. I feel that we should begin right now to treat this whole issue without violence of any kind. Then perhaps we can help to clear it up. We don't want, if we can avoid it, to stimulate proponents of Bohr to make a violent refutation of your article. Rather, it is best to cause them to think deeply about their whole position. Do you agree with this? It is not just the "neutral" reader that one has to have in mind. Can one help clear up this whole confused muddle of violence in basic physics? It might be a (very modest) beginning of a contribution to peace in the world.

NOV 24th 1967

[Date added – CT.]

Dear Jeff

I hope that by now your child is born and that your wife is well.

I shall try to give a more detailed answer to your letters.

Firstly, I am in sympathy with the general spirit and aim of your article.[11] But I believe that in its present form, it will not achieve its aim, because many points in it are not clear. In addition, its format is very inappropriate. I would suggest minimizing the math. formalism in the body of the article, leaving this for the Appendix. I picture the reader as a philosophically oriented person, interested in qu. mchs., and what it means. It seems to me that the sudden interjection of formulae into verbal text has a jarring effect on the brain, not conducive to understanding. If you must bring in formulae, explain them verbally, (especially the meaning of the symbols) in some detail. Otherwise, you tend to lose the reader very easily.

Now, about von Neumann, I am in general agreement with what you say. But I would emphasize the following:

(1) V. Neumann's axioms assume, in effect, that observable averages are linear functions of the operator matrix elements, O_{mn}.

(2) He proves that this assumption is incompatible with dispersion free ensembles – hence with hidden variables of any kind.

(3) In our paper, we give up this assumption, by assuming a non-linear connection between observable averages and the matrix elements. Thus, in effect, we make use of one of v. Neumann's results, to indicate the sort of change of theory that is needed, to bring in hidden variables.

---

[11] See p. 190, n 9 above. For Bohm on Bub's response to criticisms see C134, p. 199, n 1—CT.

(4) To recover v. Neumann's theory in a suitable limit, we introduce a randomizing process.

But more generally, we are not obliged to recover v. Neumann's theory, if we don't want to. We could instead make a theory that related to perceivable, observable fact at a much lower order of abstraction. (E.g., it would have in it abstractions corresponding to interference fringes and energy levels, but might not get these out of the same formalism as that of current qu. mchs.)

In a way, J. and P. try, as you say, to show that hidden variables are excluded, as they would imply that qu. mchs. is already objectively falsified. Of course, they are confused about this point. You would be doing the whole subject a good service if you could get to the bottom of this confusion. Assuming that they are not "raving mad lunatics", how could they come to believe that the present facts already exclude hidden variables? If you can discover this, you might also make it more clear just what, if anything, they are doing wrong.

I'll continue further later.

Best regards

David Bohm

[The following single page of type-written text has been inserted here – CT.]

DEC 10th 1967

[Date added – CT.]

THE NEGATIVE APPROACH TO THE MEANING OF LANGUAGE

D. BOHM

Words and their meanings are never more than abstractions, which cannot substitute for that to which they refer (e.g. using the word for "dinner" and thinking about what it means to us cannot provide the kind of nourishment that comes from actually eating a meal). Moreover, words cannot abstract all that is to be known about any given thing. Indeed, they do not even abstract all that is essential to the function of that thing (e.g. the word "chair" abstracts what is essential for the function of supporting a person who sits on it, but not what is essential to its functioning at the atomic or nuclear level). So, it is necessary to recognise that all language has an essentially negative and partial relationship to that to which it refers. A. Korzybski has put this relationship very succintly in the assertion:

"Whatever we say it is, it isn't."

This statement is not a metaphysical assertion about the basic nature of "what is." Rather, it is a very deep challenge to the entire structure of our communications, both external and internal (which latter are called "thought").

To understand this challenge, let us begin with the fact: "We are always talking about it" ("It" refers to anything whatsoever). When we read Korzybski's statement, our first response is to see that we have already begun to say something about "it" (whatever "it" may happen to be). And then, noticing that "it" is not what we say, and that what we say is at most incomplete abstraction even from what is to be known, we assume that "it" must be something else, as well as something more. But "something else" and "something more" are also what we say "it" is. As we do this for a while, we begin to be struck by the absurdity of the whole procedure. For whatever we say it is, it isn't.

What is the appropriate response to such a situation? Evidently, one has to stop saying anything at all, not merely outwardly but also inwardly. It is suggested here that if all the "chatter" of thought can really stop, then something new can happen. But even to say this much may be going too far. For if this means that "it" will be "something new," then the novelty that we say "it" is will be what "it" is not. The paradox with which the reader has to be left is "What is it when there is no saying at all, neither outwardly nor inwardly?"

Dec, 1967
DEC 23rd 1967

[Date added – CT.]

Dear Jeff,

I am now writing to finish up my answer to your past three letters.

Firstly, congratulations on the birth of your son. I am glad to hear that your wife, Robin, did so well with the new method of childbirth.

Secondly, when you come to England, it would be best to see me before July 15, if possible, as I shall be away in August on vacation.

Thirdly, did you get the talks by Krishnamurti?

Now, to get down to your letters. I want to discuss von Neumann again. Recall what we said:

(a) No axioms are ever really true or false, but only confirmable and falsifiable, as well as (perhaps) actually confirmed and falsified in certain ways.

(b) By saying that hidden variables are ruled out unless qu. mchs. is objectively false, v. Neumann was being very ambiguous at a crucial point.

(c) The legitimate meaning of (b) is that qu. mchs. is objectively falsifiable. The illegitimate meaning is objectively falsified.

By in effect leading the reader to confuse these two meanings, v. Neumann made it impossible to clear this question up at all. Either people accepted his statement to mean objectively falsified and thus unjustifiably rejected hidden variables. Or else, they rejected this meaning, and tried to demonstrate that hidden variables are really possible. To do this is to try to batter down an open door. Thus, the very emphasis on the possibility of hidden variables further confused the issue. For it caused people to overlook the real fact, which is that v. Neumann never really "ruled out" hidden variables at all. The solution to the confusion is to point out that "objectively false" means "objectively falsifiable" in v. Neumann's context.

But in some ways, the trouble had a deeper root. It has been an ancient Western tradition that what is true is eternally true (emphasized by Plato, among others). Thus if qu. mchs. is true, it is eternally true. Therefore it can never be falsified. And if it is falsifiable, it is not eternally true, therefore not really true, therefore already "objectively false".

This is perhaps more deeply why v. Neumann used the word "false" instead of "falsifiable". For in this tradition, "falsifiable" means "false".

Meanwhile, the tradition has been changing. So today, words such as "falsifiable" have new meanings, relative to v. Neumann's day. To overlook this is to add to the confusion.

What you say about violence is very interesting. It is certainly of crucial importance to watch out for violence in one's work. Here, one must consider not only a violent content to what is said by oneself or others, but also a violent style. Thus. one can say very peaceful things in a violent style. One also has to watch for violent meanings in one's thoughts and feelings. Violence is a total process, which destroys the proper order of functioning of the mind, in every respect.

The attempt to impose an order on the mind is violence. This is often very subtle. Thus, as I indicated in earlier letters, when a man takes drugs, he usually has some tacit end in view, which is to put his mind into a "better" or "more interesting" state. Even if he says he is only "exploring", in reality, he will find that he means by this that he is "seeking" some tacitly defined kind of satisfaction. What the drug probably does is to inhibit the critical intelligence sufficiently, so that he will fail to realize that what he has "found" is only what his own thought has put together, so that it is not what is real. Something similar can be done by alcohol, by violent excitement (e.g., primitive dances by the whirling dervishes, etc.) and by other means. When the mind deteriorates to a state of insanity, a similar breakdown of critical intelligence can produce more or less the same kinds of results.

All of this is violence. Indeed, the man "who doesn't want to know" is being inwardly very violent. For a part of him already knows (e.g., that his state of "ecstasy" is a delusion). Another part then blasts, bashes, anaesthetizes and deadens the part that "knows" so that awareness of this knowledge is destroyed. Is this not an extreme form of violence? The same happens everyday to each of us. Doesn't the White man already know deeply that human nature is one? To get the pleasure of feeling superior, he bashes, blasts and suppresses this knowledge in his own brain. Similarly the physicist knows that the other fellow's ideas may have merit, but to get a "good feeling", he suppresses this knowledge, and presents the other fellow as a kind of idiot.

I think the whole question can be presented succinctly in Korzybski's terms. What is deepest in the mind is the event level. This includes fleeting sensations of every kind, along with feelings of beauty and ugliness, anger, violence, love, hate, etc., etc.

Here, one must distinguish the above from the positivist view, which treats sensations almost as if they were identifiable things. One must instead emphasize the ineffability of the event level. There is no way to define it or deduce it analytically or to deduce something else from it analytically.

Then come the level of objects. These are perceived and felt as relatively stable entities, having qualities that are signified by the feelings and other movements at the event level. The objects form a total field, which includes a particular object that is called the subject. (That is, "me", "I", the "self", etc.). This latter point has led to endless confusion, which will be discussed more, perhaps, in later letters.

Then comes the level of verbal description. As objects are defined through abstracting from the event level, so verbal descriptions are defined mainly through abstracting from the object level (though such descriptions also abstract directly from the event level to a small extent).

Then comes the level of inference, which abstracts mainly from description. This leads to inference about inference, or abstract thought, going on to an in principle unlimited series of orders of abstraction.

Now what is "the fact". This is the name of the whole process described above, which is really "facting" or "establishing the fact". But it is also the name of that part of the process which is in the first two levels. In this meaning, the third level is a description of fact. (But, of course, such a description always contains a vast totality of tacit inferences, accepted as true and as practically the equivalent of descriptions of fact.)

Abstraction is a two-way movement. For each level conditions and shapes the level below it. Thus, to see an object in the dark is already to stabilize and shape the sensations that help reveal it to us. To describe an object verbally is already to call attention perceptually to this object, and to help shape what we see about it. To make an inference is already to alter the terms of our descriptions and thus alter their "shapes". Higher level inferences perform a similar function relative to lower level inferences.

So there is a two-way stream, abstracting upward and abstracting downward. In a way, as the words abstract descriptively from our perceptual images, so the perceptual images are abstracting from the words. Thus the observer is the observed. Each "element" in the series has both roles in addition to each other element. But one role is properly the dominant one. (The word "dominant" has the meaning it has in music – the dominant theme – and not that of domination of one man who imposes an order on another by force). So let us indicate "dominant" by a heavy arrow, "subordinate" by a lighter arrow. We then have

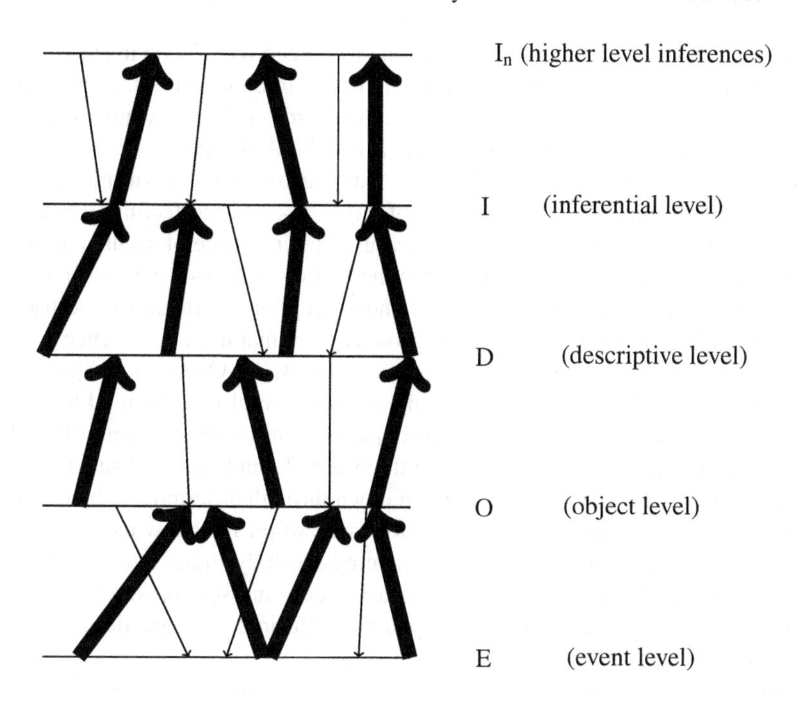

The above is the normal order of operation. To change the order of dominance in any part is to introduce delusion, confusion and violence. Thus, if an inference is confused with description of fact, we have the following:

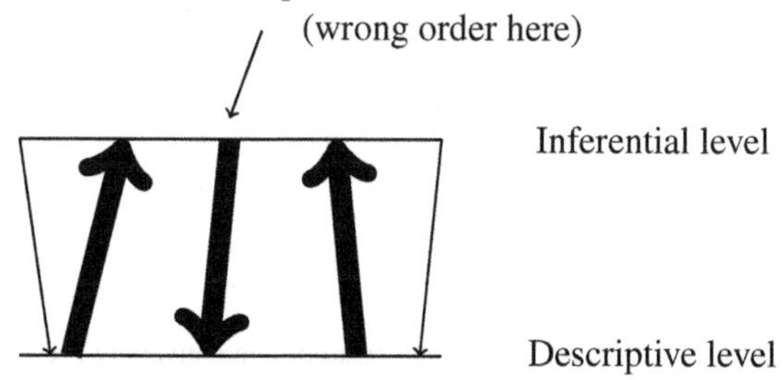

For example, the inference "The boy can be either a thief or not a thief" may dominate the descriptive level. So one's question becomes "Is the boy a thief or not?" The observed facts at the object and event level are then correlated and selected so as to provide an "answer" to this question, which would be either "Yes" or "No". Thus we get delusion and confusion.

There is already violence in this procedure. For the delicate movements of perception at the lower levels are already being blocked by the effects of the "heavy arrow" from the inferential to the descriptive level. Those perceptions that would contradict the inference are either ignored or suppressed or reinterpreted, so as to fit what is in the inferential level.

One has here a kind of "vicious circle" similar to a "feed-back" loop in electronic circuits. Descriptions of fact shaped too strongly by the inferential level, come back up, to shape the very inference that is moving downwards. Thus, the whole process builds up to a fantastic intensity, which makes it very hard to alter.

This violence gets much worse when the "heavy arrow" goes down into object and event levels. In this way, the inference of anger leads to the feeling of anger, the inference of pleasure to the feeling of pleasure, the inference of security to the feeling of security, the inference of danger to the feeling of danger (i.e., fear). One gets "feedback loops" running from feeling and perceptions to thoughts and back down again, with tremendous intensity. No wonder then, that there is violence.

Here what is important is the role of inference of a "self". One of the objects in the field of perception – i.e. the body – is already non-verbally understood to be a central source of perception and action. This object is described by the name "I" and various properties are attributed to it. When the order of emphasis in abstraction is reversed, this produces "heavy arrows" going down through descriptive, object and event levels. Because the word "I" calls attention to the whole field of what is felt and experienced, as well as to all action initiating in mind and body, these "heavy arrows" lead into key areas of brain function. "Feed-back" loops are set up, which mix up the entire process. The result is illusion, delusion, confusion, violence, destruction, suffering, sorrow, etc.

As long as there is "I" in this role, there cannot be anything other than violence and delusion. This has gone on for tens of thousands of years. Can it end? This is the key question, on which the survival of the race may well depend.

I'll write more later.

Best regards

Dave

References

Bastin, T. (Ed.). (1971). *Quantum theory and beyond*. Cambridge: Cambridge University Press.

Feyerabend, P. K. (1965). Problems of empiricism. In R. G. Colodny (Ed.), *Beyond the edge of certainty: Essays in contemporary science and philosophy* (pp. 145–260). Englewood Cliffs: Prentice-Hall.

Potter, S. (1965). *The theory and practice of gamesmanship: Or the art of winning games without actually cheating*. New York: Bantam Books.

Svetlana, A. (1967). *Twenty letters to a friend*. London: Hutchinson.

Chapter 6
Folder C134. Schumacher and Niels Bohr. January–March 1968

Jan 9, 1968

Dear Jeff

I received your two manuscripts.[1] They are, I think, much better than the original one. On the whole, I think that they could be published as they stand, with minor changes.

Firstly, the title is a bit misleading. You don't actually reconcile Bohr with Hidden Variables. A better title would be:

The Feature of Wholeness in Quantum Theory in Terms of Hidden Variables and in Terms of the Copenhagen Interpretation.

On P.6. you could perhaps consider the notion that Hilbert space with its probabilities is treated as a kind of "metalanguage". (This is due to Schumacher.) The "metalanguage" applies to the "language" of classical physics.

The feature of wholeness introduced by Bohr is to say that (because of the failure of classical mechanics) the description of experimental conditions and of experimental results is inseparably bound up in an unanalysable way. In classical physics, we could discuss, for example, the motions of planets entirely apart from a discussion of the instruments for observing these motions. In quantum theory, the conditions needed to observe momentum are inseparable from the "existence" of a state of well-defined momentum. These conditions are incompatible with those needed to define position (whereas in classical theory, no such incompatibility arises). As a result, apparatus and observed system are conceptually inseparable. Nevertheless, linguistically, we still separate them. So the separation is epistemological, rather than ontological. We do not do as Heisenberg, and portray electron and microscope as two separate objects. (I think you should emphasize in your discussion of Heisenberg that he is actually using the classical ontology in a confused way – confused

[1] Published as Bub (1968b, 1969)—CT.

C. Talbot (ed.), *David Bohm's Critique of Modern Physics*,
https://doi.org/10.1007/978-3-030-45537-8_6

because he also wants explicitly to deny it, in the very same argument in which he tacitly asserts it.) Rather, there is only one quantum object – the electron. However, it is described through the quantum metalanguage, whose "objects" are the terms of the classical "language" (whose objects are in turn the observable structure-function of the whole experiment).

One could perhaps think of the relationship between a book and a book review here. The "objects" referred to in the book are the people, events, places, described there. The book review refers to the structure of language and ideas appearing in the book. So there are two kinds of "objects": (a) People, trees, buildings, etc., along with events that describe what happens to them. (b) Novels, histories, expository books, children's books, etc., etc. These two kinds of objects really have no significant direct interaction. But to check statements made about objects (b), we have to refer to statements made about objects (a). In some peculiar way a book review is inseparable from the book that it reviews (without the latter, it would have no meaning). Moreover, the book review has some informational value about such a book. (Indeed, it may even "influence" the structure-function of subsequently written books.) So a book review is only linguistically separated from the book to which it refers. It is also only linguistically united with this book. To review a history book implies a different set of conditions than to review a novel. The form and style required for these reviews is different (but complementary in some peculiar way).

Because the quantum algorithm refers to a "metalanguage" it cannot have an unambiguous relationship to the "language" of classical physics. (e.g., a book review cannot refer unambiguously to the events described in the book that it is reviewing.) It can however refer to these events statistically (e.g., one could have a review of a whole category of books, saying that in 60% of novels of this kind, the hero triumphs over obstacles, etc.)

So let us say that the "novel" is told in the "classical" style of language, while the quantum style of "metalanguage" is able only to "review" the general statistical features of the "actual story of what happens" that has to be expressed the classical style.

I think you should somehow incorporate such notions into the introduction.

Page 9, middle paragraph. I would leave this out, as it is confusing to bring in classical physics, after we have clearly rejected it.

I would say instead that here, Bohr's linguistic thesis is plain.

Firstly, Rules Q_1 and Q_2 show that the incompatibility of different kinds of definite results in the algorithm (when two operators don't commute) corresponds to the incompatibility of experimental conditions needed to observe these results. Then Rule B merely asserts that the "metalanguage" of Hilbert space is only statistically related to the "language" of directly observable results.

In connection with Q_1 and Q_2, one must emphasize that the results of a measurement are classically describable. So Q_1 connects certain aspects of the algorithm with the unambiguously definable classical possibilities. Q_2 then shows that these possibilities cannot all be realized together (i.e., they are incompatible).

(A) refers to the relationship between the "metalanguage" and time (Dynamics).
(B) refers to the relationship between the "metalanguage" and the "language" (Linguistics).
But all this is based on the assumption of classical "language".

I think it is wrong to criticize this position (SEE POSTSCRIPT). Rather, it is necessary only to question whether or not it is inevitable. Emphasize here that another position might lead to new kinds of experimental and theoretical developments, that could be overlooked if we adopted Bohr's position without realizing that it is only one of several possible points of view (Refer here to Sect. 5).

I would also bring the point in at the end of Sect. 1.

In the hidden variable approach, then, we do not adopt this linguistic approach. Rather, we say:

(a) There are hidden variables determining actual results in each case – averaging over them is what yields the statistics.

(b) But we do not adopt the classical view that what can be observed in a measurement process is in principle separable from the total set of experimental conditions (or of the environmental conditions in general, whether part of an experiment or not). Rather we keep the feature of "wholeness" – but ontologically and not through epistemological assumptions.

So we not only introduce hidden variables, λ, but also assume that there will be a single rule for all changes in (ψ, λ) (rather than two rules, one "dynamic" and the other "linguistic"). But this rule is neither dynamical nor linguistical. Rather, it is a new kind of rule, having some of the features of both, but equivalent to neither.

It is similar to dynamics, in that there will be an equation of motion, determining changes of (ψ, λ) in each case, so that qu. mechanical results follow statistically by averaging over hidden parameters, λ.

It is trans-dynamical, in the sense that this equation of motion will depend on the whole environment, including the experimental apparatus as a special case. We assume that the "isolated" electron is treated approx. by Schrödinger's equation. But in any environment, (including that needed to make a measurement), there are additional terms, leading to a rule that will relate a definite final state to each initial state, but in a way that depends on the large-scale environment.

Refer here to Mach's principle, which relates inertial properties (differential equ. of motion of a free particle) of a micro system to cosmological environment. Our idea is in some ways similar, in that we relate the equation of motion of a quantum system to the whole environment, in such a way that an isolated particle is described by Schrödinger's equation. However, instead of bringing in only the cosmological environment, we bring in the whole environment on many levels, including that on the level of laboratory apparatus.

Now, I would give the example on P.10. I would revise the material on Pages 11 and 12, in the light of what has been said. On page 12, I would emphasize that because the micro laws depend on all the levels, the laws of the micro-level transcend the framework of dynamics, as we have known it. Because the "rule" (given on top of P.10) follows from a dependence of micro-laws on the environment determined at the level of the laboratory, it follows that we need an ordered hierarchy of inter-related concepts to describe all the levels, which are all just aspects of one total indivisible structure, which functions as a single whole. The rule on top of P.10. then relates the functioning at the microlevel to certain features of the structure at the macro-level. So in a certain way, it has a structural relationship similar to that of language-metalanguage, used in Bohr's point of view. But it is now an ontological structure, and not a structure of language.

So it retains the feature of wholeness, but does not share with Bohr the notion that only classical concepts are unambiguously definable. Rather, it uses new unambiguous concepts (ψ, λ, the state of the large-scale environment, etc.). As shown in Sec 5, these new concepts imply the possibility of new unambiguously observable results.

I would rewrite all the material up to P.13 as one new unit, starting out afresh, if I were you.

Page 22, bottom. I would emphasize here that this refutes Bohr's thesis that the only unambiguously communicable features of what we observe correspond to classical concepts or some refinement of them.

I would here the importance of considering alternative axiomatic and conceptual structures even before the old ones are definitely refuted. For these new structures suggest new experiments that we would be unlikely to consider, otherwise. They suggest new ways to falsify current theories. A theory is always improved by the existence of a counter-theory, showing what it would mean to falsify it. If it is then falsified by experiment, we have evidently learned something new. And if not, we have also learned more about the domain of validity of the older theory. So we win, no matter what happens. Failure to consider alternatives tends to lead to a tacit form of dogmatism, because the older theory now seems inevitable, in the absence of an awareness of alternative possibilities.

As far as the second article is concerned, I think it is OK as it stands.

With best regards

David Bohm

P.S. With regard to Page 9, I would say that if you want to criticize Bohr, it is best to question his assumption that the only possible unambiguous concepts are those of classical physics, or of refinements of classical physics. One could say that this seems like an arbitrary assumption. In addition, emphasize that such an assumption, if unfounded, might be very restrictive, preventing us from developing new kinds of unambiguous concepts that are potentially of experimental significance (Referring here to Sec 5).

PPS

With regard to the whole paper, it would be useful to keep in mind the following general picture.

In everyday experience and in common perception, we have learned to analyze the world as constituted of objects, each located somewhere in space, and moving in a continuous way in time. This is what Bohr calls the space-time description.

Then we have learned to suppose that objects act on each other causally, to influence and determine each other's motions. This is the causal description.

Everyday experience is based on the tacit assumption that these two descriptions are compatible. Thus, we say that causes determine later effects, through forces and their effects on movement. We see no reason why a determination of causes is ruled out by a determination of where objects are and how they move.

The ability to apply both descriptions together leads, when refined in a quantitative way to the general scheme of classical dynamics. This scheme has two roles.

(a) Description of motion, position and of forces (Space-time and causal).

(b) The use of descriptions of force to make inferences about motion and the use of descriptions of position and motion to make inferences about forces.
Evidently, the inferential possibilities of classical dynamics depend on the simultaneous applicability of space-time and causal descriptions.

Finally, there is the division of observer and observed, subject and object. In everyday life, one of the objects in the whole field is singled out and called "I", "me", "the self", "my body", "the observer", etc. Because of the assumption of classical dynamics, we suppose its interactions with other objects can be taken into account.

In particular we may also set up a special object called the "observing apparatus", which serves as an extension of the "observer". The apparatus is so related to the human body that neither the behavior of the "observed object" nor the registry of this behaviour in the apparatus is significantly influenced by what the human observer does.

Now, the implication of the "quantum" is that this whole scheme breaks down. One finds that the experimental conditions needed for a precise causal description are incompatible with those needed for a precise space-time description. Therefore inferences of unlimited precision are no longer possible (The limits of precision being given by the uncertainty principle). Heisenberg's picture of an electron as one object and a microscope as another becomes inapplicable. The appearance of statistical fluctuations in behavior of the "observed result" merely expresses the breakdown of the possibility of drawing precise inferences, due to incompatibility of space-time and causal descriptions, needed for such inferences.

The division of apparatus and electron is now gone. Instead, we have two levels of language. (a) The descriptive languages of classical physics (space-time and causal), not applicable together in a completely unambiguous way. (b) Inferences can now be made only statistically. The inferences are drawn from the quantum metalanguage, as I explained earlier. The metalanguage now describes the "quantum object", which involves electron and observing apparatus in an inseparable way.

In a way, we can regard the classical phenomena as the "subject" (These are observed directly by the human "subject"). Then the "quantum object" is abstracted from these phenomena. Just as everyday "objects" are abstracted as invariant relationships holding in the "event level" where there are many "phenomena", so the "quantum object" is abstracted from a collection of "events" (e.g., readings of Geiger counters). So there is no analytic relationship between quantum object and individual phenomena. The quantum object belongs only to the ensemble.

Suppose you look at a picture, made of dots. With a few dots you get a vague picture. With more dots, the picture grows sharper. The picture is abstracted from the dots. There is no analytic relationship between the picture and the dots. Rather, it is statistical.

You could look on the interference pattern and quantum mechanics in a similar way.

Our criticism of Bohr is that he insists on the classical language as the only unambiguous one. We suggest another language, retaining "wholeness", but with a different ontology. It has new experimental consequences worth exploring.

In addition, it has new conceptual features that seem interesting and worthy of further study.

(a) A multi-level theory, in which all reflects all.

(b) Observation is inseparable from the reflective process. So that we don't say that what is to be observed is "there", apart from the process in which it can be observed.

(c) Yet, we don't say with Bohr that the process is incapable of unambiguous description.

(d) However, to describe it, we need new non-classical concepts (e.g., multi-level theory, structure-function, reflective function, etc.).

Jan 13, 1968

Dear Jeff,

I just received your long letter.

First, as to the job situation, it is much worse here than in the USA. There is an absolute freeze. Even some new appointments that had already been scheduled for physics dep't at Birkbeck have been cut off, though the men had been informally promised jobs. University budgets are being drastically slashed. I am really sorry, but I don't see what can be done here. I will write to Eugene Gross[2] on your behalf in the next few days. I do hope that something can be found for you. I realize how difficult your problem is. Couldn't you get a year's extension where you now are?

About the contents of your letter, it is difficult to comment on them. Probably one can learn from our correspondence that when there is a failure of communication, nothing is more useless than to try to discuss this failure. This is the surest way ever discovered to "confound the confusion". Perhaps we had better leave it at the simple fact that you and I have different relevance judgements, too different to be discussed by the inadequate content afforded by letters. Maybe in direct verbal discussions, we could do better.

My own experience in formalism is that to work in it requires so much attention that little energy is left to consider new informal notions. I have never been able to

[2] A physicist who was Bohm's research student at Princeton and remained on good terms. See Talbot (2017) for more detail—CT.

develop new informal ideas while working in formalism. Probably most physicists are similar in this regard. When I look at "densely packed" formalism, such as that of J. and P., or K. and S., I feel that to understand it, I must in effect commit myself to its language. So I get caught in activity similar to trying to find out whether I used first or second grade butter on the watch[3] (i.e., K. and S. are first grade butter, but J. and P.[4] are second grade butter). However, I see that your relevance judgements in this regard are quite different.

It may well be that publication of articles, such as our discussions of the language of physics will in fact only confound the already very great confusion. Very possibly, there is no way to "clear it up" at all. Every effort to discuss it tends to make it worse. For example, your article may perhaps establish more harmony with the von Neumann school, but it tends by this very move to go out of harmony with what I am trying to say. I am not blaming anyone here. All of us are caught in this mess, and as we move, we sink deeper in it.

I'll try to comment more on your article later.

Best regards Dave

Jan 15, 1968[5]

[Postmark 14th added – CT.]

Dear Jeff

I am continuing the letter of a few days ago.

I have written to Eugene Gross and also to Dr. Low[6] in Israel. I feel that it would be very good if you could get a job in Israel, and get out of the generally unpleasant situation now prevailing in American science. I am very sorry that you have so much to worry about now in the way of getting a job. However, I can say that Dr. Low's letter to me indicated that they are favorably disposed to giving you a fellowship. Once you are there, you could probably locate some kind of more permanent position, perhaps in Tel Aviv or Haifa.

With regard to your letter, I think it is best to proceed calmly and slowly. We do have different notions of what is relevant, and these have first to be sifted out carefully, if we are not merely to add to the existing confusion. Then perhaps, both of us will be able to come to perceive new orders of relevance, in which we are in closer harmony.

---

[3]Reference to Alice in Wonderland. The March Hare had put butter on the Mad Hatter's broken watch, and defended his action by claiming it was the best grade butter. See also C135, p. 230, n 2—CT.

[4]J. and P. refers to J.M. Jauch and C. Piron (see C132, p. 99, n 1). K. and S. refers to Simon B. Kochen and Ernst Specker who proved a mathematical theorem relating to hidden variables in 1967—CT.

[5]First letter with this date—CT.

[6]William Low (Ze'ev Lev) a noted Israeli physicist—CT.

Meanwhile, it may be useful to clarify my own notions of relevance a bit more. Please try simply to look at these, without acceptance or rejection, just staying with the fact that this is the content that happens to be relevant to me at the moment (as one simply registers the fact that the clouds in the sky happen at one moment to resemble a horse, and at another moment to resemble a mountain). Actually, no specified or particular content is important enough to get into a quarrel about it. For in a moment's perception, any content may be seen to have limited relevance, and in this very perception, a creative new ordering of relevance takes place.

Perhaps I could discuss what I think concerning the work of Cauchy. In my view, the relevance of this work is much less than that attributed to it generally by mathematicians. Time and time again, mathematicians have believed that they had produced ultimate and solid foundations for mathematics, only to discover later that these "foundations" were in turn just as unclear as the original mathematics "founded" on them. Because so much energy was expended on consolidating the foundations, little energy was left for creative new discoveries. For example, mathematicians could perhaps have discovered new algorithms, more suitable for scientific investigations than the calculus. These new algorithms would have made Cauchy's investigations rather uninteresting. Indeed, some of Leibniz's ideas on the calculus could already have led to new algorithms, making the whole question of continuity of functions not very relevant. This is, in fact, a direction that I am now pursuing. However, it could easily have been done a hundred years ago, if people had not regarded the calculus as something that had to be consolidated.

In my own work, I cannot see that what Cauchy (or others did) is very helpful. On the contrary, I recall how when I first learned mathematics, I was ready to try new methods and algorithms, but the teacher said I must be "rigorous". This demand for rigor had a very negative effect on my ability to think in original ways. I am sure it has a similar effect on most young people.

To me, it appears that mathematicians (like von Neumann) are putting quantum theory on what seems to be a solid formal basis, not noticing that the formal basis in turn rest on a very confused informal basis. It is like a man who builds his house on a solid concrete foundation, which is so extensive and deep that he loses sight of the fact that nothing supports the concrete except a quagmire. And indeed, nowhere in the world is there a "solid" foundation for physics. In such a situation, what is called for is something more like an airplane, a helicopter, or a dirigible than like a skyscraper erected on a solid granite box. In other words, it is only in free movement that one can have some measure of security. To armor oneself with "solid" foundations is as irrelevant as to adopt the armored structure of a dinosaur under modern conditions.

I get the feeling that you don't appreciate fully how serious is the confusion in v. Neumann's informal approach. As indicated in the articles with Schumacher, he assumes two existentially disjoint domains, classical and quantum, in interaction. As we point out in detail in the articles, there is no way to do this that is free of confusion. And there is no way to be sure that this confusion in informal language does not confuse what is done with the formalism. (Indeed, we indicate that it does confuse what is done with the formalism.) That is why I say that such work is like erecting solid concrete "foundations" in a quagmire. In certain ways, the formal "clarification" makes the situation more dangerous than ever, because it takes attention away from the fact that all that is said and done has, after all, no foundation.

If you are really serious about putting quantum theory on a "solid" foundation, I suggest that what it entails is for you to develop a consistent (confusion free) informal language, for discussing formal theories of the von Neumann kind. Otherwise, I feel that differences such as those between K. and S., J. and P.[7] , etc. will be as irrelevant as differences between the grade of butter in the running of watches. For ultimately, all these "solid" foundations are entangled in the confusion in the informal language. So no matter how clear K. and S. may be in their mathematics, one gets lost in trying to see what it means in the context of physics. Or even worse, one may lose sight of the fact that there is an informal context that is relevant, even to what is done with the formalism.

You say that changes in the informal language are arbitrary, thereby implying that, in some way, the formal language is not arbitrary. However, in my view, the formalism is equally arbitrary (a better word would be "fortuitous").

Mathematicians and physicists tacitly tried formalisms as solid and well established structures. But actually, these are fortuitously dependent on the personalities of those who are generally recognised to be authorities and leaders in the field (which personalities in turn depend fortuitously on the social situation, the whole historical evolution, etc.). Thus, if v. Neumann, Hilbert, Cauchy and a few others had had different personalities, the whole formalism of mathematics (as used in physics) could have been radically different from what it is today. What I find so hard to communicate is the insubstantial, "vaporous" character of formal structures. They can change as easily as can the shapes seen in clouds, and have no more ultimate significance than these shapes. One man sees a horse, and another a mountain. Likewise, one man sees the differential calculus and another sees purely algebraic structures, in which continuity is irrelevant. Why do scientists take these formal structures so seriously? It is like trying to put solid foundations under the "castles in the sky", that can be seen in the clouds. Perhaps the explanation is that most people learn very early to believe that they cannot do really creative work, that they lack a special quality called "genius", that would be needed for such work. So they restrict themselves to trying to clarify (or put solid foundations under) the shapes seen in the clouds, by other men, who for fortuitous reasons, are somewhat less conditioned to believe in the limitations that they are supposed to have. Perhaps some day, people will see that "anybody can do it". It doesn't take any special "genius", whose work is then regarded as an authoritative and solid structure of ideas (or capable of being made so, by people of more limited abilities).

I hope you will remember that all that I write (and all that I have ever written) is just a set of relevance judgements, no more substantial than the shapes in the clouds. (Remember also that this holds for v. Neumann, K. and S., J. and P., Newton, Einstein, Bohr, etc.)

---

[7] See p. 205, n 4—CT.

I'll continue more later

Best regards

Dave

Jan 15 , 1968[8]

[Postmark 17th added – CT.]

Dear Jeff

This is to continue the letter of a few days ago.

As I indicated, you and I have different relevance judgements in a number of issues. Very probably, this difference has been accentuated since we have been going our separate ways, each working on his own lines. Nevertheless, it was also almost certainly always there.

Roughly, you feel that informal discussion is arbitrary, subject to fortuitous variations, dependent on subjective features (perhaps of the Ego), while the formalism is the opposite, i.e. necessary, and not dependent on subjective features (of the Ego). So it could perhaps be said that you have some sort of notion of the relevance of mathematical truth, or of truth being in some sense mathematical. On the other hand, as explained in the previous letter, I feel that the formalism has been determined largely by fortuitous features of the personalities of those who happen to be recognised as leaders in the field. So in my view, it is mainly the Ego which has been operating to determine what is to be regarded as relevant, by elaborating certain kinds of formalism. However, because of the heavy emphasis on logical deduction and proof, it has been made to appear that all is necessary and true, so that the Ego has nothing to do with it. What is overlooked here is that formalisms can change as easily as the shapes of clouds. It is the Ego which holds to one "shape" as relevant instead of another. Once a certain formal structure is thus fixed, one can be very logical and clear. Nevertheless, it may all be irrelevant, because genuine clarity often requires that the whole formalism be set aside in favor of a novel one. Thus, the emphasis on proofs and clear logical deductions can easily cover up a deep confusion around the question of why we work with one formalism all the time, instead of frequently exploring novel formalisms. So in my view, mathematics is not a kind of truth. It is a formal form of description. It can be neither true nor false. Rather, it requires perception in each situation to see its relevance and irrelevance. This is possible only if formalisms are freely alterable all the time.

It may be that Einstein had notions of mathematical truth somewhat similar to those described above. But in my view, this was a serious barrier to his progress in the later years of his life. In a few years around 1905, 1906, etc., he wrote papers on three novel and original subjects, Special Relativity, Quantum Theory and Brownian motion. If you read these [you] will be struck by the simplicity of his reasoning,

[8] Second letter with this date—CT.

by its basically informal character, by its freedom from complicated and confusing chains of formal reasoning, that are full of unsolved problems and ambiguous terms of discussion. Later, being influenced by Minkowski and Levi-Civita, he went over to formal geometrical forms, with tensor analysis, matrices, curvature, etc., etc. Although he developed an interesting theory, general relativity, in this way, he was never satisfied with it. He felt the need to go on to seek a unified field theory. But in 30 or so years of work, he went only from one formalism to another, never arriving at new discoveries. Each formalism involved non linear equations that could not be solved, and whose terms were therefore ambiguous in meaning, because one never knew in what way they could be relevant to what Bohr called the experimental conditions and the experimental results. How different this was from the simple, direct, and basically informal character of the work of the fruitful early years of his life. Perhaps if he had never been blinded by the dazzle of tensor analysis, he could have gone on making simple but deeply significant informal discoveries, instead of spending 30 years in a fruitless search for unified field theories, of a formal kind.

Perhaps you will see from this that I feel that only when you start from the informal and then feel your way slowly to the formal do you keep out of the trap of allowing the Ego to determine relevance judgements. When you start from the informal, you are sensitive to the actual situation under investigation. It is like the artist who first makes a general sketch and then articulate the details. But if you start with the formal, your choice of formalism can only be determined by the conditioning, which is the Ego.

Consider, for example, those followers of v. Neumann who want to base everything on a formal algebra or lattice of some kind, involving the possible answers to "Yes-No" questions. Why do they regard this approach as relevant? If you do an experiment, how do you observe a "Yes-No" question? Evidently, this cannot be done. The question is imposed by the theoretician. To discuss an experiment, you have to describe the experimental conditions, as well as the experimental results. Until it is possible to do this in terms of "Yes-No" questions, there is no way to connect up the formalism to experiment and observation, except by the arbitrary fiat the Ego of the theoretician, who verbally says that an experiment can be treated as a set of Yes-No questions. He says this because it makes him feel good to believe that he has solved the problem of measurement.

If the relationship of theory to the informal experimental situation is confused, this can confuse all that is done with the formalism. While one is occupied with proofs, deductions, etc., one is able to fail to notice that there is no foundation to efforts to regard what is done as more than pure mathematics. Meanwhile, attention is thus diverted from cues to changes of informal and formal languages, that are given by a consideration of the total situation, in which there is no feeling that one has to give special relevance to a particular formalism.

Perhaps I could illustrate my views here by saying that the quantum formalism is now dead, so that it is not interesting whether it is logically impeccable or not. In the early days, it was alive and full of new discoveries, though formally, it was not very "clear" what was being done. After it was "cleared up" by Dirac, v. Neumann, etc., new discoveries almost ceased. Now 40 years later, even the donors of research funds are beginning to question whether "elementary particle physics" is worth all the money that it costs. The note of progress seems to have gone down, in a kind of inverse proportion to the number of physicists working in the field, and to their technical competence in mathematical formalism. Is this nothing but an accident? In my view it is not an accident at all.

Now, it is clear that our relevance judgements are very different. What are we to do? Is it of any use for you or me to try to convince each other with proofs, arguments, refutations, etc.? Evidently not. For whatever I say has no logical weight with you at all, if my premises are not relevant to you (and of course, vice versa). As long as each of us holds to fixed relevance judgements, we cannot meet. If we do this, it is best that we simply have nothing to do with each other. For every attempt to argue or convince can do nothing but confuse the issue yet further.

This is all I meant by the statement to which you objected. "What is relevant is the irrelevance of holding to fixed judgements of relevance, especially when one wishes to communicate with others."

Of course, the phrasing of the above statement can be improved. I would rather say: "It is relevant in communication to be aware of all the judgements of relevance, one's own and those of the other, and of how fixing these tends to get in the way of communication."

However I did not mean this in the form of an edict, imposed by a mythical entity called Bohmacher. If you object to this statement, let me know what your objections are. If you feel that fixed judgements of relevance are no barrier to communication, then nobody can possibly stop you from going on with them. However, I feel that specifically in our communication at present, fixed judgements of relevance are the main barrier. I am not suggesting a rule, whereby people are compelled to change their relevance judgements all the time. Rather, I suggest only that if you see the meaning of what I say, that is all that is needed to change what you do, in our correspondence (and elsewhere, of course).

Perhaps you are taking the verbal form of my statements too literally. Very probably, this form could be improved a great deal. But I also hoped that you could "read between the lines", so that you could go beyond inadequacies in my form of communication.

More later,

Best regards

Dave

P.S. I notice you use the word "guilty" a lot in your letter. In my view, this is another irrelevant notion.

In some way, both of us are responsible for the failure of communication between us, in the sense that what we did brought it about and that what we do from now on determines whether it will continue or not. But to blame anyone, or say that he is guilty has no meaning in this context.

After all, nobody chooses his relevance judgements. They "just happen", as the rain "just falls", rather than, for example, because God chose to make it fall. Every person makes such judgements according to his total conditioning. A new understanding, a new awareness, makes him effectively a different person, who cannot do other than make correspondingly different relevance judgements.

Indeed, even in simpler contexts (e.g., getting up from a chair), one does not really choose or will to do it. If, for example, you ask yourself "Just when and how did I choose to get up?", you will see that generally speaking there was never actually a moment of choice. It "just happened", step by step. Later, we verbally say: "I made a choice to do so." This projects a little "drama" in thought of "The man who chose or willed to get up." But this drama is irrelevant, except for giving one the pleasure of believing that one is an entity who has power to exert will or choice.

So it makes no sense to say that someone is guilty or to be blamed for what he did. Whatever he did, he had the feeling of its relevance, and therefore, he could not have done otherwise. Blaming a person will never change him significantly. It is only when one deeply understands and perceives the relevance and irrelevance of his judgements of relevance that his actions can change fundamentally. And then, it will happen of its own accord, without choice or efforts of will.

Similarly, it makes no sense to praise someone for what he did. Whatever he did, it flowed from what was relevant to him at the time, and he could not have done otherwise.

Perhaps you misunderstood my letter, and thought I was blaming you for the way you wrote your article. If so, please let me correct this wrong impression right now. I meant to say that first of all, there was a lot of interesting and relevant content in your article, but that, because it brought together so many mutually irrelevant points of view, the net effect was to add to the confusion that already exists on the subject. This confusion was your responsibility, but of course, not entirely yours. For the whole subject is now so confused that any attempt to refer to it tends to fall into similar confusion. Thus, you must harmonize with whatever you refer to. But if you try to "harmonize with confusion", then you too will be confused.

When I asked you why you wanted to write an article while you are in a state of disharmony, this was not a criticism or any kind of blame, but just a simple question, that I thought you would find it interesting to consider and perhaps even to "meditate" upon. I did not wish to imply that I have done better than you in this regard. The question is relevant to me too and indeed to everybody.

I am not blaming J. and P., von Neumann, or anyone else. Each was responsible for what he did, but could not have done otherwise, without a deeper understanding of why he did what he did.

It is useful to keep in mind that deeply, no one acts from predetermined intentions (though in some superficial cases, like getting up from a chair, this notion can sometimes have a limited kind of relevance, in the sense that I can say I want to get up, and then this is followed by the action of getting up). None of us intended to adopt the views in physics or mathematics that he now has. It "just happened", because of an infinite complex of factors, mostly unknown. So even if you feel my relevance judgements to be irrelevant, please recall that it was never my intention to adopt them. Rather, they are operative, because to me, they are relevant.

In short, then, relevance is no kind of value. To give value to relevance judgements leads to confusion. If we give positive value to them, we will want to keep them, even in contexts where they are not relevant. If we give negative value to them, will want to get rid of them, even in contexts where they are relevant. So relevance judgements cannot relevantly be valued. Rather, it is relevant to be aware of the irrelevance of giving value to them.

Jan 25th 1968

[Date added – CT.]

Dear Jeff

I wonder if you received my last two letters (Dec 24 and a second sometime in January, where I gave detailed comments on your articles).

I hope your wife and son are doing well.

Enclosed is a copy of an article which I received.

Best regards

David Bohm

PS. Did you get a manuscript of mine entitled "Some Further Remarks on Order for the Bellagio Conference in Theoretical Biology"?[9]

Feb 12th 1968
Date stamp on envelope 14th
[Date and comment added – CT.]

Dear Jeff

Thanks a lot for your letter and article.[10] On the whole it seems good. I enclose some comments.

Best regards

David Bohm

Comments on Article

Page 5. Middle.

I think that the confusion in Heisenberg's analysis is worth a whole paragraph. Heisenberg treats the observing apparatus and electron as two different and separately existent objects in interaction, exactly as is done in the classical ontology. But then he uses the quantum theory to treat this interaction process. The quantum theory,

[9] Waddington (1969), pp. 41–60. See Introduction, p. 3, n 10 for more details. Also C132, p. 147, n 14—CT.

[10] Published as Bub (1969)—CT

as used by Heisenberg, is in contradiction with the diagram of the microscope and the electron, that is so familiar to everyone. Therefore, the argument is confused. At most, it is useful to illustrate the limits of applicability of classical concepts. But beyond this, it plays no real role whatsoever in the theory. Unfortunately, most texts either tacitly or explicitly assume that Heisenberg's argument is both relevant and basic, whereas its relevance is actually extremely limited, and, being in contradiction with the theory, it can hardly at the same time, be a basic aspect of the theory.

The contradiction I mentioned above is clearly asserted by Bohr. He said that the "quantum object" is on another level from the classical objects and events that we can actually observe and communicate about in an "unambiguous" way. Bohr never actually used Heisenberg's diagram, and this is probably not an accident. According to Bohr, the "quantum object" is abstracted statistically from a vast number of classically observable phenomena of a similar nature (as the picture in a book emerges as an abstraction from the printed dots). So the quantum object is inseparable from the whole set of experimental conditions.

I would say that Bohr's notion of the relationship of observed phenomena and of lower level "objects" may well be a fruitful one. But, of course, his mistake is to assert that only classical concepts can be used in describing the observed phenomena unambiguously. If Bohr gave up this latter assumption, then it is not at all clear that we would be basically in conflict with him. Perhaps it would be worthwhile for you to assert this somewhere in the article.

On Schumacher's attitude

Schumacher agrees with us that Bohr is wrong to assume that classical concepts are the only unambiguous ones.

However, he emphasizes the extremely new and creative feature of Bohr's perception, that subject and object are different levels of communication about the same field. This gives a deep insight into this inseparable relationship.

Further discussion with Schumacher brought out that he wants to emphasize, not "language" or "linguistics", but communication in general (Visual, auditory, music, language, etc., etc.). We cannot separate the question of knowledge from that of communication. Knowledge is indeed an aspect of communication. I think he is as much against "linguistic philosophy" as we are. Nevertheless, some very deep questions concerning the meaning of language are raised, when we attempt verbally to assert that "reality is basically unverbalizable". He believes that to clarify this question is the key to "substantive progress".

About the question of randomness, he emphasizes the distinction between in its creative and analytic (logico-deductive) aspects. Current probability theories attempt to to ascribe a measure (of probability) to randomness. Schumacher thanks this is a contradiction, equivalent to the term "order of disorder". He thinks that randomness is basically incomplete self-definition of an order. Thus, it is, in a deep sense, another term meaning "potentially creative". The relationship of numerical probabilities to randomness is therefore, in his view, at present ascribed in a confused way. Clarifying this question would then be very important.

If we take this meaning of the term "randomness", we see that a book review is indeed "randomly" related to the books that are reviewed in it (i.e., it does not "substantively" define the order of the books that are reviewed). This does not necessarily imply a probability measure, relating the book review to statistical distributions of words in the books that are reviewed. Similarly, the "quantum object" is "randomly" related to the perceived physical events (measurements) that take place in the laboratory. Schumacher agrees that current probability measures applied to this "randomness" are confused, and not necessarily inevitable. So, tacitly at least, he could agree with us that more general measures could be applied to the contingences (i.e., results of individual measurements). Indeed, he is more radical, and believes that something new is needed, altogether beyond the notion of applying a measure to random (i.e., creative) sets of contingencies.

In short, what he wants to keep in Bohr is what he regards as deep, radical, revolutionary, and essential. This is the notion of not regarding physical concepts (e.g. Hilbert space) as being in a correspondence with "what is". Instead, the Hilbert space generates a field of "statements" about a vast set of contingencies, providing a way of asserting the operations of necessity in this field of contingencies. (Without contingency, necessity has no field to operate). But the necessity is not absolute. The notion of "randomness" allows for the creative emergence of new orders in the field of contingencies. (Classical physics would never have allowed for this.)

Schumacher regards Bohr's use of the correspondence principle (asserting that our unambiguous concepts are only classical) as an attempt to "mitigate" the extreme novelty of his basic views about the "quantum". He thinks they are inessential, and a source of confusion, in their present form. Thus, he is basically not in disagreement with what we want to say.

Page 6. Even in our point of view, quantum objects might well have a different status from large scale objects. Indeed, the structural process point of view implies that each "object" is an abstraction of a set of other objects of a different "status". These other objects are contingently related to the abstracted object in question.

Therefore, it seems important to assert our agreement with the deeper aspect of Bohr's thesis, while we assert our disagreement with the superficial aspect, in which he equates "unambiguous" with "classical". We should also say that unfortunately, Bohr's writings have suggested to most people that what is basic to his views is the necessity for all unambiguous concepts to be classical (or refinements). If we were to take Bohr's views on the mode of abstraction as basic and revolutionary and his views about classical concepts as superficial efforts to "mitigate" this novelty, then we would even say that hidden variables are deeply in agreement with what Bohr said.

How this comes about is as follows: The wave function (the quantum algorithm) is only contingently related to the observable events. The actual events depend on the total experimental set of conditions, which determine what "observable" is being measured. We differ from Bohr mainly in saying that the description both of the results of the experiments (and ultimately of the experimental conditions) involves concepts going beyond those of classical physics.

To be sure, the present version of the theory is reminiscent of classical mechanism, insofar as we assume a hidden variable, which obeys a deterministic equation. But we could quite consistently regard this as a provisional "bridge" to a deeper "structural-process" approach. We could assert our readiness to abstract from our present theory a different approach, which drops the deterministic equation. Instead, we could (as Bohr does) define a field of contingent events or structures, observable on the large scale level. We could, however, differ from Bohr, in that this field would be unambiguously described by new non-classical concepts. We could then bring in new "micro" laws, which would assert a limited kind of necessity in this field of contingencies. This latter field would be "random" in Schumacher's sense. That is, the contingencies would not be completely self-defined in their order, so that there could be room in them for creative emergence of new orders, reflecting a "micro-level" operating in this field of contingencies.

To do this, we need to develop new non-classical concepts. The value of hidden variable theories is that they serve as a "bridge" to help indicate what these new concepts might be. This is in addition to their value in helping to show what it means to falsify the orthodox theories.

To be sure, it would be difficult to put this into your article. But if you could indicate something of this nature, the article would gain force and value.

Probably, it would be inappropriate to emphasize this question too much. But if you could, in different places, indicate that all sorts of theoretical ideas have to be considered, to relate to each other and to criticize each other, you could, without committing yourself to hidden variables, assert that consideration of them has a role to play in the overall dialectic of development of new physical theories. You could assert that Bohr's deeper views are not at all incompatible with the deep intention behind the development of hidden variable theories. Indeed, in the long run, the structural process point of view aims to realize a theory that is in essence compatible with Bohr's deeper intentions. What is necessary is to point out is how superficial is Bohr's emphasis on classical concepts. Don't fall into the trap of interpreting Bohr as being in opposition to you. You yourself said that "Bohr's philosophy does not really exist". So why don't you too "play the game" and re-interpret Bohr so as to agree with you. In my view, you are closer to the essence of Bohr than Heisenberg and most physicists are. If you were to say that the essence of Bohr is compatible with the deep intention behind hidden variables (and that he is only superficially against it) this might generate so much surprise that the reader's attention would be awakened, so as to question the whole story. This is really what is to be desired.

Page 30 (Conclusion).

If we take Bohr's "deeper" thesis, it implies neither "rockbound indeterminism" nor determinism. The very word "random" now means "creative". We all agree that every law must leave room for creative emergence of what is new. Bohr's notions make room for this in a radically new way. A given law does not determine a certain field of contingencies, but merely asserts some predispositions (or "propensities" as Popper would say) in the field of contingencies. (A "delta function" is the limiting case of complete necessity.) Therefore, to say that Hilbert space determines propensities in contingent sets of events leaves room for entirely different orders in these contingencies (e.g., those implied by hidden variables).

The "orthodox" view of Bohr's thesis is then that "randomness" implies current notions of probability measure – therefore "rockbound indeterminism". Notice that the algorithm of Hilbert space is only contingently related to the notion of probability measure. It could mean something else, but still a relationship in a large set (ensemble) of events. Our own view of Bohr's thesis is that randomness does not always mean probability measure, as defined in the orthodox view.

Rosenfeld and DLP have come out in favor of this "orthodox" view on probability measure. At the same time, they seem to wish to assert that these views on probability measure are alone enough to justify the scheme of quantum mechanics without radically new epistemological views (such as those of Bohr). This position is, of course, confused. But your last paragraph (P. 31) is also confused.

If you accept all of Bohr's theses, then as far as one can tell, the result is consistent. So although his theses are not inevitable, they are possible. And there is no logical reason compelling one to look for a new theses. As far as I can see, these reasons are tactical and methodological. We showed that Bohr's thesis is not inevitable. In view of the need to criticize older concepts with the aid of newer ones, why not explore our thesis (without commitment) as part of the dialectic of development of scientific theories? This exploration may well have the same deep intention as is behind Bohr's deep thesis (about "objects" of one level having a different status from "objects" of another level, in the sense that one is an abstraction applying "randomly" in the other, which latter is a field of contingencies relative to the former). On the other hand, the study of hidden variables does drop Bohr's superficial thesis about the classical concepts being the only unambiguous concepts.

The notion that classical concepts are an approximation to a quantum theory of measurements directly contradicts the whole of Bohr's thesis. But I wonder if it is of any use to say this here. Why not leave out altogether all mention of DLP in this article? After all, it will all come out in the next article. What you say here about DLP can only lead to confusion.

More remarks on Schumacher's ideas.

Schumacher regards the question of division of subject and object as basically a formal one, arising in the sphere of communications (including language as a special case). In other words, the notion of subject and object have no content. That is to say, there is no factual, objective significance to this division. Rather, it is a form imposed by our mode of communication.

This form is conventional. Historically, certain conventions have arisen. (e.g., the subject is "inside the skin" while the object is "outside the skin".) But entirely different conventions are possible, and indeed advantageous. Bohr would, for example, take the configuration of observable events as a kind of "subject", while the object is the "quantum system" abstracted from these. In the customary historical convention, subject and object are taken effectively as separate and distinct objects in interaction. In Bohr's view, subject and object interpenetrate inseparably, in a kind of "wholeness". The division is a formal one, for the purposes of communication.

You say that Bohr's views were physicalistic and not linguistic. But what is physics? Physics is defined by a set of conventions (largely tacit and historically conditioned).

Why are biology and psychology not taken to be part of physics? Evidently, their subject matter is interwoven inseparably. But people thought it convenient to abstract physics as a "distinct" subject.

It is important to expose these conventions to our view, because they could be getting in the way of clear and free thinking. Now, basically it is through communication (mainly by language) that physics is defined in a tacit manner. The division of subject and object is an essential aspect of this definition. The world is called the "object" and the "thinking subject" is conventionally said to be "observing" this world. Evidently, whatever is said to be examined by the subject is, by definition, some kind of object. By definition, the "real subject" can never examine himself, in his very act of carrying out the examination in question.

Thus the linguistic and conventional division of subject and object has defined what we mean by the words "scientific knowledge" and ipso facto, what we mean by the terms "physicalistic" and "linguistic". Bohr was, according to Schumacher, tacitly touching on these questions (which of course have nothing to do with linguistic philosophy).

It is important to note that Schumacher also regards the definition of words like "randomness" and "statistical" as historical conditional conventions, that could be altered, with advantage.

Questions of language can according to Schumacher, be important. The reason is that "what is the case" is asserted through linguistic forms, conventionally defined. When these forms are taken as universal, we are committed to analyses, whose structure is really conventional, but which seem to be inevitable. Nobody is primarily concerned with language. Yet, whether we like it or not, we will be primarily concerned with linguistic puzzles, unless we are clearly aware of the vast role of language, in what is called "fact" (e.g., the musician is not primarily concerned with notes, but if he ignores the effects of their structure, he will become the prisoner of this very structure).

You are quite right (and Schumacher agrees) to say that Bohr confused the issue by trying to salvage what was true of the classical conceptual structure and to build on it. But let us now say that this was a "superficial" part of Bohr's theses. The deeply revolutionary essence was in the need for a hierarchy of concepts, with the notion of a random (creative) relationship between different levels of the hierarchy (Probability measure is only a trivially restricted case of randomness). Other hierarchical views overlooked the creative (random) relationship of levels leading only to a more complex form of "ironbound determinism" or "ironbound indeterminism." The hierarchy of creatively related metaconcepts is not in itself restrictive. Rather, it opens up vast possibilities for development.

Bohr did lack a clear notion of creativity (The latter is too vast to be called a "concept"). But he had something like it implicit in some of his views.

The important point is not only that a conceptual framework can evolve creatively. Much deeper is the need to have a language structure that leaves room for such an evolution, without the need for a tremendous "crisis". The notion that "metaconcepts" apply "randomly" (creatively) helps provide such a structure.

General speaking, fluctuation phenomena are the field exhibiting the possibility of creation. If the metaconcepts are recognised explicitly to apply in the general statistical sense, this does not leave room for creation in each level. This is extremely important. For example, let us consider human perception. Here, the "subject" can usefully be defined by the convention that the subject is some set of mental "events" perceived at a very low level of abstraction. The "object" is than abstracted from these (as we can see very clearly on a dark night, when vaguely defined "objects" are always forming and dissolving). Evidently the "object" interpenetrates the "subject" in an inseparable way.

What is crucial here is to notice the "random" relation between object and subject. The "object" in no way provides a strict determination of the order of the "subjective" perceptual events. Rather, these are essentially free and creative. But in following their free creative mode of creation and annihilation, they subordinate themselves in certain very general and almost unspecifiable ways to the general pattern of the "object". In addition, they contain vast further orders, called "beauty", "harmony", "love", etc., etc. By convention, these have been said to belong to the "subject" and not to the "object". But it is entirely a matter of what is useful or convenient whether we say the beauty is in the object or in the subject. For after all, all subject and object are one, interpenetrating and inseparable.

Similarly, as Schumacher points out, in special relativity, the events observed in a certain frame of reference are the "subject", while the "object" is the set of "invariant relationships", abstracted from these events, and independent of Lorentz frame. Relativity theory defines the "object", but leaves the details of what happens in any frame to be completely free and potentially creative.

Likewise, in quantum theory, what happens at our perceptual level is free and creative. We can abstract all sorts of "objects" from this process, as perceived. The "quantum object" is one of these.

Don't you think this opens up a tremendous field for structural process? (It is really an extension of the notion of referential and inferential structures, to cover the whole of what we know and perceive.)

Feb 14, 1968

Dear Jeff

I just got a letter from Mary Hesse,[11] indicating that she is ready to publish your first article, but feels that the second is too mathematical for their journal. She suggests publishing the second article in another journal, but says it could perhaps be published along with the first, (though she evidently greatly prefers to do otherwise).

---

[11] Mary Hesse was a British philosopher of science, then editor of The British Journal for the Philosophy of Science (BJPS), to which Bub had submitted his articles—CT.

What do you think of publishing the second article in a more technical journal. You could add a paragraph or so of further introduction, to relate it to the first one.[12]

Best regards

David Bohm

PS. Enclosed is a notice about Krishnamurti that you may find interesting.

Feb 20, 1968

Dear Jeff

Thanks for your very interesting letter about Bohr and your various "experiences". When I falsely recalled Bohr's statement as "trapped in language" instead of "suspended in language", this shows that, like you, I felt that Bohr's views imply a kind of imprisonment or fettering of the intelligence. However, as I indicated in my last letter to you, perhaps we can divide his views into his "deeper thesis" (subject and object as interpenetrating aspects of one set of creatively contingent phenomena) and his "superficial thesis" (only classical concepts are unambiguously communicable). If we reject his "superficial thesis", then perhaps your objections to Bohr would drop away. What do you say to this?

Of course, you are right to say that we can always change the language in terms of which phenomenon are to be described (so that we have a new kind of "plain language"). Nevertheless, the "language and metalanguage" question will arise once again, whenever we try to consider the conventions that distinguish subject and object. Of course, physics is by definition a field in which this distinction should not be relevant. Therefore, you are right to say that our language for observable phenomenon should be such that we do not need a metalanguage to talk about it. Rather, all these questions "beyond the language" are properly to be regarded as tacit rather than explicit. (Here, I am sure that Schumacher would agree too.) However, it is not at all certain that even in our "language" there could be no breakdown of Boolean logic. Isn't Boolean logic itself an idealization, based on our "plain language"? Your position amounts to the assumption that the foundation of physical language is in what is explicit and precisely definable. My view is that it is in the depths of the mind that are tacit and unspecifiable. The explicit emerges out of the tacit, as the precise image in the centre of vision emerges out of the vague background in the corner of the eye (as in the "psychedelic" image emerges out of a background of flux). In this regard, physics is not different from any other field of perception.

So in the last analysis, the specification of structural process may be tacit rather than explicit. It may well be a mistaken ideal to aim to associate each object (thing) with a Boolean lattice of its attributes, which are in principle measurable. Indeed, the whole notion of regarding measurement as basic to physics could be called into question. In a fundamental sense, measurement is very superficial, perhaps even a triviality.

---

[12]The first article was in fact published in the BJPS, Bub (1968b), the second in the International Journal for Theoretical Physics, Bub (1969)—CT.

In other words, Bohr may be wrong in emphasizing what is unambiguously communicable in the foundation of physics. Rather, physics may be more like art, in that unambiguous communication emerges out of a vast background that is not precisely specifiable or measurable.

Now, about LSD and other such drugs, your letter was very interesting. It is hard to comment on your specific experiences, which I have never shared. However, I can say something on more general lines, that we do probably have in common.

Firstly, you question my view that such drugs "inhibit the inhibitions" due to thought, and thus "free" some other aspects of the functioning of the mind (such as the free flow of images). Here, let me call attention to the fact that studies of brain function have established the generality of this pattern, by which the various functions inhibit each other.

For example, we generally have only a relatively weak awareness of the sensations of our clothing, of sitting in the chair, etc. Some people, with certain brain damage in the thalamus, feel their clothing and such sensations with a maddening and intolerable intensity. It has been shown that these damaged sections of the thalamus normally inhibit critical awareness of such sensations. Indeed, by cutting out certain parts of the thalamus of animals, similar symptoms are produced.

All this is reminiscent of your descriptions of your own experiences. If you assume that the drug further inhibits your awareness of sensations of contact of the body with the ground or with the chair, this would explain the sensations of "floating in space" that you describe.

Although I haven't experimented with drugs, I do not find your description totally foreign to my experience. I see, in awareness, roughly three levels of perception.

1. Sense perception.
2. Emotional perception.
3. Intellectual perception.

Of the three, the last (intellectual) is clearly the slowest. I would say that emotional perception (beauty) is the fastest, while sense perception (form, color, etc.) is somewhere in between.

Perception is all one process. But its "spectrum" of frequencies is roughly analyzable in the above order. When awareness "emphasizes" one part of the spectrum, then attention is directed mainly to that part.

For example, when perception is mainly intellectual, awareness tends to focus on the "lower frequency" part of the "spectrum" of perception. Emotional perception (beauty) focuses mainly on the "highest frequency" parts of the "spectrum".

Therefore, when you are thinking, you are "soft-pedalling" all the faster processes. They are still there, but they are in the "background" of awareness. Likewise, there is a state of emotional perception, in which intellect is soft-pedalled, and there is a state of sensual awareness, in which both intellect and emotion are soft-pedalled.

I am led to assume that drugs like LSD and marijuana work by inhibiting the intellect, so that awareness focusses on the senses (and possibly on the emotions). You are right to distinguish them from alcohol, which probably inhibits almost everything. With these drugs, perception of process speeds up. This is quite inevitable, because the intellect, which works so slowly, has been inhibited.

There is another aspect of this process that is significant. We must distinguish, at sensual and emotional levels, between conditioned perception and unconditioned perception. Conditioned perception is structured by word and idea, as laid down in the past. Unconditioned perception is direct response of senses, emotion (and intellect) to the unknown, without the mediation of words or ideas, laid down in the past. Conditioned perception can still be very fast, even though the intellect is slow. We can compare the situation to a drama. The dramatist writes down a few words, quite slowly. From these words, the actors can elaborate a vast and complex set of actions, that take much less time to carry out than the dramatist took to think these words, and write them down. Yet, the actions are not spontaneous, free and independent. They are conditioned by the words of the dramatist. Similarly a few thoughts, very slow in themselves, can lead to the elaboration of a vast "dramatic" response at the sensual and emotional level. It is crucial to distinguish this "dramatic" response from the true unconditional perception that is a direct (unmediated) response to the unknown.

In reading four or five books written by people who have taken LSD, I have noted that one can trace most of the "visions" described by these people to verbal conditioning, which is is apparent from what they themselves say in their books. The "experiences" themselves are very fast, vital, colorful, and convincing. But I cannot escape the suspicion that they are mainly a "dramatization" of verbal conditioning rather than a direct contact with the unknown (e.g., of the kind that Krishnamurti talks about).

Some of what you describe strikes me as possibly being of a similar nature. For example, you became aware of the "Greekness" of the friend whom you mentioned. Isn't "Greekness" a purely verbal concept, based on past conditioning at an intellectual level? Wasn't your experience of "Greekness" perhaps a "dramatization" of this concept, at the sensual and emotional levels? Were you really in direct and immediate contact with this man as he really was, in his vast and unknown totality of eternal movement?

When the intellect is inhibited, one is able to see other aspects of the mind more clearly. So, in a way, something is gained. Yet, in the long run, perception conditioned by a drug is just as limited and meaningless as is perception conditioned by habits of thought. It doesn't reach out to totality, to the unknown depths that are immense and immeasurable.

One of the difficulties with such perception may well be that it is limited by the relative absence of intellect, so that it is unbalanced and incomplete. In particular, when thought does come in, one cannot avoid confusion.

For example, you describe how after a while, thoughts come in. "That was good", "I feel my clothes", "I must remember what it felt like", etc., etc. You tried to fight them off, but they got worse, and soon you come "down" from being "high".

Because your intellect was not very awake, it probably could not really understand that when you said "This effort to stop thought is just another stupid thought", this was a value judgement. Such a value judgement acts like a blast, like a bomb exploding in the middle of the the brain, to introduce utter confusion. The word "stupid" applied to anything is in itself a tremendous blow at the stem of the brain, which leaves one stupefied and confused. The effort of thought to stop thought is in itself a contradiction.

But you were treating it as a problem. Now a problem can be solved, so that to think of a problem is to force the brain to continue the thought of the problem, until a solution emerges. But a contradiction cannot be solved. It has to be dropped. However, by treating a contradiction as a problem, the brain is caught in going on forever with the contradiction.

Had you been able to see this, you wouldn't have had to come "down". But in this case, perhaps you wouldn't have needed the drug in the first place. Basically, the "self" is a contradiction. But we are conditioned to treat it as a problem. So we try to "set it right". Since this is impossible, we are eternally trapped in thought at a very deep level. This very emphasis on thought in contradiction is a soft-pedalling of the faster emotional and sensual sides of awareness. (It also prevents real intellectual perception, which would be free of contradiction.) Therefore, what is called for is the ending of this contradiction, called the "self", which is mistaken for a "problem". The drug doesn't really do this. It apparently gives moments when the thought of the "self" is "soft-pedalled". But it is still there, in abeyance. It has not been dropped forever as the meaningless contradiction that it is. Unless this happens, playing with drugs seems to be just another "gimmick".

For this reason, I am inclined to doubt that these drugs can really teach us very much about the mind. Try to understand what is involved here. The mind, in each one of us, is deeply conditioned to contradiction, confusion, violence, and delusion. The drug alters the effects of this conditioning, without (as far as one can tell) altering the basic conditioning itself. How is it possible for a confused mind to know what it is really "learning" from the drug? Recall also that the "research worker" who "observes" someone else taking the drug is himself deeply, basically, and hopelessly confused. Whether he takes the drug himself or observes another who takes it, what is to be expected, starting from such confusion?

Doesn't it seem, rather, that what is called for is that each man shall begin to inquire into his own confusion? I suggest that when this confusion is gone, each man will be in immediate contact with something immensely beyond the "highest" experiences made available by any drug.

Best regards,

David Bohm

PS. I am looking forward to seeing you this summer.

Feb 26, 1968

Dear Jeff

Just received your letter of Feb 21. Briefly, I suggest that you modify your article[13] on the lines you propose. Change the title to include the notion of reconciliation. Put as much of the "new wine" in as you can, without making too much work for yourself.

[13] i.e. Bub (1968b)—CT.

The DLP issue is much less significant. Why not publish your reply to DLP in the same journal as DLP published their articles?[14] Begin it by summarizing your article on Bohr, and by emphasizing how DLP differ from Bohr, while they claim to be supporting him.

I'll reply to the rest of your letter later.

Best regards

David Bohm

PS. The new ideas could well be the basis of a further article, as you suggest.

March 8, 1968

Dear Jeff

Enclosed is a reply of mine to a contributor to the Bellagio Conference, which you may find interesting.

Best regards

D Bohm

March 4th, 1968

[Date added – CT.]

[There follows a type-written document: "Addendum to Remarks on Order for the Bellagio Conference on Theoretical Biology", which was published as "Addendum on Order and Neo-Darwinism" in Waddington (1969) pp 90-93. See Introduction, p 3, n 10 – CT.]

March 1, 1968

Dear Jeff

In your letter you asked what I thought about your notions of randomness as applied to the structural process underlying quantum theory.

On the whole, I would say that I agree with what you write on the subject. I would add a few points, however.

Perhaps we could regard the wave function as belonging to the quantum level of S-P (structural process). The dual function, ξ, may then be a kind of "image" of the quantum system in the scale large scale environment (including the apparatus). This is a bit like the idea that an electric charge produces an "image" in a metal sheet. If the sheet is a perfectly homogeneous conductor, the image is uniquely and necessarily related to

[14]This was published as Bub (1968a)—CT.

the charge. If it has inhomogeneities, these introduce a feature of contingency. As the "real" charge moves, its image jumps around in a somewhat fortuitous way (because the inhomogeneities are fortuitously related to the movement of the "real" charge).

Similarly, we may assume that ξ is an "image" of ψ that tends to move somewhat fortuitously in relationship to the latter. Current quantum mechanics assumes this relationship to be completely fortuitous or random. We can propose new theories, in which the "degree of fortuity" is limited, eventually approaching a relationship of simple necessity, as an extreme limit. Therefore, as you say, the role of ξ is to enable us to treat quantum theoretical randomness as potentially creative – i.e., capable of new orders emerging. (Read the Further Remarks on Order for the Bellagio Conference[15] for a discussion of how the fortuitous becoming the necessary is the basis of creativity).

If this is a correct view, then the temperature of the environment is what is crucial. At low temperatures, the fortuitous element in the environment is reduced. But of course, to get beyond qu. mchal randomness, we may need spatially homogeneous environment, as well as temporarily homogeneous process.

Best regards

David Bohm

References

Bub, J. (1968a). The Daneri-Loinger-Prosperi quantum theory of measurement. *Nuovo Cimento B, LVI, I*(2), 503–520.

Bub, J. (1968b). Hidden Variables and the Copenhagen Interpretation-A Reconciliation. *The British Journal for the Philosophy of Science*, *19*(3), 185–210.

Bub, J. (1969). What is a hidden variable theory of quantum phenomena? *International Journal of Theoretical Physics*, *2*(2), 101–123.

Talbot, C. (Ed.). (2017). *David Bohm: Causality and chance, letters to three women*. Berlin Heidelberg: Springer.

Waddington, C. H. (Ed.). (1969). *Towards a theoretical biology 2*. Edinburgh: Edinburgh University Press.

[15] Waddington (1969) pp. 41–60. See Introduction, p. 3, n 10—CT.

Chapter 7
Folder C135. Bohr Versus von Neumann. September–December 1968

Sept 24, 1968

Dear Jeff,

I received your two letters, and have sent off the recommendation to MIT as well as notified the College about the need for a PhD certificate.

The money situation, as described by you, does not look promising. I don't feel, at the moment, that the set up in Toronto would be appropriate. In any case, they haven't contacted me, as yet. To try to agree on a programme of research, so as to combine my interests, with those of the people who determine the grants, would probably result in a kind of self-cancellation, in which no interests would be realized. The very notion of having to report on "progress" would invalidate the basic form, needed for exploring the unknown.

Could you perhaps get some money to come here for a while (to London)?

Now, now about the contingent parameters, I want to make a slightly different proposal.

Let ψ_{ij} represent a wave function, with i indices corresponding to operations going on in a certain relatively localized region, while j indices correspond to operations going on in other regions (i.e., generally more than one region).

We now consider relatively localized operations, described in terms of the i index. It is important not to call them "measurement operations", as you do in your letter. My view is that these are "natural operations", which take place of their own accord. "Measurements" refer to special cases of these "natural operations", taking place under selected conditions, with arrangements of matter called "physical instruments", and with a further selection of those operations that are regarded as relevant. Thus, in a laboratory, a Geiger counter may click, a truck may pass by, and someone may switch on a light, all at more or less the same time. All of these processes are attended by vast acts of "natural operations". But of all these, only the results of operations in

C. Talbot (ed.), *David Bohm's Critique of Modern Physics*,
https://doi.org/10.1007/978-3-030-45537-8_7

the Geiger counter are regarded as relevant. And indeed, even in the Geiger counter, only one result, (i.e., the click) is regarded as relevant.

So the word "measurement" is not relevant to the fundamental descriptions of physics. It is of a status rather like that of the word "money". Money is needed to do experiments, but the word "money" has no place in the descriptions of physical phenomena or in the laws of physics. Similarly, measurements may (or may not) be important in scientific activity, but the term "measurement" has no place in the basic descriptions and laws of science.

Now I propose that in a localized operation, the contingent parameters depend only on the i index, and not on other indices. Thus I write

$\xi_i{}^R$

representing the contingent parameters that are relevant to localized operations belonging to the region R.

I now propose the following equations

$$\frac{d\psi_{ij}}{d\tau_R} = \lambda\psi_{ij} \sum_n J_n \left(\frac{J_i}{K_i} - \frac{J_n}{K_n} \right)$$

where $K_n = \xi_n^{*R} \xi_n^R$ and τ_R is "proper time" for the Rth region.

$$J_n = \sum_j \psi^*_{nj} \psi_{nj}$$

You will readily verify that

$$\frac{dJ_i}{d\tau_R} = \lambda J_i \sum_n J_n \left(\frac{J_i}{K_i} - \frac{J_n}{K_n} \right)$$

It follows that an algorithm will go through. If the $\xi_i{}^R$ have a random distribution, it follows that the probability of the ith result of a localized operation is

$$P_i = \sum_j \psi^*_{ij} \psi_{ij}$$

This operation will leave the wave function ψ_{ni}, and 0 for $n \neq i$.

If you work it out, you will see that for two regions, R and R', the results of the EPR experiment are consistently described.

An important question is then that of whether the parameters $\xi_i{}^R$ can be described as ordered in a way that is relevant to the general experimental conditions. For example, a particular $\xi_i{}^R$ may either be constant or change in a specifiable way for some period of time, τ_R, after which it becomes subject to contingencies that are the fortuitous results of what is outside the present field of discussion. If so, it becomes possible to reveal the parameters $\xi_i{}^R$, in terms of the order of results of successive operations. But there may be other ways to reveal the new order that is implicit in this description. For example, one may do the EPR experiment under conditions in which there would be no time for a "signal" to pass from region R_1 to region R_2. If the statistical distribution changed, this too would reveal a new order, beyond that of the "quantum" description. So it is important not to overemphasize the significance of any one way of revealing new orders, implicit in new descriptions.

When we come to the question of a "relativistic" theory, we are faced with a certain confusion. Bohr's consistent quantum description is admitted to be relevant only to non-relativistic theories (e.g., it depends on the smallness of $\frac{e^2}{c\hbar}$). Relativistic algorithms that are called quantum theoretical have been developed. But these are not consistent. This inconsistency is not merely in the infinities, etc., to which such theories give rise formally. Even more, there is no consistent description of the informal experimental conditions, similar to what Bohr gives for non relativistic quantum theory.

When Einstein criticized Bohr in terms of the EPR experiment, he began with the non relativistic formal algorithm. But in the middle, he brought in the idea of a light signal as a limiting velocity. This is an informal motion that is essentially relativistic. Thus, Einstein's discussion was not relevant to Bohr's statements, and Bohr's answer was not relevant to Einstein (because Bohr's notions are relevant only non relativistically). Bohr and Einstein did not notice that they had no way to meet. Thus, their discussion was a source of confusion to both of them.

Since relativity and quantum theory have not really even "met" thus far, one may suspect that a new language is needed, in which both "quantum" and "signal" are not fundamentally relevant terms of description. Rather, these will have to be abstracted as secondary features, relevant in some limited and contingent domain (as Newtonian conceptions are relevant in the domain of small $\frac{v}{c}$).

Now, one thing that is fundamental to a quantum description is the notion of a single wave function (inappropriately called the "quantum state"). A basic notion in relativity is the local character of all basic descriptive terms. But the wave function belongs to the totality of what is under discussion (in principle covering the whole universe). So we have a tremendous difference in basic descriptive orders.

How can this difference be brought into harmony? I suggest that to each relatively localizable region R, there is a relevant wave function, ψ_{ij}^R. But because ψ_{ij}^R depends on indices j as well as i, this means that the whole "universe" is potentially relevant to the "full" description of each localizable region. This is what we in fact find in common perceptual experience. But there is a special mechanical or dynamical mode of description, in which each element is said to be describable in terms that do not involve other elements. We are now considering that such dynamical descriptive orders have limited areas of relevance, not extending to the area now under discussion.

This means that the use of Schrodinger's equation as a kind of dynamical law for ψ_{ij}^R is also no longer relevant. What shall we do instead? I propose that it is necessary to perceive new orders of relevance.

If we go back to general relativity, we see a similar situation. Consider certain kinds of field quantities, such as the symbols $\Gamma_{\nu\alpha}^{\mu}$, describing a parallel displacement. When these displacements are integrated around an infinitesimal circuit, we are led to a new quantity $R^{\mu}_{\nu\alpha\beta}$, which is the curvature tensor. The assumption that the laws of physics are to be expressed in terms of $R^{\mu}_{\nu\alpha\beta}$ is then the assertion of a new order of relevance (i.e., of integrals of $\Gamma^{\mu}_{\nu\alpha}$ over infinitesimal circuits).

Now, now in our situation, we don't want to refer to infinitesimal circuits, because the whole notion of continuity may also be irrelevant. But we can still refer to finite circuits.

To do this, let us introduce the notion of a transformation $\Gamma^{R,R'}_{ij,kl}$, which plays a role analogous to a generalization of the notion of parallel triangles. That is, we assert that the relevant comparison of wave functions in different regions is through the "invariant difference operation"

$$\delta\psi^R_{ij} = d\psi^R_{ij} + \sum_{kl} \Gamma^{RR'}_{ij,kl}\psi_{kl}$$

where $d\psi^R_{ij} = \psi^R_{ij} - \psi^{R'}_{ij}$, which is the "simple difference".
Then, we consider "circuit integrals" of such displacements, giving rise to a "curvature", associated with a circuit, C

$R^C_{ij,kl}$

When we assert the relevance of the quantity $R^C_{ij,kl}$ to the laws of physics, we are bringing in a new order of relevance, in terms of which the ψ^R_{ij} for different R can be related. But these relationships will be of new kinds. Schrodinger's equation can only be a special contingent relationship, relevant in limited areas.
Now, in Einsteinian theories, there is a further relationship, such as $R_{\mu\nu} = T_{\mu\nu}$ (where $T_{\mu\nu}$ is the energy-momentum tensor) which connects geometrical relevancies ($R_{\mu\nu}$) with dynamical relevances ($T_{\mu\nu}$). In the theory of contingent parameters, we could have a corresponding relationship connecting $R^C_{ij,kl}$ with some corresponding function of the contingent parameters. This would perhaps be the means by which the contingent parameters would be revealed, by their implications for the directly descriptive parameters ψ^R_{ij}.

All this is very sketchy, of course. It still has to be developed. And anyway, it is only heuristic, at best. We may well be led to very different modes of description, in the long run. But it is perhaps useful to consider it as a content that gives one some notion of what are the real questions involved here. In other words, we have to be free to set our thinking loose from habitual "moorings", and to experiment with new forms and orders of relevance, more in an "artistic" fashion, than in the fashion of what is now generally called scientific (but which is really technological).

Now, to come to your comparison of logic and geometry. Logic: quantum theory :: Geometry: Einstein's relativity theory. First, let me say something about geometry. I propose that geometry is mainly a language – a mode of description and inference – and therefore cannot be regarded as "empirical". Rather, the real question is: "Is a particular geometry relevant or not?"

Consider the postulate of parallels, for example. It is well known that this has little meaning, apart from further assumptions on how light and matter are assumed to move, in relationship to what are called parallel lines. Thus, any facts explainable by non-Euclidean geometry plus the assumption that light rays follow a system of "parallel" lines can also be explained by Euclidean geometry plus the assumption that light rays follow suitably "curved" paths. To judge between these theories depends on a perception of their respective relevancies. And this judgement is more akin to aesthetics than to technology (i.e., it cannot be based on the results of "measurements" or other "empirical" factors).

A geometrical language is useful in helping us to perceive relevance in a visualizable form. But because geometry means "measurement of the Earth", it also tends to lead to an unconscious metaphysics of a very harmful kind. For the "geometrical objects" tend to be thought of as "actually existent", as if one could touch them and see them. In fact, of course, they are only abstractions. But sooner or later, one begins to think in terms of geometry. In other words, "What is" is Euclidean space, Minkowski space-time, curved space, Hilbert space, or whatever else one is led to consider, by current scientific theories. As a result, one loses the necessary freedom to change the description in a fundamental way.

Now, Einsteinian relativity is primarily a form of language: i.e., the assertion of the irrelevance of "absolute" frames of reference. Minkowski geometry is another language, which enables us to visualize much of the content of Einstein's language. The two language forms are relevant to each other. But if we say that Minkowski geometry is "empirical", it is like saying that the subject-verb-object relationship is "empirical". What is appropriate here is only the relevance of a language form, and not its "empirical" verification or falsification.

Much the same sort of remarks can be made about logic. Quantum theory is, like Einstein's theory, a form of language. Logic is yet another form of language. That is to say, we feel the necessity for our communications to be in a certain order that is called logical. We have tried to describe this order by means of what are called rules of logic. But recall that the exception is the rule. That is to say, the relevance of a rule is in those situations that can clearly be seen to be either exceptions or non exceptions to the rule. Thus, the rule "Keep to the right" has no relevance, either as exception or non exception, when one is eating dinner. So the relevance of our so called rules of logic is shown just as much by those areas where they fail to hold as by the areas in which they are valid. And, of course, one must consider areas where our rules of logic are irrelevant.

As soon as we say "Logic is empirical" we are irrelevantly treating logic as an object of discourse, like a chair or a table. For, we fail to notice that what we say "about" logic immediately becomes an inseparable aspect of the logical rules that we are tacitly and informally following. In other words, it is irrelevant to talk "about" quantum theory, because this very talk is quantum theory. Similarly, it is irrelevant to try to talk "about" logic. This very talk is the logic that we want to talk "about". So, we can "talk quantum theory" and we can "talk logic". Our only remaining question is: "Is there any relevance to what we are saying?" Relevance cannot be analysed, and cannot be confirmed or falsified "empirically" (i.e., by technological means).

Just as we can have alternative geometrical languages, and no "empirical" way to choose between them, so we might have alternative descriptions of what we wish to call the rules of logic, without any "empirical" way of deciding between them. But what is more significant is that the really destructive factor is the attempt to treat logic and geometry as metaphysical objects of discourse, rather than as forms of language. It is this factor that has led to the modern tendency to take the forms of inference (i.e., logical rules) as having primary relevance, while the informal description is regarded as a secondary feature, which is just taken for granted. That is to say, high-powered mathematicians (like v. Neumann and his followers) have given tremendous attention to very sophisticated forms of inference, while they just accept common everyday informal descriptions, of incredibly naive character. The naive character of their informal language is not in harmony with the extreme sophistication of their formal language of inference. But this does not disturb them, because they regard the informal language as not very relevant. Or else, if they admit its relevance, they feel that it is a fixed thing that can never be changed basically, so that all the attention is directed only to changing the formal language of inference. But what is relevant now is to change the informal language, so that informal and formal languages can be in harmony. You will never get this harmony, in my view, if you try to start with the formal language of inference, while you tacitly accept most of the extremely naive and crude informal language to which physics is now committed.

With regard to the paper of Clark and Turner,[1] I have talked with Clark (who is now a part-time research associate here, working in Birmingham one day a week, and living in London, coming to Birkbeck as "honorary" unpaid research associate). He accepts your criticisms, but points out that the words "dual space" are what confused him and Turner. They treated the contingent parameters as literally a dual space. The use of "bra" and "ket" notation encouraged them here. It would be better to say simply that the "spaces" are different, without referring to "duality" at all. (Don't use Dirac's notation, but just ψ_i and ξ_i.)

We all send you and your wife our best regards. Will write more later.

Dave.

P.S. To sum up the situation, a language form or a set of rules has areas of relevance, outside of which it is irrelevant.

In its area of relevance, it has sub-areas of non-validity (or exceptions) and sub-areas of validity (or non-exceptions). The only "empirical" questions are those that have to do with areas of exception and areas of non-exception to the rules. The deeper question of what is the area of relevance is not in essence "empirical" (though it also does not exclude the "empirical", as potentially relevant to the question of relevance). Consider for example, the Mad Hatter[2] in Alice in Wonderland saying:

"I don't know why this watch doesn't run, although I used the best butter".

One could imagine an answer saying:

"The trouble is that you actually used second grade butter."

One could then try to settle the issue "empirically", by looking at the packet of butter, and seeing whether it were "first grade" or "second grade". But there is no

[1] See Clark and Turner (1968) (information from Jeffrey Bub)—CT.

[2] See also C134, p. 205, n 3—CT.

"empirical" way of deciding whether the grade of butter used is relevant to the running of watches. It is more a question of general harmony of concepts and percepts, which is basically "aesthetic" in nature. Similarly, given a certain theoretical content, there is no "empirical" way to determine its relevance. But unless it is relevant, "empirical" observations referring to this content will be largely sources of confusion.

Therefore, one may well ask: "Is the proposal of new rules of logic relevant to the physical situations indicated by the term 'quantum'?" Unless it is relevant, attempts to test these rules "empirically" will be confused.

One interesting point is that those who propose new rules of logic are still informally using the old rules of logic, to describe their new rules of logic. This is a disharmony between content (new rules of logic) and form (old rules of logic). It is similar to the statement: "Never use a preposition to end a sentence with".

The separation between the formally prescribed rules of inference and the informal rules applying to this prescription is itself an arbitrary and irrelevant act, a kind of disharmony. For the formal rules are the informal rules, and vice-versa (as the modulation of a radio wave is the radio wave, and vice-versa).

What is needed, in my view, is a new informal language that discusses the relevance and irrelevance of formal statements, in particular contexts and areas of experience.

Nov 5, 1968

Dear Jeff

Received your article[3] and letter. I apologise for the delay in answering, as I have been rather busy. Meanwhile, I have written references for you to Kansas, Boston, and Los Angeles. I hope you get a suitable job soon.

Your article seems most interesting. It clarifies the question quite a bit. It is useful to relate our own work to my 1951 papers,[4] as you do. Indeed, one can say that in my 1951 papers, I use contingent parameters that are restricted to being delta functions in configuration space. But the probability doesn't come out naturally any more. Rather, it has, in some way to be imposed, as $P(X) = \psi^*\psi$.

Perhaps the words "hidden parameters" or "hidden variables" should now be dropped altogether. Instead, we could talk of an "extension of a given description, to include further parameters that are contingent in the context of the original description". For short, we could say "an extension to contingent parameters" or "description in terms of an extension to contingent parameters".

I would suggest that you remove the word "neo Copenhagen" from your paper. There is no Copenhagen point of view. Bohr, Heisenberg, Rosenfeld, Pauli, etc., etc., all had different and mutually incompatible ideas on the subject. Rather, I would say that my 1951 papers were "an extension of Bohr's wholeness of description to include contingent parameters".

---

[3]Published as Bub (1969)—CT.

[4]Bohm (1952a, b)—CT.

Firstly, because of historical reasons, I formulated the 1951 papers in terms of measurements. Now, I would say that each set of parameters (e.g, for an "electron") had to be complemented by parameters corresponding to the rest of the universe. This may include what is called an "observing apparatus". But more generally, the latter is irrelevant. Every such description (in the 1951 papers) contains a formal and contingent distinction between foreground (the electron) and background (the rest of the universe). From one instance to the next, this distinction may vary (as does that between contingency and necessity). But what is universal to this form is that somewhere it has such a distinction. However form and content are an unanalyzable whole. So without a specific content in which this form is relevant, the form would be "empty". It is our work to try to develop such a content, that could be relevant in perception beyond the word. As yet, our work has only gone a limited way toward this aim.

Would you send us the references to the work of Kochen and Specker?[5] One can say, from their work, that extensions of quantum theory to contingent parameters cannot work, if the form of the description is the classical dynamical one, of disjunction between "observed system" and "rest of universe".

This terminology is superior to saying that there are "no hidden variable theories".

Best regards

Dave

Nov 19, 1968

Dear Jeff

Thanks very much for your article[6] and for your letter. I have sent a letter to York, as you requested. I hope that by now you will have some definite offer, from one of the places at which you applied. Things seem to be getting more difficult now, with regard to jobs. But there seems to be a good chance that among all these applications, you will get one or more offers of positions.

I hope you received my previous letter, in which I gave some responses to your article. In particular, I would like to repeat that phrases like "Copenhagen" or "Neo Copenhagen" point of view don't really mean anything, because there never was an agreed "Copenhagen interpretation" (e.g., Heisenberg, Pauli, etc., all had points of view different from Bohr's and from each other).

More generally, I would say that the content of your article is a valuable contribution to the subject, but that its form is not in harmony with the content. Perhaps the enclosed article[7] will help explain what I mean by this in more detail.

---

[5] See the references given in Bub (1969)—CT.

[6] See p. 231, n 3 above—CT.

[7] Probably Bohm and Schumacher,"On the Failure of Communication between Bohr and Einstein", included here as Appendix D—CT.

First of all, by emphasizing "hidden variables" so much and at the same time asserting that the term is meaningless or otherwise inappropriate, one tends to confuse the reader. For informally, "hidden variables" are being treated as very relevant in your article, while formally and explicitly, you assert their irrelevance. (This is rather like what Bohr did in his answer to Einstein, as described in the enclosed article). In particular, you say also that "hidden variables" constitute an appropriate description for theories that would fit the criteria of Kochen and Specker. But actually these are no more "hidden" than are those of our own Rev. Mod. Physics paper,[8] or my 1951 papers. Thus, in classical statistical mechanics, atomic variables are not "hidden", though they do correspond to the general requirements of Kochen and Specker.

It may be that your difficulties stem from the adoption of an "informal form" of argumentation and proof for your paper. Is this a relevant form to state what is novel? It seems to me that the argumentative form always tends informally to adopt the existent older modes of thought and to criticize them. Criticism may be positive or negative. That is, you may agree with certain points and disagree with others. But this approach begins by accepting the relevance of the older modes, so that it is not appropriate for saying something new, that implies their irrelevance. To say something new, you have to set the older forms aside, so that you cease to adopt a critical attitude to them, either positively or negatively. Likewise, to attempt proofs in terms of older theories also interferes with saying something new, which requires a different language form.

If I may make a suggestion here, I would say that what is missing in your article is a discussion of necessity and contingency. What really motivated my 1951 papers was informally and tacitly the question of expressing the contingency of individual events. For accidental and fortuitous reasons having to do with the historical development of the subject, I inappropriately tried to express contingency in terms of "hidden variables". Likewise, I was led to talk of the wholeness of observed system and measuring apparatus, when it would have been relevant to talk in terms of those parameters that one specified explicitly in the foreground of the discourse and those that are only implicit and that describe a background (usually called the "rest of the world"). In certain cases, this background may be regarded as functioning like a measuring instrument, but this is, of course, only a limited and special kind of context.

We can now say that any statistical theory can be extended by the introduction of parameters that are contingent in the context of the theory in question. One mode of extension is to describe these parameters informally as existentially disjoint from each other and from the background, or general context of discourse. This is the classical dynamical mode of description (e.g., atomic variables to explain thermodynamics). Another mode of informal description is to regard foreground and background as a whole, without existential disjunction of the kind described above. This is what I did in my 1951 papers and what we did in our Rev. Mod. Phys. papers.

In such a paper, "hidden variables" should appropriately be mentioned rather unobtrusively, perhaps in a footnote, to call the reader's attention to the historical

[8] Bohm and Bub (1966a, b)—CT.

context, and to the inappropriateness of this terminology. In this way, the form of saying it would agree with the content, i.e., that "hidden variables" are irrelevant, and therefore of no significance.

It would be useful to go over briefly the discussion of Bohr's deep philosophical insight into wholeness (the term "thesis" is however perhaps too argumentative in its form). This insight is highly implicit, tacit, and informal, but it is really what is most novel in Bohr's point of view. The basic point is the inseparability of form and content. That is to say (as in a game), the form reveals itself as a set of rules that are working in each concrete instance that constitutes the content, so that the form is the content. (See the enclosed paper for a more extensive discussion of this point.)

Now, we come to more particular modes of realizing this wholeness of form and content. Thus, Bohr considered the content to be a statistical set of experimental conditions and experimental results. These were described informally in terms of classical language. The informal incompatibility of different experimental arrangements corresponds to the non commutativity of certain operators in the formal algorithm. Your discussion (particularly of Kochen and Specker) makes it clear that no extension of this kind of theory in terms of contingent parameters can have the classical form of existential disjointedness. So, if one wishes to extend the description to include individual events as contingencies, one has to have a form of wholeness, rather than existential disjointedness. Thus, Bohr's deep and general insight of wholeness is still relevant. But his more particular insight of complementarity (described in terms of classical language) may cease to be relevant.

In particular, in my 1951 paper, it was implicit (but never stated unfortunately) that the "background" is not to be described classically. Rather the "apparatus" always had a wave packet, with a "particle" inside it. The "packet containing the particle" was the only relevant packet. But this "packet" with a "particle" inside is like a combination of Hamilton–Jacobi theory (a wave) with a particular ray (the particle trajectory). It is really a different classical description, with new potential content (e.g., a statistical "jumping" of the "particle" between packets to describe a state of finite temperature, and to include statistical mechanics as basic to the large-scale description).

One could combine the 1951 theory with certain aspects of our own theory: For example, we could add non-linear terms to the 1951 paper tending to cause the "wave function" to go to zero, except in the "packet" containing the "particle". These terms could be significant only in the "multi-parameter thermodynamic type background".

I hope this will help you to orient your modifications to the proposed article.

Best regards

Dave

Dec 5, 1968

Dear Jeff,

Thanks very much for your letter of Dec 2. I do hope that the question of your getting a job will soon be settled favorably. It must be very worrying to be subject to such uncertainties.

It is very hard by correspondence to discuss issues as subtle as those raised in your letter. I know by experience how easy it is to convey a wrong meaning in a letter, one that is not intended. In conversation, all this can be detected and corrected in a few seconds, but by mail, it leads to a growing structure of confusion and failure of communication. I shall therefore try to be as clear and straightforward as possible, and hope that you will "read between the lines" where necessary, if you should get a feeling of lack of understanding, or irrelevance, or any other sense of failure of communication (not that I actually expect this to happen, of course!)

Firstly, I was very much struck by your statement that you are the disharmony of opposing judgements of relevance (i.e., those of the group following von Neumann and those that you find in my work). Of course, there is no meaning to your trying to reduce this disharmony. But perhaps we can discuss these questions, and see more deeply what is implied in them.

Whenever there is a disharmony of form and content, communication is almost certain to be faulty. For the mind takes in the content overtly, explicitly and consciously, while it takes in the form tacitly, implicitly and sub-liminally (or unconsciously). So your reader takes in mutually irrelevant judgements as to what is relevant. And just because a great deal of this is "unconscious", he cannot help getting caught in confusion (after all, confusion is, in essence, just the result of the working out of mutually irrelevant judgements of relevance). So the almost certain result of publication of your paper will be the "confounding of confusion".

Given that all this is the case, I am led to ask: "Why do you want to publish a paper, written while you still are a disharmony of form and content? Why can't you wait until (at least as far as you can perceive) you are such a harmony? Then your paper will be a contribution to harmony, peace and understanding, rather than to disharmony, contention and confusion."

Now, let me suggest some of the main reasons why you are a disharmony at this moment. In my view, the main reason is that, in some sense, you are trying to "reconcile" my relevance judgements with those of v. Neumann, Jauch and Piron, Gudder, et al. Since this cannot be done, you are the "problem" of trying to achieve the impossible. You will inevitably continue to be this "problem" unless and until you see its irrelevance, in which case you will lose interest in the "problem" and cease to be it (i.e., interest and relevance are one whole, whatever is relevant to you will interest you and what interests you is what is relevant to you).

In this connection, it would be appropriate to say something about v. Neumann, et al. It is clear that v. Neumann and his followers rule out Bohr as irrelevant, not by saying so explicitly, but rather, by simply not mentioning him at all. Nevertheless, if v. Neumann had been genuinely serious on this point, he would have discussed the issue and stated why he regarded Bohr's views as irrelevant. By saying nothing at all on this score, he informally implied agreement between him and Bohr on all that was essential. But since their views are so different (as is evident at least sub-liminally to almost every reader), the reader is led to try unconsciously to find what assumption gets rid of the confusion. The assumption that apparently does this is: "Only the formalism is relevant". This is of course very probably just what v. Neumann wished to convey, in some deep sense. But then, this led to the "measurement problem"; i.e., how can the formalism be related to experience: And from this flowed endless confusion.

Now, I feel that because v. Neumann wrote as if Bohr never said anything at all, one is led to doubt v. Neumann's seriousness on this point, and therefore the relevance of what he says. It is only on the assumption that Bohr's work is actually irrelevant that one can justify the notion that v. Neumann was serious in these discussions. Indeed, the followers of v. Neumann do tend to imply the irrelevance of Bohr's work, as you yourself point out. But then, since you evidently feel Bohr's work to be relevant, this leads me to another question: "How is it possible for you to regard both Bohr and v. Neumann as relevant?" To me, it seems that if Bohr is relevant, v. Neumann's work is almost pointless. And if v. Neumanns work is relevant, then Bohr is pointless. Any effort to reconcile these two judgements of relevance would then have to lead to disharmony. Perhaps this would explain (at least in part) why you are at present a disharmony of form and content. For you are adopting both Bohr's and v. Neumann's mutually irrelevant judgements of relevance.

In a similar vein, let me raise a closely related question. Of course, you are right to emphasize the need to discuss the logic of J. and P., in terms of their own views as to what is a relevant aim of research in physics. And of course, you note that you have to discuss my views in terms of my judgments of relevance on this score. Here, you would probably agree with me that if Gudder compares my formal equations with those of others in terms of his own purely formal criteria of relevance, he is not discussing my views at all. He is tacitly ruling them out as irrelevant. But by making a "discussion of my theories" the form of his discourse, he also implies that he is discussing my views. Thus, he is "confounding confusion", so that what he says is worse than useless. However, in your discussion of my views, placed in parallel with those of v. Neumann, J. and P., Gudder, et al., you are forced to include all sorts of mutually irrelevant judgements of relevance, and asking the reader to accept the content of these judgements; i.e., to think in terms of the forms implied by these judgements. How is the reader to do this? Will he not become, as you have become, the disharmony of form and content? Will this not confuse the reader, rather than help him to clarity?

My question is then: "Will you continue to try to discuss mutually irrelevant judgments of relevance in the same paper, with the implication that the reader is asked to subscribe to all of these judgements, or to try to "choose" between them?" (If you haven't been able to make such a choice, how could the average reader, who is much less well informed and has thought a lot less on the subject?)

Is there any way for you to be creative, without adopting my views, or Gudder's views, or Schumacher's views, or someone else's views? I feel that the answer to this question is "staring you in the face". What you can do is to recognise explicitly what your paper already is implicitly, i.e., a new kind of historical discussion, showing how different physicists had different relevance judgements, and how their failure to take this into account properly led to a breakdown of communication, with its resultant confusion. All sorts of questions are actually raised by you, that are germane to this point. What did various people, such as v. Neumann and myself mean by the term "hidden variables"? In terms of each person's meaning, did that person actually accomplish what he claimed to accomplish? If not, why not? What probably confused him? How could his objectives have properly been achieved?

In this way, you will no longer be asking the reader to accept, or choose between, mutually irrelevant judgements of relevance. Rather, he will be learning that what is relevant is the irrelevance of holding to fixed judgements of relevance, especially when one wishes to communicate with others.

Now for some more technical points. Firstly, you say that "hidden" means "operation behind the scenes". Let us provisionally accept this definition. But then, this is just what I meant by the word "hidden" as applied to the parameters of my 1951 paper. And it is an equally relevant meaning of the term as it appears in our Rev. Mod. Phys. paper.

However, you suggest that it is "natural" to take v. Neumann's meaning for the word "hidden variables". I wonder what it means to say there is a "natural" problem of hidden variables. Usually, when people say that something is "natural", what is meant is "habitual", in the sense that it seems easier to think that way and harder to think in another way. But the word "natural" also tacitly and informally implies a kind of necessity, and thus denies the contingent character of all such meanings (i.e., they are contingent on a host of factors, historical, environmental, psychological, etc., rather than "built into" the very structure of the subject under investigation.) So I would say that habitually physicists had come to think that all descriptions had to have the character of being "potentially disjoint from the rest of the world, belonging to the observed system alone". However, I discovered in 1951 that one could have a description which did not have this character, in terms of parameters that were "hidden" in the same sense as v. Neumann could say that the parameters he talked about were "hidden".

Very probably, v. Neumann would have regarded such a form of description as irrelevant. But I could now ask a further question: "Do you also adopt v. Neumann's position that it was "natural" (and therefore in some sense inevitable) to restrict the term "hidden parameters" to mean parameters describing disjoint parts of the world, considered as objects potentially available for observation?"

I feel that if your answer to this is in the affirmative, it will be difficult for you to give a clear presentation of my views, and that you will thus enjoin on me the arduous task of distinguishing what you say about my views from what I feel that these views actually are. Indeed, in my view, it would be more appropriate to say that v. Neumann's meaning of the term "hidden variables" was the "unnatural" one. For by considering what Bohr had to say, in an informal way, it became almost trivially obvious that no "disjoint hidden parameters" are possible. The mystery is then why so much effort has gone into proving the obvious. (This is, in fact, as I recall, how I felt intuitively about this problem in 1951.) In other words I felt that it was "natural" to regard v. Neumanns kind of "hidden variables" as irrelevant to the actual nature of the quantum theory.

In my view, it would be useful to suggest, at the very outset of your article, that the confusion arose, in large measure, because the scientific community was not generally able to consider the relevance of descriptions going beyond the classical disjunction of the observed object from the rest of the world. My own confused response to this confusion was then to take this behavior of the scientific community as relevant, at least tacitly, by adopting the informal form of trying to "refute" von Neumann's "proof". However, it was hard to avoid doing this, because everybody was always tending to raise the question. It would have been better if I had started by firmly asserting the irrelevance of what von Neumann had done to what I was saying, rather than to do this in a rather implicit form, towards the end of two long articles, at a place where it could easily escape the notice of most physicists.

I feel that you give the impression that whereas I had no clear idea of what I was "up to", those who followed v. Neumann were by contrast quite clear in this regard and were indeed following "natural" lines of development (implying, that in some sense, my lines were "unnatural" or "less natural"). Your main criticism of von Neumann et al. is that they misled themselves on how successful they were in attaining their otherwise valid and clearly defined objectives. But in my view, this puts things upside down. The entire v. Neumann line was deeply confused and therefore worse than useless. The mere fact that v. Neumann ignored Bohr and put his whole "authority" behind this, did irreparable harm to physics. And if von Neumann had listened to Bohr, he could never have written as he did. He would have seen that his whole approach was a combination of mutually irrelevant notions, and therefore, a "confounding of confusion". Because you do not mention any of this in your article, then you are "willy nilly" tacitly subscribing (at least in the minds of most readers) to the current view that regards v. Neumann as relevant and Bohr as irrelevant. (Recall that the situation is such that both cannot be relevant together, in this context.)

Next, I think that there was a very significant difference between Bohr and Heisenberg. Certainly, I could never–never–never agree to having my views called "Neo Heisenberg", whereas "Extension of Bohr's views" is, in my opinion, not an entirely inappropriate way of referring to them.

Heisenberg's term "uncertainty" or "indeterminacy" or "indefiniteness" implies an informal form that is a disharmony of form and content. First, you assert "definiteness" and then deny it by saying "indefiniteness". But to deny definiteness is to make definiteness relevant. So you are making relevant just that which you wish to assert to be irrelevant. In other words, the deep meaning of quantum theory is not the absence of definiteness, but rather the irrelevance of definiteness. Indeed, to use the word "indefinite" means implicitly that lack of definiteness is only a contingency (e.g., contingent on the quantum algorithm) and that, in the informal language, room is left for definiteness in principle to be relevant. It is implicit in Bohr's statements that definiteness is irrelevant, and that words like "uncertainty" are therefore a disharmony of form and content.

Heisenberg's work is characterized by an overall insensitivity to the harmony of form and content. His "microscopic experiment" is an example of this (as explained in our paper). In this, he is far from Bohr, in whom one sees everywhere a recognition of the relevance of harmony of form and content. Even when he doesn't achieve this harmony, he shows an understanding of its importance. But Heisenberg often seems to aim purposely at such disharmony, or in some sense, to prefer disharmonious forms of communication.

It seems to me that Heisenberg was influenced by Bohr to talk in terms of complementarity, while deeply, his philosophy of "uncertainty" (i.e., disharmony) remained in the informal form of his discourse. Vice versa, Bohr was influenced by Heisenberg to talk in terms of "uncertainty relationships". However, in my view, this term is totally out of harmony with the informal form of Bohr's discourse, and has served greatly to confuse the meaning of Bohr for most physicists.

Finally, I do not think it would help for me to write a comment on your article. It would probably lead people to say: "After all, it must be a confused subject, because even those who work on it together cannot avoid getting into unresolvable arguments." In addition, let me say that the whole form of argumentation, proof, refutation, etc., is not relevant in this context. To "refute" is usually to pick on certain points of someone's discourse, regarded as relevant in your own view, and then to show that the other fellow's notions fail to harmonize with your own. Our own title for the J. and P. article should have been "On the Irrelevance of what J. and P. have to say for physics." But this would have been called "impolite" (E.g., as politeness prevented Einstein from entitling his article "On the Irrelevance of Traditional Notions of Space and Time".)

I shall end up with some more questions. Are you trying to make comments "about" the work of other people, such as v. Neumann, J. and P., Gudder, and myself? Or are you trying to say something new? If so, are you regarding the technicalities of current efforts to put the "axiomatic" foundations of quantum theory into better order as a possible source of novel discoveries?

In my view, the key to something new is in changing the informal language first. To go on trying to repair the present formalism is like a person who lives in a condemned building, and spends a lot of money repairing it, decorating it, etc. I would ask him why he does this. Why doesn't he look for (or try to build) a sound new building, into which he can put his energies. He may have to live in the old building for a while yet, but he need not think in terms of putting the old building into perfect order.

Regards

Dave

Dec 10, 1968

Dear Jeff

I just received your second letter (Dec 5). I hope you received my answer to yours of the previous day. Meanwhile, I think it worthwhile to add a few further comments.

First, I am glad to see how you are beginning to understand the irrelevant way in which words are often used, to create what appears to be "good feelings", rather than to communicate concerning their supposed content.

Of course, you are right to cease to use words like structure, function, wholeness, etc., in this way. As I indicated in my previous letter, your latest article does have a potentially relevant content of a novel kind. However as long as its form and content are not in harmony, it will not actually communicate its content in a useful way.

You ask "What is quantum theory anyway?" This is indeed a very good question. In the article with Schumacher, it was suggested that there is no such "thing" as quantum theory. That is, quantum theory is not an object of relevant discourse. To talk "about" the quantum theory is actually to change its content in a generally irrelevant way; i.e., without our realizing that we are changing quantum theory when we think we are only "reporting factually" what the theory is "about".

Thus, when v. Neumann, J. and P., Gudder, et al. say something like "Quantum theory is a non Boolean lattice", they are not only adding something to quantum theory (i.e., a new formal description saying something "about" the earlier formalism), but they are also subtracting a great deal (i.e., the essential features that gave the theory content and made it fruitful). For example, most of the ability of quantum theory to be useful and relevant in connection with actual physical content depends intimately on largely informal requirements of continuity and single valuedness of the "wave function" (e.g., without it, energy levels of atoms, explanation of superconductivity and superfluidity, etc., would fall to the ground). It is almost impossible to indicate these essential features, in terms of a description, such as "non Boolean lattice". I would be ready to challenge anyone to make the notions of J. and P. or Gudder relevant to any actual physical content. I think you will see it can't be done. Rather, this stuff is just like the Medieval discussion of the number of angels that can dance on the head of a pin. However because of modern conditioning of thought, whenever physicists see "densely packed" formalism, they have a pleasing emotional reaction, equivalent in meaning to "This is real, solid stuff, not just empty philosophical verbiage".

Wouldn't you say that to go on making "commentaries about" theories that other people gave physical content to is neither very relevant nor very useful? When one understands that in any case, to "make a commentary about" a theory is actually to produce a different theory (but one that is different in a trivial and irrelevant way), one sees that he might as well admit, at the very outset, that all he can really do is to "make something new and different". To make something different (as v. Neumann, J. and P., et al. do) and then to use the same name (i.e., "quantum theory") is to confuse the issue. At the very least, it could be called "v. Neumann's, J. and P.'s, Gudder's theories of lattices, propositions, etc.", which are intended ultimately to be relevant to what is studied in physics (though as yet, they have not been). They could then note that in their theories are aspects that are vaguely similar to aspects of what has been called "quantum theory". They could then assert their belief that these aspects are the "essence" of what has been called "quantum theory" and that has been left out (e.g., single valuedness of wave function, Bohr's discussion of wholeness, etc.) will be ultimately seen to be irrelevant.

If they did this, then they would show, at least, that they are in some way, serious. Otherwise, one suspects that what is mainly behind their work is to use "emotionally charged" words that give a certain feeling of satisfaction. In modern physics, one of the most "charged" set of words is to say: "We are hard headed, clear thinking scientists, whose work is based on precisely defined formulae, and not on "wooly" or "fluffy" informal "philosophical" notions." (I talked to Jauch and saw that this is just the sort of language he takes pleasure in using.) This way of talking gives one a "solid and satisfying" state of feeling.

Best regards

Dave

Dec 11, 1968

Dear Jeff

Enclosed is a paper that will help explain what is meant by communication.[9]

I feel that your own paper is potentially relevant to communication. Its essential content is:

(1) Bohr and von Neumann could never have communicated, without basic changes in their points of view.

(2) This failure of communication was never perceived, and was indeed ignored.

(3) This led to irrelevant discourse, in which people talked without actually communicating, but believed that they were communicating.

(4) Attempts to "disprove" hidden variables of a disjoint nature were irrelevant, because there was no relevant informal language, that would allow one to discuss disjoint variables consistently in a "quantum" context. (For this reason, I feel more that there is latent confusion, even in the work of Kochen and Specker). Attempts to "disprove" disjoint variables in a quantum context are like attempts to "disprove" quadrangular triangles. These attempts are in some sense, worse than useless, because they introduce confusions into our thinking.

Either Kochen and Specker have no informal context at all in which case the proof is a piece of "pure mathematics", irrelevant to physics. Or else, it has, in some way, the informal context given in Bohr's language, except that this is described in a less consistent way than Bohr did. Now, in Bohr's language, the notion of a "disjoint variable" has neither meaning nor relevance. How is it possible for K. and S. to give it enough meaning and relevance, so that "disproving it" is more than mere confusion?

---

[9]See p. 232, n 7 above—CT.

In other words, I wonder if any attempt to "disprove" disjoint variables is not, in itself, necessarily a source of confusion and break down of communication. Once you adopt Bohr's context, even tacitly and implicitly, (as K. and S. probably do), then there is no point in even mentioning the words "disjoint variables".

On the other hand, as I showed in the 1951 papers, it is coherent to have "hidden" parameters ("operating behind the scenes"), provided that these are not disjoint.

V. Neumann (and others) introduced confusion by tacitly equating the words "hidden" and "disjoint". If they had paid attention to Bohr, they would never even have considered trying to do this. V. Neumann tried to rule Bohr out by bringing in his "classical observables", but of course, these lead to the "confounding of confusion" (as discussed in articles by me and by me and Schumacher). How do Kochen and Specker deal with this informal context, without at least tacitly basing their work on confusion?

Best regards

Dave

Dec 17, 1968

Dear Jeff

This letter is mainly to comment on your proposal concerning the Stern Gerlach experiment.

Detection apparatus

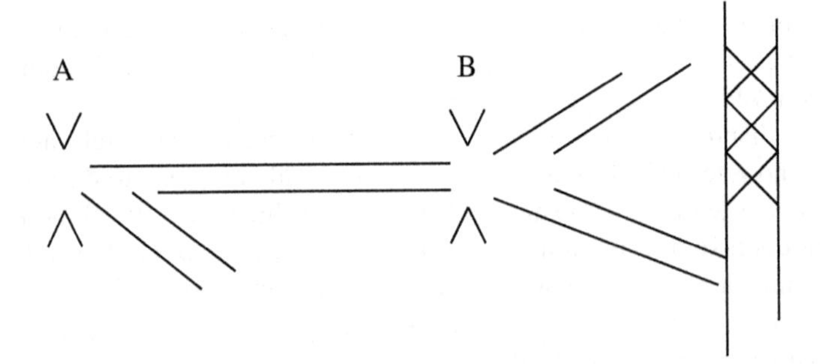

Suppose that with the aid of [the] inhomogeneous magnetic field A, we split a beam of molecules and choose a particular spin, say $\frac{1}{2}$. Then we orient B at a small angle relative to A, so that the component of the hilbert space vector in the measured direction ($\bar{n}'$) is very small. That is, with regard to $\bar{n}'$, we can write

$$c_1 = 1 - O(\epsilon^2)$$
$$c_2 = O(\epsilon)$$

where $O(\epsilon)$ means "of the order of ϵ".

You then propose that it may take a long time, before "projection along the direction of $\bar{n}'$" is completed, for those cases, in which the ultimate result is $|c_2| = 1$, $|c_1| = 0$. You ask whether one could not carry out some kind of experiment, to "detect" this time lag.

The difficulty with your proposal is that our informal language is ambiguous, with regard to the meaning of those cases, in which projection is "incomplete". It gave clear meaning only to those cases in which λ is so large that projection could be regarded as "complete".

There are several kinds of ambiguity. Firstly, it is not clear how many degrees of freedom are relevant in the description that is the context of this phenomenon. If you will read my book, Quantum Theory, Chap. 23,[10] you will see that there is an ambiguity in what is to be called the "coordinates of the observing apparatus". For example, one could say that the position and the momentum of the atom are the "coordinates of the apparatus that measures the spin". Or one could say instead that they are among the coordinates of the "observed system". One must note that the division between "observed system" and "observing apparatus" is thus a purely formal one (which is however of course part of the informal form), and that it has no physical content at all. (It is only a feature of our mode of description, like the distinction between a signal and its meaning. A signal is its meaning, and the observing apparatus is the observed system.) However the von Neumann approach talks in terms of an actually irrelevant existential disjunction between "observing apparatus" and "observed system". It is as if one said that "signal" and "meaning" were two existentially disjoint "systems" that would be in "interaction".

When one wishes to apply our approach (in the Rev. Mod. Phys. papers) in this context, what is not clear is, first of all, how many degrees of freedom one should include. On the face of the matter, one would be inclined to include the coordinates (or momenta) of the atom, along with the spin. When the magnetic field "interacts" with the atom, it gives the latter "momentum" that "depends" on its "spin". I would say that it is at this moment that the "measurement" takes place (and not later, when the atoms separate in space, because of their different momenta). So, we write for the wave function

$$\psi = \sum_{in} c_{in} \psi_i \phi_n$$

where ψ is the wave function of the spin ($i = 1, 2$) and ϕ_n is the wave function of an atom with a fairly well defined momentum. Then we should have the contingent parameters, K_{in}.

Our equation becomes:

[10] Chapter 22 of the Dover, 1989, edition Bohm (1989)—CT.

$$\frac{dc_{in}}{dt} = \lambda c_{in} \sum_{jm} J_{jm} \left(\frac{J_{in}}{K_{in}} - \frac{J_{jm}}{K_{jm}} \right)$$

λ would be large only while the atom was passing through the magnetic field.

This procedure would be ambiguous in the following respects.

1. It does not determine the hilbert space vectors, ϕ_n completely. Presumably they would correspond to some partial definition of the momentum and some partial definition of the position. But there is nothing in our informal language which indicates the features of the experimental context that would actually determine ϕ_n completely.
2. It is not clear whether the coordinates of the detecting apparatus are not relevant in this context.
3. The ambiguity described in (2) becomes significant when you consider the case where λ is not large enough to bring about a "complete projection". In this situation, the "detecting apparatus" must be involved in the further process that "completes the projection". But our informal language has no terms in it to describe such a situation.

Our theory is then limited in its context of relevance, to situations in which "projection is complete in all cases". If we wish to go into broader contexts, we need a novel informal description. This cannot be deduced, inferred or implied, from the formalisms of von Neumann, J. and P., Gudder, Kochen and Specker, or anyone else. Nor can it be deduced or implied from Bohr's informal discussions. It requires a creative new step.

What I am led to ask is, however, whether this is indeed a relevant direction for further progress. The possibility of "incomplete projection" comes from our formal description in terms of a differential equation. This is one of the ways in which we are continuing the forms of classical dynamics. Do we really wish to commit ourselves to the assumption that the form of a differential equation is the most relevant feature of physics, and that all our informal language must be adapted to fit this form? Or is it not the case that the informal form is free and creative, and that the formal forms of equations must change, to adapt to new informal forms? No doubt, von Neumann and those who followed him see the creative possibilities solely in what is done with the formal forms of equations. But how do you see it?

Finally, with regard to your article, I would like to sum up the situation once again.

1. Von Neumann and his followers are deeply confused, because they assume that the "observing apparatus" and "observed system" are existentially disjoint and in interaction. Their assumption that "hidden variables" must likewise be existentially disjoint merely "confounds the confusion". Efforts to "disprove" hidden variables, based on such an assumption, are worse than useless. It is like a man who says "all triangles are quadrangular" and then tries to "disprove" the possibility of existence of triangles.
2. Others like Heisenberg and the majority of modern physicists tend to adopt a mixture (or "linear combination") of von Neumann's approach and Bohr's approach. This too is confused. Insofar as they adopt Bohr's approach, to "disprove hidden variables" is an irrelevant and meaningless effort. Insofar as they adopt von Neumann's approach, they are merely uttering irrelevancies, from the very outset.

The only relevant content of your paper is to try to reveal and explicate all this confusion, and general failure of communication. Otherwise, you will merely be combining, elaborating, and contributing to the breakdown of communication.

Best regards

Dave

Dec 19, 1968

Dear Jeff

This is just a brief addition to the letter of a few days ago.

If you seriously want to discuss the meaning of the work of Kochen and Specker (K. and S.), it is necessary to give an informal description appropriate to their formalism. Your suggestion that this is just "von Neumann's book," (which would be what one is supposed to mean by "quantum theory") is not adequate. For as pointed out in my previous letters (and in various preprints), von Neumann's book is informally a morass of confusion, on which nothing can properly be based. Nor would it make sense to use Bohr's informal language for this purpose, because in terms of the latter, the words "hidden variables" are totally irrelevant and meaningless from the very outset. Nor can we say that von Neumann's treatment is in any sense equivalent to Bohr's principle of complementarity. Indeed, the two points of view are irreconcilable, in the sense that each of them implies the complete pointlessness of the other.

What is needed in your article is first of all a thorough informal discussion of a theory, in which the term "phase space of classical dynamics" is replaced by "Boolean algebra" or "commuting algebra". To the best of my knowledge, this has never been done. In effect, mathematicians "wave their hands" implying that all this is obvious and trivial, and that only subtleties of the formalism are worth discussing. But I am willing to bet you that you will find that it is not at all easy to do this (although I think it is probably possible).

The next step is to take the non commuting algebra of quantum theory and give a corresponding informal discussion, enabling one to see that this algebra gives a relevant discussion of physical situations, and is not just a piece of pure mathematics. Once again, mathematicians seem to believe that by bringing in a few of the physicist's words (e.g., "quantum theory", "observable", "charge", "mass", etc.) they will ensure that they are "talking about physics". They "wave their hands" and piously hope that these words will make their conclusions relevant to physics. But as Bohr showed by example, it takes a serious discussion of the informal language to do this. It was done tacitly and implicitly for classical theory over the centuries. And a new theory requires new work on the informal language.

At this stage, I find myself puzzled. Without bringing in something like Bohr's language (which makes K. and S. just as irrelevant as is von Neumann) how will you do this? And if you start with von Neumann, you are bogged down in hopeless confusion from the very first step.

Assuming that you manage to pass this second hurdle (i.e., to give a consistent informal discussion of quantum theory that allows the term "hidden variable" at least potentially to have more meaning than "quadrangular triangle"), then you should discuss informally what it means to say that the embedding of the non commuting algebra of quantum theory in a commuting algebra is equivalent to what one means in classical physics by the term "hidden variable". It seems to me that this might be done by considering special cases in which non commuting algebras can be embedded in commuting algebras. (I am told that these exist, but that they violate certain further assumptions made by Kochen and Specker about the "quantum algebra".) Perhaps in this way, one could see that the notion of "embedding quantum algebra in classical algebra" is (in the physicists context) more than sheer confusion, and has at least potential meaning.

The situation here is that there is a danger in proofs of impossibility that one is just lost in confusion. In order to prove impossibility, in a coherent and rational way, it is necessary first to begin with a case where the property in question is in general possible. Then one shows that certain further contingent assumptions are not compatible with this result, so that it is generally possible but contingently impossible. The danger is that one will start with an absurdity that has no meaning, and is therefore already "impossible". To try then to prove the impossibility of such an "impossible" is confusion. Thus, if one starts with the tacit assumption that a "quadrangular triangle" is generally possible, and tries to "prove" that this assumption is not compatible with the contingent hypothesis of Euclidean geometry, such an effort is evidently worse than useless.

Similarly, it is very probable that the term "Hidden variables of a classically disjoint kind" may be just as confused as the term "quadrangular triangle". Mathematicians have great faith that this is not the case, but this faith is no substitute for a serious inquiry into the question. After all, in Bohr's language form, which is the only known consistent description of the "quantum" context, the notion of "disjoint observable" is of the same status as "quadrangular triangle". Mathematicians have tacitly ruled Bohr out as irrelevant. But this doesn't necessarily prevent them from uttering confused absurdities all the same (at least insofar as they imply that their formalisms are relevant in physics).

Best regards

Dave

Dec 27, 1968

Dear Jeff

It occurred to me that some of my previous letters may have been a bit too strong, with regard to v. Neumann and those who followed his work.

It remains true that this whole line is informally confused. However, the other side of the picture is that Bohr's coherent informal account of the quantum theory was very difficult to understand, at the time it was put out. This is probably why people like v. Neumann essentially ignored it and tried to develop their own accounts of the subject instead.

Now, if you write your own article by tacitly equating quantum theory with "what is in v. Neumann's book", you will find yourself entangled in some elaboration of v. Neumann's confusion. What you could do instead is perhaps to begin with what physicists were doing informally in quantum theory (i.e., using operators, eigenvalues, a certain probability interpretation, etc.). You could point out that there was no adequate informal theoretical discussion of what they were doing. The majority of physicists (who effectively ignored Bohr) were therefore treating quantum theory as an elaborate set of formal mathematical rules along with another elaborate set of rules for applying the formalism to experiment. (Thus, it was like a very elaborate version of the Rydberg–Ritz rules in spectroscopy, before the days of Bohr theory.) Your contribution would then be to show (using the formal arguments of Kochen and Specker) that this set of rules of computation plus rules of application was not compatible with an explanation in terms of hidden variables that are potentially disjoint from the rest of the universe.

You could also point out that v. Neumann and his followers (up to Kochen and Specker) were first of all, entangled in informal confusion, and then secondly were confused, in that even formally, they did not actually prove what they had set out to prove.

It would clarify your article if you made it clear that between Bohr's informal account and that of v. Neumann, there was no possible contact, but that most physicists tacitly believed that such informal considerations were of secondary or tertiary relevance, so that the question was never seriously considered, to any significant extent.

Finally, you could point out that my 1951 papers (and those of ours in Rev. Mod. Phys.) have no relevant relationships to the work of v. Neumann at all (including Kochen and Specker). So the main "theme" of your paper would be to reveal how little communication there has been between physicists who thought they were talking to each other. Best regards, Dave.

References

Bohm, D. (1952a). A suggested interpretation of the quantum theory in terms of hidden variables I. *Physical Review*, *85*(2), 166–179.

Bohm, D. (1952b). A suggested interpretation of the quantum theory in terms of hidden variables II. *Physical Review*, *85*(2), 180–193.

Bohm, D. (1989). *Quantum theory*. Mineola: Dover. Reprint of New Jersey original (1951)

Bohm, D., & Bub, J. (1966a). A proposed solution of the measurement problem in quantum mechanics by a hidden variable theory. *Reviews of Modern Physics*, *38*(3), 453–469.

Bohm, D., & Bub, J. (1966b). A refutation of proof by Jauch and Piron that hidden variables can be excluded in quantum mechanics. *Reviews of Modern Physics*, *38*(3), 470–475.

Bub, J. (1969). What is a hidden variable theory of quantum phenomena? *International Journal of Theoretical Physics*, *2*(2), 101–123.

Clark, P. M., & Turner, J. E. (1968). Experimental tests of quantum mechanics. *Physics Letters*, *26A*(10), 447.

Chapter 8
Folder C136. Differing "Relevance Judgements". January–March 1969

Jan 20, 1969 Postmark 18th

[Comment re postmark added – CT.]

Dear Jeff

Perhaps in this letter, I can more or less finish the reply to your long letter. As I re read it, I see again and again that our relevance judgements, are in certain ways, very different. It distresses me to see how far we are from communicating, even though we worked together and discussed so much together. How much further from communication I must be from others, with whom I never even had any common basis! Yet, I cannot choose my relevance judgements simply to put me in greater harmony with you or with other people. It would be absurd for anyone try to do this. Nor does it have meaning to tolerate other people's points of view. Indeed, this is almost an insult; E.g., as a white man, I tolerate your black colour. I find it unpleasant, but I suppress these feelings, and somehow manage to get along with you. This too is a sterile lack of communication. What we have to do is to listen deeply to each other's relevance judgements, without choice or value judgement or blaming or praising. Then perhaps communication may arise.

It is in the relevance of the informal and formal languages that our greater difference lies. You regard the informal as a kind of starting point, whose ultimate purpose is to arrive at new formalisms. You believe you can, by means of a formalism, transcend informal phrases such as "observer and observed". For you, the ultimate understanding is in terms of a formalism. As you say, eventually the informal statement will be dropped (though in some sort of way, we will still need informal statements).

Now, for me, I feel that I understand something mainly when it is clear what the informal situation is. I feel that at best, formal statements are cues to changes in the informal language. I would never think of starting with the informal and then dropping it when I arrived at a good formalism. Rather, I would say that a given informal description might, in a short time, give rise to 20 or 30 or 100 formalisms. Formalisms are like "shapes" seen in the clouds. They are like the hundreds of

C. Talbot (ed.), *David Bohm's Critique of Modern Physics*,
https://doi.org/10.1007/978-3-030-45537-8_8

sketches that the artist will make as he paints a picture. What you are doing is discarding the picture as "vague", and keeping one of the sketches on which people fortuitously happened to have latched.

For example, I feel that in the long article I wrote for the Cambridge conference,[1] I showed the irrelevance of the notion of "hidden variables", to the situation in physics which has developed as a result of the application of the quantum theory. Indeed, I proposed that "contingent parameters" provide a more relevant set of terms of discussion of this context. However, you in your letter say that you regard hidden variables as a key notion, to show what the quantum theory means. From the fact that you did not comment very extensively on this article, I feel that you regard it as irrelevant. Very probably, you feel that at best, such an informal article could serve as a starting point for the discovery of new formalisms. Since it shows no signs of doing this, you do not feel it to be worthy of a great deal of attention. On the other hand, I feel that it is a starting point to discover something new that is informal, and therefore it is something that begins to make the vast mass of formalism relevant.

Of course, it is no use for me to blame you because your relevance judgements are different from mine. That is just a simple fact.

Here, let me say that I have always felt this way. When I was a student, I was shocked and distressed to discover how almost everyone gave such relevance to the formalism, and regarded the informal at best as a secondary aid to get better formalisms, and at worst, as useless and meaningless. I often thought of leaving physics, as a graduate student, and when I got to teach at Princeton. However I saw that the situation was the same in other fields, so I did not leave. In the whole of life, men are putting the formal as primarily relevant, so that the informal situation has become chaotic and even dangerous. But I find no way to communicate this to most people.

The way I have always started is to change the informal language by seeing a new relevance of what can be observed in everyday perception to the situation in physics. It is not at all arbitrary, no more so than is the action of the artist who slowly feels out how to paint a picture. On the contrary, the artist who began with elaborate formal details could only be imposing his past ideas on the present situation in front of his eyes.

To me, it seems that all knowledge, including physics, stems from such perception. As indicated in the article with Schumacher, I feel that our present "instrumental" modes of perception are mainly elaborations of the formal language of physics and therefore inadequate, as well as arbitrary. It is necessary to start once again from fresh perception. This <u>may</u> mean that most of the present formalisms are not very relevant, except perhaps as rough sketches made by previous artists, which can be considered, provided that one doesn't give them too much significance. What I am trying to do is to develop a new informal language that is directly relevant, both to

[1]Bastin (1971) – see C133, p. 184, n 6. Bohm's contributions were "On Bohr's Views Concerning the Quantum Theory", pp. 33–40 and "On the Role of Hidden Variables in the Fundamental Structure of Physics," pp. 95–116. The latter is the "long article" referred to. Bub's contribution to the conference was "Comment on the Daneri-Loinger-Prosperi Quantum Theory of Measurement", pp. 65–70 – CT.

ordinary perception and to the situation in physics. In doing this, I am prepared to change the formalism from one day to another, to help improve the overall harmony of the description. And it goes without saying that the informal is always changing in this way.

I think you may see that a great many confusing details of argumentation between us arose out of this difference of relevance judgements. It wouldn't be worth while to go into all the details. The main thing to remember is that you feel that the situation is clarified when informal questions are replaced by precise formal questions, subject to proofs, deductions, etc. I feel that this approach gives excessive relevance to particular formalisms, which are fortuitous, accidental, and arbitrary. Rather, I feel that from a slow and careful change of the informal language will flow a kaleidoscopically changing series of formalisms. Thus proofs and deductions are not very relevant, because as a rule, what is called for is to change the formal descriptive language so that all becomes clear directly without proofs and formal arguments.

I find that I haven't quite finished the answer to you yet, because I must discuss the question of the "about", which I shall do in the next letter.

Best regards

Dave

Jan 18, 1969
Postmark 21st

[Comment re postmark added – CT.]

Dear Jeff,

Perhaps now it will be possible to finish the answer to your long letter (though whether this answer will communicate something to you, I have no way of knowing, until I hear your reply).

I feel that you really didn't understand the notion of "talking about", as discussed in the articles by Schumacher and me. Perhaps it will help to go into it some more.

Now, as indicated in the articles, there is a certain limited kind of relevance to talking about an object that is potentially or actually within perceptions going beyond the word. Thus, I can look at the table I am writing on, and I describe it as hard, wooden, green, of such and such a size, having atomic constitution, etc. One can go on indefinitely, adding properties, qualities, structures, etc. These are all united because they are "about" the perceived table. The unity has genuine content because what it means is revealed in direct perception. It is not merely a game of manipulating words.

But let us go to an opposite extreme of abstraction. Is it relevant, for example to talk about love, as if love were an object of discourse, similar to the table? Evidently, this is a very different case. Firstly, love cannot be identified or pointed to or characterized in any relevant way at all. Whatever you say about love, this will generally produce feelings that could perhaps be called "what one is talking about". But these feelings are actually only produced by words and thoughts. For example, many parents say: "I love my children", getting a pleasant glow as they say it, and identifying this glow with "the love that they are talking about." But actually, they never take the trouble to understand their children, and instead impose a pattern on them which is really an extension of their own Ego gratifications. As a result, most children grow up confused and stunted in a psychological sense, and this helps maintain the general chaos in the world today, as these children in turn fail to understand what they are doing with their own progeny. So what did the parents mean in "talking about how they loved their children"?

What we are suggesting is that love cannot relevantly be talked about, in the same sort of way as one talks about a table. That is to say, the word, love, does not properly have object relevance. At most, it may have what I call flashing indicator relevance. Consider the flashing indicators on a motor car. It indicates something very dynamic and momentary, i.e., action. If people thought that the indicator itself was what was relevant to look at (i.e., the object of discourse) the result on the roads would be terrible chaos, confusion and destruction, as cars plowed into each other in vast multiple collisions.

So the word "love" can have flashing indicator relevance, if it is indicating what is beyond the word in feeling, thought and action. Unfortunately, however, people have gotten into the habit of saying "I love you" and then "looking" at the feelings and euphoric glows that these words can produce, as if they were permanent objects of discourse. They then describe these feelings, and thus get more euphoric glows. So they begin, in effect, to act as if there were a "thing" or an object called love. But actually, this "love object" is maintained by words and thoughts. Later, when it can no longer be maintained, its illusory character may reveal itself, producing tremendous shocks and unhappiness. For example, a parent may say "I love my children", but force them to live a life that has no meaning to them. He then is able verbally to identify his love feelings with his "sternness," i.e., his ability to impose his views on them "for their own good." He gets a terrific shock when they later "turn against him" and follow their own inclinations.

It is not generally realized to that to treat love, hate, fear, anger, aggression, ambition, etc., as objects of discourse is just as destructive as to treat the flashing indicator on the car being in itself a relevant object. A large part of the vast misery, destruction, fear, and anguish that grips the human race today is just a kind of "multiple collision process" that results because people have mistaken the feelings produced by words for a reality that is independent of words (E.g., one can see how patriotism is "whipped up" in this way).

Actually, the word and its total meaning, including the feelings to which it gives rise, are a whole, in which analysis is of little or no relevance. After all, if words had no meanings, they would be just noises. The word is its meaning. (It is the meaning that makes it a word and not just a noise.) But if we separate off a part of the meaning of the word (e.g., feelings to which words give rise) as if it were a distinct and separate object of discourse, we will introduce hopeless confusion into our thinking and action. It will not be the kind of confusion that arises as we explore a new subject and find ourselves unclear. For this tends to clear up as we work and explore. Rather, it will be a kind of confusion that grows without limit. The reason is that basically, when we talk for example "about" love, it is the fortuitous conditioning of the Ego that determines the content of the "about", (as whatever happens to feel pleasing, gratifying, envy, convenient, secure, etc.). Such fortuitous conditioning is of course irrelevant. But because the Ego is always "defended", we cannot see its irrelevance. Thus, the more we talk "about" love as an object of discourse, the more we fill our minds with irrelevant feelings, euphoric glows, self pity that the "loved one" does not return our "love" etc. We get the impression that we are "looking at" as "perceiving" our feeling of love, just as we can do with the table. But these feelings have been manufactured, rather than perceived. Because they are not genuine, we may be sure that when people have begun to talk a great deal "about" love, that love has already vanished (if indeed, it was ever actually present). What they are talking about is actually the outcome of the Ego. More accurately, it is the Ego, because the Ego is one whole, with its form and content. That is what is meant by saying "The 'about' is us". Both the content that is ascribed to the "love that one is talking about" and its form are one whole with the Ego of the person who is talking. Beyond that, it has no reality or actuality.

Let us now go on to discuss a theory in science, such as relativity or quantum theory. Can we talk "about" this as if it were a perceivable object, or is the theory not more like love than like such an object?

First of all, what is a scientific theory? For example, Einstein communicated his theory in certain published papers. But Einstein is dead, so that the theory can no longer relevantly be regarded as one of Einstein's properties or qualities. Insofar as the theory still "lives", it is not an attribute of an entity called Einstein. Indeed, even while Einstein was alive, nobody else was able to "look into his mind". Whether Einstein is dead or alive, what happens is that we can read his papers or talk with him, and in this action, something emerges that is us, not Einstein. Each person "makes" of relativity something that is "his own", a bit different from what other people "make" of it.

Is relativity to be in identified with the printed words and formulae in articles and books? Evidently not. For if all physicists in the world were suddenly to die, there would be nobody left to whom these words had meaning. And of course, relativity is its meaning. There is no relevance even to consider the notion of a difference between the theory relativity and what it means.

Many physicists behave as if there were some fixed and well defined mathematical truth about the theory of relativity, which truth is its meaning. Understanding relativity is for them a process of coming closer to this truth, which is thus regarded as a kind of "mental object." But is truth an object? Or isn't truth in essence just like love which can relevantly be indicated in an oblique flashing way, but which cannot be made an object of discourse, to be "looked at", described, manipulated, etc.

What happens when one reads a paper on relativity is an infinitely complex response, in which there are all sorts of intimations, sentiments, fears, pleasures, pains, anticipations, etc., of a nature similar to what happens when one plays a game. This complex unspecifiable movement is one's "understanding of relativity".

It includes the possibility of verbal extension and further formal abstractions from initially informal verbal descriptions and communications. The formal verbal aspect is only a tiny part of what is going on in one's "understanding of relativity". But most physicists seem to feel that it is all that is really significant. What they do is like saying: "The essence of an iceberg is the top part. I am interested in the bottom part of an iceberg mainly insofar as it leads me to the top. After I get there, I dropped the bottom, and build all my ice structures solely from the top of the iceberg (though in some ways, it seems to be still necessary to mention the bottoms of icebergs from time to time)".

So in communications concerning relativity, it is not relevant to talk "about" the content of this theory, as if it were a distinct and separate object that we can look at, describe, make inferences about, etc. The content is one whole with the informal form as described above, so that communications in relativity should properly have flashing indicator relevance rather than object relevance. The flashing indications are relevant to the vast indescribable movement that is one's understanding of relativity, as it is at a particular moment.

You have asked yourself questions, such as: "How can I understand what Bohr or v. Neumann or Einstein did?" The important point to note is that such understanding is assimilation; i.e., however you came to understand the theory as it originally presented itself to you, this is now changed, so that it has become a harmony with you. Indeed, your "understanding of relativity" is you. And your "non understanding or relativity" is also you, because when you "don't understand", what this means is that you feel a disharmonious response, as you read or consider what has been written or said.

The difficulty is that our language confuses us here. It says explicitly and formally that a theory is a "mental object" external to us, and that we have to "look at it" and to "understand what it really is". But as I indicated, nowhere is there a perceived object that could relevantly be called a theory. Whenever I look at "relativity theory", it is always me that I am looking at (i.e., what I have "made" of it this far). If I once see this, I do not fall into confusion, because I see that "my understanding of relativity" is not the same as yours or Einstein's. It is different. In a way we could call it a different theory, with certain similarities to what other people "make" of it, or to what I "made" of it yesterday. I do not therefore imagine that I am coming clearer to the "truth about relativity." For example, at a certain point, I may understand that relativity is of very limited relevance, so that in the context of interest to me, relativity is "dead". So Einstein understood relativity as "alive and kicking", where I understand it as moribund, if not already dead. Each of us may have been "right" in his own context.

If scientists realized that they were changing theories when they think that they are only talking "about" them, then they would start to look for new consequences, rather than act as if all were the same as before, except that it is now on a "solid and clear logical foundation". Cauchy changed what Newton and Leibniz did, in a number of fundamental ways. V. Neumann changed what Dirac, Heisenberg, et al. did, equally fundamentally. If these changes had been admitted, and their consequences gone into, the result might have been further progress. But because they said "There has been no change in essence", they introduced confusion. That is, they equated or identified two theories that were fundamentally different in certain ways. To equate things that are different is confusion. And because they held on to the view that nothing essential had changed, this confusion could only increase. It was not like the simple confusion that arises when one faces the unknown. This creative confusion moves spontaneously toward clarification. But once you have fixed the identification of what is different, there is no way for this to happen, so that confusion has to go on increasing. That is why I said that the effect of v. Neumann's theory was "worse than useless".

Here, it will be useful to bring in what you say about v. Neumann. You say that Heisenberg, Jordan, Dirac etc. gave a wide variety of formalisms, none logically satisfactory. V. Neumann set out to develop a precise formalism that would be in harmony with what he thought to be the essential informal situation regarding the quantum theory. Here, I think that the word "relevant" would be more appropriate than the word "essential". In other words, v. Neumann made certain relevance judgements concerning the informal situation regarding quantum theory. These were in harmony with the sort of mathematical language in which he was trained and in which he was so proficient (as in fact, so you go on to say a bit further down on that page of your letter).

Now, the interesting point is that other people had made other relevance judgements. Bohr had relevated the informal language, while later, in my 1951 papers,[2] I relevated something else again. No doubt everyone would make different relevance judgement concerning this informal situation. But for certain reasons, v. Neumann's judgements were given greater and greater weight by most physicists, while other judgements tended to be dismissed, at least tacitly, as irrelevant.

The key point here is that v. Neumann thought that he was not changing the theory essentially, but that he was merely giving it a more satisfactory formulation, bringing it closer to the ideal of mathematical truth that he and most physicists of his time shared. But the fact is that his theory was, in certain respects like that of Dirac, Heisenberg, Schrodinger, etc. It was just another formalism. It was merely v. Neumann's personality that led him to relevate the formal details that he did as essential, while other questions considered by Bohr, Heisenberg, Dirac were regarded as inessential. If v. Neumann had merely said "Here is another formalism, more logical than the others, but not necessarily more essential", then he would not have initiated confusion. However, this would have given v. Neumann very little satisfaction. For v. Neumann was his ideal of mathematical truth, which was also one whole with his technical proficiency and inventiveness in mathematics. He would have felt it to be uninteresting (i.e., irrelevant) merely to propose another formalism, with some advantages and some disadvantages worthy perhaps of exploration. Rather, he had to be the one who gave a logical and clear foundation to the very essence of the truth, concerning the quantum theory.

If it had been seen that v. Neumann just produced yet another formalism, then it is possible that certain crucial questions that were implicit in the "messier" work of

[2] Bohm (1952a, b) – CT.

earlier physicists would not have been dropped as "inessential". So physics might have taken a very different and more fruitful course.

Of course, v. Neumann's formalism also turned out to [be] unsatisfactory. It was followed by algebras, lattice theories, etc. All those who followed insisted that their formalisms were yet more "essential". It seemed to them that they were working toward the ultimate mathematical truth about a certain stable and well defined mental object called quantum theory. Actually these theories are all radically different. For example, to replace classical mechanics by a Boolean lattice opens up a tremendous range of new questions, relevant in this framework, but not when we treat it as a set of orbits or functions in phase space. But all this is tacitly dismissed as irrelevant. For they all believe that they are approaching some fixed mathematical truth, so that these differences are inessential. The whole object of what is done is thus to try to prevent change and progress (which latter would be a challenge to the very basis of the personalities).

But to add to all this, there has been a kaleidoscopic succession of basically different formalisms (including quantum field theories, renormalizations, S-matrices, current algebras, etc., etc.) As far as I can see, none of these is particularly relevant. Very probably, no formalism is much more relevant than a viewpoint or a kind of mirror, that gives us a flashing insight into the informal (i.e., the unknown). But if you were to try to communicate this notion seriously to a mathematician, he would be deeply and genuinely hurt. For he is his notion of mathematical and formal truth, so that to say the above is, in effect, to say that his whole personality and Ego are not very relevant, after all. At best, he can give us momentary and flashing insights, whereas his whole personality is such as to demand that we regard his ultimate goal as formal and mathematical truth, with some kind of eternal value and validity. What all this means is that it is not relevant to talk "about" theories as if they were mental objects. The content of our discourse can be relevant to theoretical activity (i.e., the activity of theorists) as a kind of flashing indicator, that is momentarily significant, as a rule.

When Schumacher and I discussed Einstein's and Bohr's work, this is the sort of relevance we had in mind. But of course, we did not manage to communicate this to you, at least not in our papers.

After all, Bohr and Einstein are dead. Whatever we say "about" Einstein and Bohr, its form and content is always us (i.e., our response to whatever it is that they did). So Bohr and Einstein are really a kind of allegory, whose real content is our own failure of communication, at this very moment. What we hoped to do in this discussion is to flash an indication as to what is actually happening in the reader at the moment that he reads the papers. Bohr and Einstein, as individual entities, are irrelevant in this context. We are not trying to talk "about" them. But our whole heritage of language is such as to throw us into the form of the "about". So we rather depended on the reader seeing the need to allow a kind of "poetic licence"; i.e., not to take the form of our discourse too literally.

To sum up, then, it is actually impossible to talk "about" a theory as an object of discourse. Every discussion produces something different and novel, in some ways. However if we try to talk "about" a theory, this leads to an ever increasing confusion, because we are tacitly or explicitly treating relevant differences as irrelevant. Whenever we do this, the "about" is actually us (i.e., our personalities, our Egos) both in content and in form. For it is only because certain relevance judgements are identified with "one's very self" that they are "defended", and do not change freely, even when there are cues or indications of their actual irrelevance. As soon as a "fixed mental object" forms in awareness, it is the Ego (i.e., the conditioning) that determines its form, its content, and its tendency to continue irrelevantly. And thus, our thinking and perception are distorted by what is arbitrary. We are able to move away from the arbitrary only when the irrelevance of the "about" is deeply perceived.

Finally, it is necessary to mention that you and I are different people, and that we are different relevance judgements. You are the judgement that the informal is relevant mainly insofar as it leads to satisfactory formalisms, which alone permit a genuine understanding of a situation. I am the judgement that understanding is basically the assimilation of a situation in the informal, and that the formalism is like "a facet of a facet on top of the iceberg". Perhaps there is no way for us to meet at all. In this case, you could dismiss me as irrelevant, except insofar as I might accidentally say something that would liven up the at present "dead" situation in the formalisms of physics. And of course, to me, it would be dull and uninteresting merely to try to produce a new formalism, which would keep other physicists occupied with applications, proofs, criticisms, justifications, refutations, theorems, etc. Other physicists tacitly dismiss the very essence and core of my work as irrelevant, but some of them don't yet want to dismiss me as irrelevant, because of the "off chance" that I could produce something that gets them out of the current doldrums. But I feel that if I tried mainly to produce a formalism, I would become as dull as the others. In other words, deeply, there is no difference between me and them. Anyone who relevates formalism to be the essence of physics must get caught in dullness, because to do this is to talk "about" physics, and thus to overlook change, in the interests of some kind of fixed mathematical truth.

However, I have hoped that perhaps something different could happen. Perhaps you could seriously entertain my notions of relevance without choice or criticism or praise or blame, just neutrally and factually. Then you might begin to get a glimmering of new orders of relevance. If you communicated this to me, I too could change, and thus perhaps something novel, interesting, and worthwhile could take place.

Best regards

Dave

P.S. Perhaps the first step on your side toward communication between us might be to read my long paper on hidden variables.[3] Here, I explicitly start to change the informal language in an orderly way – something that you feel can be done only in an arbitrary fashion. Then you could look at the papers with Schumacher[4] in a new light, because these too are aimed at changing the informal language.

[3] See p. 250, n 1 above – CT.

[4] I.e. Appendix D and E – CT.

What struck me is that you tacitly dismissed all this work, which represents my deepest feelings on the subject, as v. Neumann and his fellow mathematicians tacitly dismissed Bohr as "inessential" when developing their formalisms which were aimed at capturing what was essential in the informal situation regarding quantum theory. If you had at least discussed the papers carefully and shown why you regard them as irrelevant, this would have made sense. In this connection, let me say that I have gone into v. Neumann carefully over the years, and do have a notion of what he did. V. Neumann's work is discussed in some detail in my papers.

Let me say it again that it is not a question of blame or praise. All this is irrelevant. I understand that your whole background has been such as to make you believe that changes in the informal language can only be arbitrary. With this background, it was natural and inevitable that you would dismiss these papers as irrelevant. But how are we going to communicate if you regard my deepest thoughts and feelings as arbitrary? Is it not necessary to begin by entertaining these thoughts, if your intention is communication?

Jan 21, 1969 Postmark 22nd

[Comment re postmark added – CT.]

Dear Jeff,

Although I knew from general information available to me that it was extremely unlikely to help, I called up Popper yesterday, to ask if there was any possibility for a job for you at LSE, perhaps a joint appointment with Birkbeck, and if not, whether he could suggest anything else. He said that he has 5 good students and several former students who are unsuccessfully looking for employment. Moreover, he says he has little influence. Generally speaking, in England, Oxford or Cambridge graduates are favored for posts involving philosophy. Posts involving philosophy and science together would be an innovation, not readily accepted. In any case, what is accepted as relevant is philosophy along with history of science.

In addition, appointments at Birkbeck have been frozen. The explicit policy is no new posts, plus careful scrutiny even of appointments filling vacated posts, to see if these cannot be left vacant, in order to save money.

So all in all, the outlook does not look promising in England. I very much hope that you locate something in America or in Israel. It must be very hard for you to have to face all this uncertainty, especially while your wife is expecting another baby. I would very much like to help you, but just don't see what I can do. People have begun to believe in the relevance of saving money at all costs, so that they will be ready to destroy everything, if a few dollars are saved, which can then be burned up or blown up in Viet Nam, also judged to be so relevant that everything else can go down the drain as long as the fight in Viet Nam can continue. We can only hope that these judgements of relevance will soon change. Otherwise all of us will fall into the soup, in one way or another.

Meanwhile, I hope that all these letters (especially the long letter I wrote yesterday) can help establish a new communication between us. Perhaps it would help to recall that I regard formalisms as ephemeral and momentary flashes of insight, having having indicatory significance for the vast informal and unknown movement that we call understanding. Thus, it may perhaps be, as you say, that von Neumann's or J. and P.'s or K. and S.'s formalisms are not branches of classical statistical mechanics. But then I would say that even if this could really be proved in an ironclad way (of which I am still a bit sceptical) it means very little. To me, Heisenberg, Dirac, Bohr, S matrix theory, renormalization theory, may also contain flashing insights, just as relevant as those of all these mathematicians. How do I know that these theories are also not branches of classical statistical mechanics? Then, as I am beginning to see that one can develop a fantastic variety of new formalisms. How do I know that these are not branches of classical statistical mechanics? And in any case, does it really matter very much? As long as the formalism is different, it has novel characteristics worth exploring. I certainly wouldn't exclude a formalism from consideration just because some mathematician had proved it a branch of classical statistical mechanics. For that matter, what is classical statistical mechanics? Once it was said to be a theory of orbits, then a theory of functions in phase space, and now it is a theory of Boolean algebras on lattices. As I indicated in earlier letters, these are all radically different structures of ideas, in certain respects. Mathematicians have judged these differences to be irrelevant and said that they are all "essentially" the same theory. However, in my view, what happens is that a mathematician arbitrary fixes on a certain formalism that he calls quantum theory and another that he calls classical statistical mechanics. He then proves (or hopes to prove) that one is not a branch of the other. After he has done so, then where are we? I can easily give you 20 or 30 formalisms that I could call classical statistical mechanics, all different, in certain key ways, from those that were called classical statistical mechanics by v Neumann, J. and P., K. and S., etc. One could spend a lifetime investigating whether all that I happen to call quantum theory is or is not a branch of what I happen to call classical statistical mechanics. It is a game that is enjoyed by pure mathematicians. But is it relevant to physics?

For example, you say that an adequate formalism for quantum theory should incorporate an analogue for Gleason's theorem in Hilbert space. I ask why you regard this as relevant. I am working now on new notions, in which Gleason's theorem simply has no place, because Hilbert space has no place (except as a crude simplification of another theory). So why should I spend a great deal of time and energy trying to incorporate an analogue of Gleason's theorem? To me, it is largely a piece of pure mathematics, relevant in a limited set of concepts chosen by certain mathematicians as an interesting field to work in.

While I am wasting energy proving all sorts of theorems about an arbitrarily fixed formalism, I could more fruitfully be exploring new formalisms. Actually, it is more interesting, easier, livelier and more relevant to do this than to do the other.[5]

What is the essence of quantum theory? You mention this idea a number of times. (E.g., the non existence of hidden variables is the essence of quantum theory). Does quantum theory have an essence? Or isn't this essence just a relevance judgement of the person who is talking? For example, in some of my books, I said that the essence

[5]This paragraph has double vertical lines down each side and N.B. written in the margin – CT.

of quantum theory was discreteness of transitions and the wave particle duality (both of which are foreign to classical statistical mechanics). You are not satisfied with this, because your personality is such that you say "I understand" only when you have put it in a precise formalism. My personality is such that I say "I don't understand" when it is formal, and "I understand" only when the main line of reasoning is informal. So wouldn't it be better to drop the idea of a unique essence of a theory? This is confusion, because it equates so many things that are different.

Perhaps it would help clear things up between us if I said that we are faced with a situation similar to that faced by Galileo. Very new ways of thinking are required. As you say, most people understand by "observed object" and "rest of the world" something that makes my attempts at communication meaningless to them. But in Galileo's time, most people also had concepts that prevented them from seeing the relevance of what Galileo was trying to say. This did not mean, however, that Galileo should have adopted the views (e.g., Ptolemaic) of other people, merely in order to be able to talk to them. For by doing this, he would be destroying what he had to say, thus making the communication useless (and worse, because it would confuse old and new ideas). Rather, what was needed was to experiment and discover new ways of communicating what was novel. That is what is needed today too.

Of course, you may say that I am presumptuous to say that I have to do what Galileo did. But please recall that in my view we are all basically the same. Everyone has to do what Galileo did. Those who do not do so are engaged in irrelevant discussions "about" the content of other men's work, in which they actually change this content fundamentally, while they say it is still "essentially the same". If each man realises he can do nothing but change the form and content of science, so that he is responsible for orderly creation rather than chaotic and confused creation then the whole situation in science will be enormously better than it is now.

You may feel that my views are nonsensical. If so then you can dismiss me, perhaps in a kindly way, as a bit crazy, and go on with your interest in all these mathematical theorems. But if you seriously wish to communicate with me, I see nothing for it but that you must at least entertain what I have to say, "play with the content of it", and see whether it is perhaps not entirely crazy.

If you would read the article I wrote with Schumacher again, you will perhaps see that the question of "observer and observed" is very relevant to what v. Neumann does with his formalism (See especially the dialogue on p.8. [See p. 358 – CT.]). V. Neumann begins with the notion that there is a clear distinction between object and observing apparatus. Nevertheless, even if we leave Bohr's work out of consideration as irrelevant, because it is vague and informal, Heisenberg had shown that no such distinction can coherently be made (e.g., in his microscope experiment). So v. Neumann was also ruling Heisenberg's work to be irrelevant. In what way could it then be said that v. Neumann's theory contained the essentials of the informal situation concerning the quantum theory? If someone made a theory in which differential equations were tacitly treated as irrelevant, and claimed this was the "essence" of Newton's work, one would be justified in feeling at least surprised, if not also somewhat confused. Would it not have been clearer for v. Neumann to say: "I think all this informal discussion by you physicists is a lot of hot air. Let us throw it all down the drain, and begin with my formalism, which is clear, logical, true, and relevant! This would at least have made it evident that v. Neumann was not discussing the same informal situation as is generally referred to by the words "quantum", when this word has been used by physicists. Rather, he had proposed both a different formalism and a very radically different informal description.

If this had once been seen clearly, the next step would have been to ask: "Is this new theory of v. Neumann relevant, either to the "quantum" situation or to some other situation?" Beginning with the "quantum" situation, one could see that in the "quantum" situation, there is no way clearly to distinguish between "observing apparatus" and "observed object". As Heisenberg's analysis shows (leaving out Bohr as irrelevant), the "quantum" situation does not harmonize or cohere with a situation in which "apparatus" and "object" can be treated as distinct. In other words, the "quantum" situation is different from the one discussed in v. Neumann's theory. But he says tacitly that his theory is "essentially the same" as the "quantum" situation. So he is ignoring relevant differences. And as I pointed out, confusion is just the action of ignoring relevant differences.

Now, as long as v. Neumann holds to the view that his theory is in essence the same as the "quantum" situation, he must continue the action of confusion. This confusion is not the creative kind, that tends to resolve itself as you work. Rather, it tends to spread and entangle, because the relevant differences that are persistently ignored have significant implications in wider and wider fields, as you keep on working under the assumption that there are "no differences in essence".

Only by admitting that he had proposed a very new and different theory could v. Neumann have gotten out of this kind of persistent spreading confusion. But to do this, he would have had to admit that his theory was not relevant to the "quantum" situation. It would be up to him to find what situation, if any, provided a context in which his theory was relevant. Perhaps he could have shown how to use his theory without raising questions like those raised by Heisenberg. But this is very doubtful. For example, in talking to Loinger and Prosperi last summer, I found that they had certain serious and confusing problems, to which they called my attention. In any situation of appreciable complexity, (e.g., a two slit experiment) they did not know how to determine the right boundary conditions for the "wave function of the electron in relationship to the apparatus". This was highly ambiguous, in their theory. And so it should be, because the distinction between apparatus and observed system denies the whole informal form of the quantum theory. V. Neumann would have found similar confusing ambiguities, if he had tried to discuss an actual experiment in detail. However, by talking "about" experiments (i.e., replacing them by idealized descriptions that did not raise these troublesome questions) he could fail to be aware of the fact that his theory was too different from the "quantum" context to be relevant there.

Bohr did give a basically coherent and consistent description of this "quantum" situation, though it was phrased in a way that was very hard for most physicists to understand, and mixed up a bit by bringing in certain irrelevant features of Heisenberg's language forms. This doesn't mean that Bohr's views contain the "essence" of the situation. Rather, the notion of essence is irrelevant in this context. All that anyone can do is to give a coherent account, formal and informal, that is relevant to a situation. The very giving of this account makes it a new situation. In other words, a situation is one whole with the description of that situation. A different description leads to different actions and thus constitutes a new situation.

So when Bohr thus created a new situation, leading to new ways of doing work in physics, what was called for was a novel coherent account of this new situation, and not just an attempt to talk "about" the content of Bohr's work, to "clarify it" or to "put it on a solid and clear logical foundation." And this in turn would have called again for novel and radically different forms of description, leading again to novel situations. In other words, there was nothing for it but unending creation of an orderly kind, in which each man realises that he is always doing something new.

I hope that all this will establish a new situation between us (rather than to "clarify" or "improve" the old situation).

Best regards to you and your wife

Dave

P.S. It may help if I say that when I see a formalism, my response is to try to abstract something alive and informal from the dross in which it is buried. (This is, of course more or less the opposite of your wish to begin with some informal suggestion, and replace it with a well defined formalism as soon as you can). With some formal work, (such as that of Hodge in topology), I have spent a long time, because I sensed that something worth while is buried in the mass of irrelevant equations that fill the book. With v. Neumann and his followers, I don't get that feeling at all. Rather, for reasons that I gave earlier, I feel that, for the most part, they are confusing the issue.

Generally speaking, I have found that few physicist can do other than treat my informal understanding as, at best, something of secondary relevance. So I gradually fell into the habit of "bending over backwards" against all my natural inclinations, to force myself to read formalisms, listen to formalisms, etc. All this was done merely so that I could communicate with other physicists. But now, I begin to see that this was largely a waste of energy, as well as a source of confusion in my own mind. I now see that communication is actually impossible, with a person who rules out my views as irrelevant. To cover up my feelings and accommodate to his leads to a very superficial and sterile kind of contact. Unless we both begin by perceiving all the relevance judgements involved, it means very little. Do you see my feeling that the formal is useful, mainly if it leads to something novel in the informal, and that formalisms are ephemeral structures of ideas, that can change as easily as shapes in clouds?

Feb 10, 1969

Dear Jeff

Just received your letter of Feb 4. I will answer a few of the main points now, and more later.

Firstly, I am not actually coming to America. I am sending a tape of my talk to Illinois.[6] Incidentally this talk is relevant to our correspondence, so I'll send you a copy when it is ready (in a week or so).

Your account of why you are interested in theories is very helpful, for establishing communication between us.

Of course, we both agree that a theory is not a simple reflection of what is called reality. Rather, reality is the informal, the unspecifiable content that "slips through" all forms, as defined verbally or otherwise. But I can't make sense of the idea that the physicist communicates via his formal theories. This seems to say that he does not communicate informally, to any significant extent. In other words, you seem to maintain a sharp dichotomy: The formal is the means of communication, the content (i.e. the informal) is the end toward which communication is directed.

But I have tried to discuss the informal form. In other words, we have to be careful not to confuse the totality of all form with what is commonly meant by the word "formal" or "formalism". To me, the main act of communication is through the informal form. The formalism is only an elaboration and articulation of this informal form, to make it relatively precisely defined in certain ways. But every formal form contains an unstated division between the tacit and the explicit. The more precisely the explicit is defined, the vaguer becomes its tacit significance. It is only in the informal form of language that this vagueness can be taken into account, so that it does not lead to confusion.

In other words, just as there is a vast set of theories of varying levels of abstraction that are relevant to each other in certain ways, so these theories shade imperceptibly to merge and amalgamate with the informal form of language in general. The over-all situation could be described through the metaphor of the iceberg. The informal form is the vast "submerged" part of the iceberg, and mathematical formalisms are tiny facets on the peaks of the iceberg. But the spirit of the age is to talk only about the formalisms, as if the rest of the iceberg did not exist. In other words, one tries to make the facets of ice directly relevant to the immeasurable ocean of the unknown. In fact, we cannot do this. Even the mathematician is forced to use informal language in talking about his formalism. But he can talk as if it were not relevant to pay careful attention to this informal language and as if all the communication were by means of formalisms. However this must lead to confusion.

Basically this arises because the mathematician still brings in the informal form of language "by the back door". And just because of this he is not clear on how he brings it in. All the while, his eyes are on the front door where the "white invited guests" are coming in, resplendent in their beautiful clothing (i.e. the formalisms), while he only occasionally allows a sneaking glance at the back door, where the "uninvited black servants who actually do the work" are coming in (i.e., the informal forms of language).

[6] A symposium on the structure of scientific theories held in Urbana, Illinois, March 26 to 29, 1969. The proceedings, extended by a lengthy introduction, were edited by Frederick Suppe in Suppe (1974). See C137, p. 301, n 1 – CT.

It may be true that Einstein did "paste together" a number of old formalisms. I did this too, and so did most other physicists. But I think that this is the result of our fortuitous heritage from the past, and not really an inevitable necessity. What is called for is to start to talk in a really new way, informally and formally. Bohr actually began to do this, though he blocked himself at a certain point, by saying that all unambiguous communication had to be in terms of classical concepts (suitably refined, where necessary). One can not only ask whether this statement is true (i.e., are there not unambiguous means of communication in terms of non classical concepts?) More deeply one can ask whether it is relevant. In other words, is physics necessarily and inevitably committed to whatever could be meant by the words "unambiguous communication". Or couldn't physics be more like poetry, music, and art?

Bohr was tacitly attached to a certain view of scientific truth in physics, i.e., unambiguous communication. Einstein was attached to another view of scientific truth, i.e., the world is a union of distinct elements, this union being described in terms of certain universal field equations. Because these two views of scientific truth were mutually irrelevant, Bohr and Einstein could not communicate. And because all of us are the views of Bohr and Einstein, this contributes to a general break down of communication. What is called for is that we understand the irrelevance of fixed views as to what constitutes scientific truth. And this implies also the irrelevance of fixed views as to what science has to have in it (e.g., unambiguous communication, or also, precisely defined field equations).

You say that our notion of an objective world is the result of "cap. for rat."[7] In a way, I agree with you. I agree with you that deeply the table is like love, in that it cannot relevantly be "talked about". But there is more to the situation than this bald statement. Let us say that reality is the unknown, which cannot be the object of relevant discourse. But let us make a metaphor, with "flashing indicator relevance" rather than "lasting object relevance". In other words, as the poet can flash indications on love, which is the unknown and the unspecifiable, so the scientist can flash indications on reality. So the essence of science is deeply similar to the essence of art. The scientist is not trying to communicate "unambiguously" about reality.

To begin my metaphor let me imagine an infinity of moving aspects in a vast ocean. Let me imagine a man, looking at these aspects, and relevating certain of these general features. He might discover, for example, that certain forms or patterns were invariant under space displacement, time displacement, rotation, or deformations of various kinds. He could then describe this invariance by inventing the concept of a permanent object moving through space, rotating, deforming, etc. He would not say this "object" actually exists. Rather, he would say that it is a way of thinking that can be relevant, in man's perception and action. For example, he might have a helicopter. He could imagine this as yet another object. He can imagine this helicopter crossing a "patch of ocean" with certain relatively invariant forms in it, and first encountering one of these forms, then another, etc. When he took action and directed his helicopter accordingly he would discover that the order in his concepts was relevant to the order he actually encountered.

[7] cap. for rat. = capacity for ratiocination. Information provided by Professor Bub – CT.

Of course, this does not explain why such conceptual orders are relevant. The reason may be infinitely complex, in the sense that it goes beyond all our ideas on space, time, object, separation and union, etc. The same holds for the table in front of me. I cannot explain why the concept of object has a certain limited relevance in this context (e.g., it is limited in time, or the table will eventually disintegrate, break up, burn up, etc.). But it is a fact that it is relevant in this way, because I can participate in action, in the context of what I call perception of a table, in a way that is not confused. But if I try to imagine a "thing called love", there is no context at all, in which action directed by such a concept will be free of confusion (i.e., disharmony).

What do I mean by confusion (or disharmony)? This cannot be defined. Can you define what you mean by "cap. for rat."? This has a significance for you, which is immediate and direct, but there is no "natural definition of the essential content of ratiocination." So likewise, there is no "harmonious definition of harmony." All of us are in the same boat. Every statement has an explicit meaning and a tacit meaning. The most precise mathematical statement rests on unspecifiable and tacit meanings, as vague as the term "harmony". In the last analysis, it is only "harmony" that distinguishes acceptable mathematics from unacceptable mathematics (i.e., as mathematicians are now beginning to see, the notion of a solid logical foundation for mathematics is not relevant.)

To me, "cap. for rat." means to regard the formal as an ephemeral and relatively minor elaboration or aspect of the informal form.

It is very surprising for me to see you regarding von Neumann's work as an example of "cap. for rat." When I try to read von Neumann, my head starts to swim. I ask "What in the world can he be trying to say?" It is like the delusions of a man who thinks that he is Napoleon. Every step is very natural and logical. But the whole informal form is wildly confused and illogical. This informal form has no rational necessity in it at all. It is the arbitrary and fortuitous form of von Neumann's personality. And because of the spirit of the age, it is also a common form of personality, of a great many scientific workers. You say that you do not find the notions of necessity and contingency sufficiently intuitive and simple to provide a criterion for the adequacy of theories. I am again surprised by what you say. For tacitly, by what you write about, you imply that the von Neumann type of theory is intuitive enough and simple enough to allow room for a criterion for the adequacy of theories. To me, theories of the v. Neumann type are purely formal abstractions, with no relevant intuitive content at all. Indeed, when one reads v. Neumann to see how he proposes to discuss his theory intuitively and informally, one discovers an incredibly mixed up boiling mass of confusion. You cannot have made such an examination of v. Neumann's work seriously, or else your head would also spin, as mine does, when I try to think of the meaning of v. Neumann's theory.

You say that without formal discussion, an idea itself is empty. In my view, this is not so. It is only in certain rather limited contexts that an informal discussion requires a formal articulation. For example, the poet can discuss the beauties of nature, without formal articulations, sketches, diagrams, measurements, etc. Your statement implies that science is basically different, in that it requires unambiguous communication, which in turn would require formalism. I would say that science is basically the informal and that the formal will arise in certain limited contexts only, as a further articulation and definition of the content relevated in informal discussion.

The scientist should be compared more to the poet than to the architect. In Greek "poesis" means "to make" i.e., in some sense, "to create". The architect makes only certain limited categories of forms. The "poetic spirit" envisages new orders of relevance and new forms, leading to the "making" of what has never been made before.

Then we return to the question: "What does it mean to talk "about" poesis?" Isn't this similar to trying to talk "about" love? At most, one can make "flashing indications" that are relevant to poesis. But to do this is itself poesis.

So whether he likes it or not, every "critic" can do nothing but "write poetry"– i.e., to make something novel. The question is: "Does he make something new that is a harmony of content and form, or is it a disharmony?"

If he believes that he is merely writing about the "old poem" when he is, willy nilly, writing a "new poem", his poesis will be confused, a disharmony of form and content.

I want to suggest that the genuine poet, the genuine artist is not merely making a pastiche of known forms. Similarly genuine communication is art – i.e., poesis. It involves the forming of novel forms, and the experiencing of novel feelings. Bohr's work is not basically a pastiche – it is actually a novel harmony of form and content, (as holds for every genuine work of art).

You write as if involvement in communication and language also involved a necessary commitment to make pastiches of old forms. You might as well say that the visual artist's involvement in space, form and color meant that he is also committed to doing nothing but making pastiches of what other artists have done. After all, haven't all artists used the "language" of paint, lines, shading, and solid forms, such as sculptures patterns and structures? Does this mean that Cezanne is making a pastiche of the work of earlier artists? To learn from someone else's work doesn't commit you to doing pastiches, does it?

You ask me to say precisely what I think is wrong with quantum theory: Perhaps you will see my answer now – i.e., I regard this as an irrelevant question. All I can really say is that what is called quantum theory is perceived by me as a disharmony of form and content – and this is what I feel to be wrong with it.

In elaboration of this point, let me say that it is already confusing to use the word "quantum theory" to cover the work from Schrodinger, Heisenberg, Bohr, through Dirac, v. Neumann, J. and P., K. and S., as well as field theory, renormalization theory, S matrix theory, current algebras, etc. These are all different theories, different in content and radically different in form. What they have in common is mostly the tendency to repeat certain "sacred incantations" such as "vector in hilbert space", "unitary transformation", etc.

The first step towards harmony would be to say explicitly that quantum theory does not actually exist. So nothing is "sacred". One can drop words like "hilbert space" as freely as S matrix theorists have dropped words like "a wave equation with an interaction potential."

If quantum theory is admitted to be an irrelevant term of discourse, then of course, it is not relevant for me to try to say just what is wrong with quantum theory. Rather, I say the question now is to go on with new modes of discourse, informal and formal. Theorists have actually been doing this surreptitiously over the past 30 years or so. Let us just do so in an open and direct way. "Quantum theory" is no longer relevant. "Uncertainty principle" is not relevant. Let us in the spirit of poesis be sensitive to discover new orders of relevance.

More later.

Regards

Dave

P.S. It occurred to me that the Greek word "poesis" means "to make", while the Latin word "fact" means "what has been made". This is significant in the sense that it is the fact is made, through perception leading into rational ordering and communication. The question always arises: "Is what has been made relevant to the further content of perception-action?" For example, I may say: "The fact is that I see a chair in front of me." If this fact is relevant, I can sit on the chair, move it, carry it, change it, break it up, burn it, etc. etc. If I discover that the overall order of experiencing this context of perception is in harmony with that implied by the fact, I say that my statement is relevant, and then perhaps true or false. E.g., if I say, the chair is green, this can be a relevant form, with non valid content (since the chair may actually be brown). But if I say, "Psychological tests show that a certain man's intelligence is green", this is, from the outset not a relevant form, so that the question of truth or falsity of content does not arise. How do I know the irrelevance of the form: "Intelligence is green"? This is a vast question that cannot be given a well defined answer. What happens is that when I try to act (or imagine acting) in an order relevated by that statement, I get into confusion. So I say: The form "intelligence is green" is not in harmony with any content that has been or can be relevated in perception-action.

It is clear that the proper "making" of the fact involves a continual relevation of different forms, different orders, etc. and the continual perception of the harmony-disharmony between these orders and the orders arising in further perception-action.

Now, if we are to see the fact in a novel context, we need to be sensitive to new orders of relevance. This is where "poesis" comes in. Without "poetry" we fall into the habit of "making" routinized mechanical kinds of facts, and overlooking what is novel and different from what has been known. That is why science is basically poesis and not unambiguous communication. That is why it is not basically the making of "pastiches". To see a new kind of fact requires "poetic vision" and novel forms of communication. Without these latter, poetic vision becomes confused, as we try to put novel content into old forms that are no longer relevant.

To sum up then, I would say: "The fact is made basically through poesis", so that the man who tries "only to speak prose" is actually just using very bad poetry, of a highly mediocre character.

It is in this sense that I would say "J. and P. are writing some pretty lousy poems."

Feb 12, 1969 Postmark Feb 11

[Comment re postmark added – CT.]

Dear Jeff

This is a continuation of yesterday's letter.

I see now that you and von Neumann have a close affinity in points of view; and this explains why our communication has been rather limited, thus far at least.

To clarify the differences between us, let me say that where you talk of capacity for ratiocination, I would talk of the ability to reason. Tacitly, in the context you use the term, ratiocination means formal logical reasoning. I regard this as not very significant. It can sometimes be useful, but more often, it is a source of confusion, for it is like the man with systematic delusions. The more you concentrate on formal arguments, the more your attention is diverted away from the fact that in the informal form, all is wild irrationality and arbitrariness. This is characteristic, for example, of modern technology. Everything is rationalized and ordered by computers, but the ultimate purpose is atomic and bacteriological destruction of every living thing on earth, as well as maintaining some people in wealth, others in dismal slums, and meaningless empty lives for all, whether poor or wealthy. A similar formal rationality and informal absurdity is what characterizes a good deal of modern physics, as well as most other aspects of modern life.

Now in my view, informal reasoning is the essential aspect of rationality, in science as in the whole of life. The division between formal and informal is at present much too sharp. This sharpness is artificial. What is called for is to realize that the formal is a special case of the informal. In other words, the informal, suitably defined and limited becomes what is called the formal. Any particular formalization is relevant in a very limited context, perhaps for a moment, and then it can be discarded, after it has served its purpose. I emphasize the comparison to the sketches of an artist and not of an architect. That is, one may find it useful to make sketches; these are of little value in themselves, but go into the "back of the mind", where they may serve in the informal action, which is the creation of something new, not a pastiche of other people's work, and not basically of the nature of "formal" sketches.

We agree that in some sense, reasoning is behind the experiencing of objects. However, I emphasise informal reasoning and you emphasize formal ratiocination. I emphasise that this informal reasoning operates in a broader context of perception and action. You seem to imply that it operates abstractly and formally, "almost in a world of its own." That is to say, one manipulates the symbols and pushes the pencil according to the rules; and out of this come the "objects" of physics. Or else one thinks informally "about" the formal symbols, and out of this comes the "objects" in question.

It is implicit in your view (as in v. Neumann's) that you, the thinker are thinking "about" a world that is external to you. You aim to make your thinking the "truth about the world". In a similar way, you will have to think that the observing instrument has "true" information "about" the observed object, and that you, the thinker, similarly have "true" information "about" the relationship between instrument and observed object. Then you can think that you the thinker, on the next level, have "true" information "about" how you are related to your thought about the instrument and the observed object. And so it must go. This confused regression is inherent in the v. Neumann approach, which cannot consider the wholeness of observer and observed. And the v. Neumann approach is inherent in any procedure which starts with a formalism. You have to imagine or assume that the formalism is "about" something else, i.e., an external reality. Or else it will be just "pure mathematics".

So I don't see how you are going to understand the wholeness of things by beginning with the informal, mainly in order to bring you to a formalism. It is only in the informal that one can coherently and relevantly discuss wholeness. Even the formal is then still the informal; i.e., a momentary sharpening up of the informal language, for a limited purpose. But if you regard the formal as the main point of the description, you have no choice but to say it is "about" something external. And thus you introduce a disjunction, which can never be overcome by union. For when you try to think of the union of the content of your formalism with the external world, you tacitly disjoin yourself, the thinker who is thinking "about" a union of external world and content of formalism that is external to him. And so it goes ad infinitum. That is to say, you are caught in what I would like to call disease of "v. Neumannitis". This is the unending need to disjoin things and then to try to unite them, but in this very effort at union, to achieve only further disjunction.

All this is not merely "philosophy". It confuses what we do "in physics" with the formalism (as will be shown in more detail in a paper that I shall send you soon).

So you see, there is still a big gap between our relevance judgements.

Best regards
Dave

Feb 13, 1969

Dear Jeff

To continue the previous two letters, let me sum up the situation in this way: The difference between our relevance judgements is not in our views about the unreality or reality of objects. We both see as relevant the question "How can anything at all exist?" The key difference is that you feel that to inquire into such questions, what is relevant is somehow to arrive at precise formal statements. You are willing to consider the informal, but largely as a means of arriving at such formal statements. You also admit that what is deeply relevant is the informal, the unspecifiable, the undefinable, but you feel that to come into proper relationship with this, we need above all a precise, logical, clear, solidly based 'mathematical formalism'. Then you may feel that you "understand".

On the other hand, I feel that such formal aspects have been heavily overemphasized in modern science. I feel the informal form of ordinary language is the context in which creative new changes (and not merely pastiches) can take place. To illustrate my way of perceiving, I can say that when I wrote my book "Quantum Theory", I found the need to use mathematical formalism very irksome. As a rule, I could perceive informally without intermediate steps just how one expression implies another. But I realized, for example, that in a certain kind of case, something like six or eight intermediate steps are canonically required. So I put in the requisite number of steps into the "proof". Naturally, the steps were frequently wrong though the end result was practically always right. I did go over the book three times, and my assistant in Princeton spend a year going over it to remove such "errors". Nevertheless, several hundred (or thousand) such "errors" still remain. Something similar happened in my Relativity book.

The fact is that I experience formalism as a barrier to understanding and a block of communication. Behind its logical formal form is generally a highly confused and irrational informal form. Since I read an article by first picking up the informal form and then going on to the formal details, I find that it is very painful, as a rule, to go on with it, because the typical article is in its informal form a boiling cauldron of confusion. This I feel especially strongly in the articles and books written by von Neumann and those who follow on lines initiated by him.

So I say that to "make" something new, what is needed is "poesis," rather than "architecture". The metaphor of the poet is therefore what I wish to call attention to. It is only through "poetry" that what is novel emerges into form.

Best regards Dave

Feb 14, 1969[8]

Dear Jeff

You asked me in your letter just what I think is wrong with quantum theory. I answered that any attempt to define this question precisely is not relevant. Broadly speaking, what is called quantum theory suffers from an extreme disharmony of form and content. Part of this disharmony is just that many disparate and mutually incoherent points of view are called by the same name (e.g. Heisenberg, Schrödinger, Bohr, Born, Pauli, Dirac, v. Neumann, J. and P., K. and S., field theory, normalisation theory, S matrix theory, group theory (SU3 et al) current algebras, etc., etc.). Merely to realize that the differences between these theories are very relevant while the similarities (e.g. common use of words like "hilbert space" and "unitary transformation") are rather superficial, would help greatly to establish a more harmonious overall situation. For example, you would then have been led to define your question more carefully. You could have asked "Just what is wrong with Bohr's account of the "quantum" situation?" or "Just what is wrong with v. Neumann's account of this situation?" These two questions are very different and almost unrelated, because Bohr's account and v. Neumann's account are so disparate.

With regard to Bohr, I feel that a promising start toward new modes of informal discourse has been blocked by certain fixed notions of scientific truth (i.e. unambiguous communication) which led Bohr to insist that the informal language had

[8]First letter with this date – CT.

to be that of what he called "everyday life" or "common sense", suitably refined where necessary to become the language of classical physics. I would say instead that one can start with complete "poetic" freedom in the informal language, and make "unambiguous communication" a special case, relevant in certain limited contexts of instrumentation. But if one accepts Bohr's point of view, one is led to a tremendous disparity between the informal language of "common sense" and the formal language, including phrases such as "an observable is an operator, or a matrix", "two observables do not commute, and their commutator is $\sqrt{-1}$"; "a system is described by a vector in hilbert space, undergoing unitary transformation", etc. It is implicit in this view that it is no use to try to change the informal language, and that the way to more progress is to try to develop new formalisms. This is in many ways similar to your own view, i.e., changes in the informal language are "vague", so that your ultimate aim is always to produce a new formalism (which has, however to be a "pastiche" of older formalisms). Thus Bohr's view is characterized by extreme disharmony between informal and formal forms of language. Also, by insisting on "unambiguous communication" he forces us effectively to fix the informal language, and this is not in harmony with his evident aim of changing the informal language.

To me, it seems necessary to start now to experiment with changes of the informal language, which would remove the disparity of formal and informal forms. The formal forms would then be special cases of the informal form, momentarily "sharpened up" for special purposes and then dropped when these purposes are accomplished. So I would give formalism a much less dominant role than it has now in physics.

With regard to von Neumann, my criticism is that I simply cannot make sense what he is doing. Of course, I am ready to admit that he is a good mathematician. So I will take his word for it that if I push the pencil according to the rules, and if I begin where he began, I will end up where he ended up. But as I explained in yesterday's letter, I feel that the right way to read a paper is to get a feeling for the informal form first and in this context to understand the formal aspects. When I try to do this with v Neumann, I find that the experience is very painful. My head starts to spin, as I get the impression of boiling confusion and wild irrationality, behind the rational form in which one equation follows "logically" from another. It is similar to the feeling that I get by considering how our society is setting up rationally ordered computers for irrational ends. Thus, we have a computer system to direct the ballistic missiles that can only result in our being torn to pieces and blasted to smithereens, if it is ever used. Likewise when one applies for a job in a large corporation, there is a computer to decide whether you are suited for the job. All this 'rational' computation takes place in a context in which people are full of jealousy, envy, fear, greed, hatred, etc., as they work toward meaningless ends (such as persuading people to consume more Kellogg's Corn Flakes).

I cannot avoid the impression that this is just the sort of thing v. Neumann and his school are doing. Indeed, it is, in my view, not an accident that v. Neumann was so interested in computers. For this is the paradigm case of a field where one accepts all the irrationalities of the informal situation, and tries to work out a rational "programme" for more "efficiently" accomplishing these irrational ends. This is an extremely dangerous sort of thing to do.

It is indeed not an accident that people like v. Neumann and Wigner are proponents of the "cold war game", in which one "rationally" computes how to respond to the moves of one's opponent. This is coherent with the approach which sets up formal rules, without attempting to inquire into the informal situation in which these rules are to operate. To fail to make such an inquiry is equivalent to accepting this situation as it is, and to treat it as basically unchangeable.

I hope this answers your question. I hope also it helps make clear the difference in our relevance judgements that still remain.

Best regards

Dave

P.S. The situation with regard to von Neumann's theory and with regard to computers is similar to what you describe in your efforts to get a job at York University. Formally, the form of your discourse and that of the people at York is rational and logical. But informally, each person has his own peculiar and fortuitously determined preferences and desires, to which he holds irrationally, even when there is evidence of confusion in them and conflict with the preferences of others. Everyone talks as if it were only a matter of finding a "logical solution", when in fact, the whole situation is irrational in structure, and can therefore have no rational solution. The first step to cleaning all this up is this to realise that it is the informal solution that is irrational. We need a new mode of discourse to clear it up, because all the old terminology (e.g., "interests of the department") is hopelessly entangled with the irrational informal solutions.

Similarly the questions raised by von Neumann are informally entangled in hopeless irrationality. As long as one uses terms like "quantum state" or "state vector in hilbert space" or anything similar, one can be ever so rational in formal logical reasoning, but confusion will flood in on the informal side. (Just as confusion is inevitable as long as the people at York talk in terms of what has no meaning; e.g. "interests of the department" or something similar.)

Feb 14, 1969[9]

Dear Jeff

On rereading your letter, I see that I haven't answered your questions about my attitude to hidden variables.

As I said in the last few letters, my objection to "quantum theory" is its general disharmony between form and content. In particular, the formal language has all sorts of radical new implications of wholeness, while the informal language is still the old one of "particles", "fields", "measurements", "charge", "mass", etc. So what we need is a kind of "poetry" that develops a new informal language, rather than a further elaboration of the formal "architecture".

[9]Second letter with this date – CT.

Now with regard to hidden variables, all that I have done on the subject is confused. This confusion was inevitable, once I tried to relate my work (in 1951)[10] to what von Neumann had done. For as I indicated in my previous letters, von Neumann's work is informally a boiling cauldron of confusion and irrationality (though formally it adopts a precisely logical mode of discourse). Therefore, any attempt to relate one's work to that of von Neumann must involve one in confusion. If you "harmonize with confusion", you are confused. And if you oppose or criticize confusion, you are also confused. The only thing to do about confusion is to see its irrelevance, and keep well away from it. This is what I should have done with von Neumann.

Unfortunately, I conceived the notion in 1951 of trying to show by means of a counterexample that "hidden variables" were possible, in spite of von Neumann's "proof" to the contrary. But since v. Neumann hadn't proved anything, my "counterexample" had nothing to "counter". So it moved around in wild confusion, as happens when you deliver a blow against what appears to be an object, but what is actually a structure of smoke, fog and mists.

I should have realized then that von Neumann's work had no meaning. Of course, as a formal means of manipulating symbols, it was perfectly logical. But the informal discussion that related it to the "quantum" situation in physics was a quagmire of twisted complexities. If you think that necessity and contingency are too complex for you, I can't imagine your reactions when you really begin to understand what von Neumann is up to.

Aside from the misconceived wish to counter what v. Neumann had said, my deeper interest in "hidden variables" always had to do with the question of contingency. Actually it isn't all that complicated. Consider an ensemble of uranium nuclei, undergoing radioactive decay. The whole ensemble is described formally by a certain wave function.

$$\psi = \psi(\bar{x})e^{-\frac{\lambda t}{t_0}}$$

The exponential decay time is roughly 10^9 years. This wave function doesn't change appreciably over the first million years or so. Yet, a considerable number of individual atoms will decay during that time. If we say that the wave function is the most detailed formal description available, we must admit that a lot of things happen (i.e., individual decays) which are just not describable in terms of this wave function. As far as the theory is concerned they are contingencies. That is, nothing describable in terms of the theory could raise the question of why a particular atom decays at a particular time. The theory does not attempt to discuss such questions. It discusses only statistical ensembles. In this regard it is like insurance statistics. The death of an individual human being is likewise a contingency in the context of such statistical information.

Now, in all other cases, contingencies have proved to be describable as necessities, when the context is suitably broadened. Thus, if an individual human being steps in front of a speeding car, death is inevitable. But, of course, in the context of insurance statistics, there is no place to discuss such questions. One would have thought, on

[10] See p. 255, n 2 – CT.

the face of the matter, that the quantum theory is similar. That is to say, the wave function has no place to discuss the question of why an individual nucleus decays at a particular time, but in a broader context, such questions may well be discussable.

Here, we come up against the extraordinary insistence by practically all physicists that the quantum situation was different, in the sense that it would be illogical or meaningless to consider such a broader context, in which contingencies could be seen to become necessities. Just simply to show that this view is false, I proposed the hidden variable theory of 1951. I still believe that it did show the falsity of this view. But my mistake was to relate it to von Neumann's work. I should also have said that von Neumann's work was totally irrelevant to the question, and simply refrained from bringing it in any further.

I can say that our Rev. Mod. Phys. papers[11] showed in yet another way how contingencies could be described in the "quantum" context. It had the advantage of suggesting a possible experiment, for defining the contingent parameters.

I think you are aware that I never regarded this paper as a serious proposal for an actual theory. Evidently, it is a pastiche of other people's formalisms, and was never intended to be otherwise. For its main purpose was to illustrate how contingency is to be described in this context, and not to develop a new informal and formal language.

The situation between us is now this. You regard contingency and necessity as incredibly complex notions, while (at least tacitly) you regard the approach of von Neumann, J. and P., K. and S., etc., as manageably simple. I regard contingency and necessity as relatively simple, and the v. Neumann line as not only incredibly complex, but also incredibly confused. Thus, when you cite a lot of formulae from K. and S. in your paper, I can't really understand what you are up to. I can probably show that one step follows from another. But what I don't see is how any of the symbols are at all related to the "quantum" situation except by means that are hopelessly entangled in v. Neumann's confusion.

So it seems that we are not able to communicate, until we can get clear on our relevance judgements on this point

Perhaps it would help if I said that deeply, I never regarded measurement as a problem. To a certain extent, I became involved in such language, because of my wish to find a suitable Ph.D. thesis for you (i.e., as you were caught in the wish to get a thesis, I was caught in the wish to fulfill my function of helping you get one). So I too was not as critical of what you wrote as I might have been. I too did not wish to delay your thesis indefinitely by raising all sorts of delicate questions, especially since you had no means of supporting yourself without a degree.

It is only in the v. Neumann approach that measurement is a "problem". This is because v. Neumann says that there is an "observed system that exists in a well-defined quantum state". It is up to the observer, with the aid of his instrument, to obtain knowledge as to what this state is, and then to assign it an appropriate wave function. When he does the next observation, he is forced to say that the "wave packet is reduced" or "collapses". And he thus has "problems", because elsewhere, he has said that the laws of physics allow only for unitary transformations. Thus,

[11] Bohm and Bub (1966a, b) – CT.

measurements have to be assigned a special "law of their own" and the "problem" is: "How can we know what this law is ?"

My view has always been that measurements are side issues, and do not have a special law of their own. Rather the physicist arranges a simple situation and calls this a measurement (though why he does so is not very clear). Our Rev. Mod. Phys. paper brought out this point implicitly, and in my own paper, I made it more explicitly. I never wished to claim that our Rev. Mod. Phys. papers were satisfactory, in all respects. But in my view, they were not actually directed at the "measurement problem". Rather, their deep intention was to discuss contingency in the "quantum" context. As I said before, it was your phraseology (and interests) that led to such heavy emphasis on the measurement problem.

In my view, any attempt to discuss measurement or observation as part of a fundamental theory must lead to confusion. For then, you will have to ask: "What is the law of the measurement process?" To discover this, you need another "metameasurement process". And so it goes, until you get to the mind of the observer, as v. Neumann inevitably does.

But when you get there, your troubles (or "problems") have hardly begun. v Neumann invokes the "principle of psycho-physical parallelism" to assure us that what is in his "mind" is also in the "physical world". This line of reasoning implies first of all the disjunction between v. Neumann's mind and the physical world, and then, an attempt at union through the assumption: "I, von Neumann, know the truth about the physical world." But then, if you ask him "How do you know that what you know is the truth?" he must answer "I know that my mind and the physical world are related by parallelism." Then you can ask "How do you know that?" If he were honest, he would say "This is an assumption that I find convenient, but it may well be a figment of my imagination." But he never says this. Instead, he leaves you to accept his assumption, as if this were some sort of "common knowledge".

In a recent talk at Trieste, Wigner[12] (who follows the v. Neumann approach) said something very revealing. He said: "I do not like to say that my friends are mere machines. Therefore I admit that like me, they are capable of free choice." What is interesting is that Wigner never asked if he was a mere machine. He "knew" that this was not so, but out of the "goodness of his heart", he allowed for the possibility that his friends were also not mere machines. (What about his enemies? He never discussed this question.) Such an approach always starts out with the assumption: "I exist as an entity who is capable of rational thought." But when I look at Wigner, I see that his thought is irrational, and mechanically conditioned to be what gives him pleasure. The question is then: "How are Wigner and I to communicate?" I regard him as trapped in mechanism, and (since I am probably not one of his friends), the goodness of his heart doesn't extend to admitting that I am more than a machine. Perhaps we are both machines? Can two machines communicate?

What all this is driving at is to hint to you that when you ask "How can anything exist?", you have to extend the question to "How can I exist?" Perhaps as long as

[12]Not available, but presumably the basis for Wigner (1969).

there is belief in the entity called "I", rational thought is not possible, so that it is no use to ask how such thought can give rise to objects.

I hope you see that by direct perception, I do not mean that the objects in front of us actually exist as such. To indicate what I mean, imagine placing several hundred physicists in compatible diving suits in tanks of lukewarm water, so that they would neither see nor hear nor feel anything. Let them be put in communication, through microphones and earphones. Let them discuss nothing but physical theories. Let them do this day after day, week after week, month after month. What do you think would come out of all this?

What this is meant to show is that somehow, something more than words has to enter the content of our communications. And here I do not refer mainly to the results of experiments. I refer instead to a vast background of informal experience, both perception and action. It is from this background that our more formal concepts are abstracted. Even the most abstract mathematical logic "swims" in this informal context, which is relevant whenever one uses informal language to talk about the formalism. For example, one uses words like "this", "that", "not", etc. Wherein lies the "thisness of this" and the "thatness of that", as well as the "notness of what is not true"? All this is informal and unspecifiable. If we are suspended in language, not knowing which way is up or down, how much more are we suspended in immediate perception-action, which is what is meant by experiencing from moment to moment. Without this, language can have neither relevance nor meaning.

Finally, with regard to the "measurement problem", what I say about it is that there can be no such a "problem". If one seems to have such a "problem", this is a sure sign that one is engaged in the irrelevant. Thus, Bohr never regarded measurement as a "problem", because he did not say that there is an "external object in a quantum state that has to be known by an observer with the aid of an instrument." Rather, object, instrument, experimental results, and experimental conditions are a whole, in which there can be no "problem" of describing the "process of measurement" in terms, such as the "collapse of the wave function".

My advice to you with regard to the measurement problem is very simple. It is just: "Drop it".

Best regards

Dave

Feb 16, 1969

Dear Jeff

I understand how difficult it is for you to discuss these questions clearly while you are still uncertain about your job, and while your wife is expecting another child. I do hope that by now, you will have had a suitable offer from one place or another, and that all goes well with your wife (and child).

It occurred to me that what we are discussing is still not entirely clear. The way in which confusion becomes entangled is really annoying, but it is necessary to see if all these "tangled threads and webs" that we have woven can be straightened out.

With regard to hidden variables, my deep intention has always been to say that the "measurement problem" is irrelevant, in the sense that there is no meaning to the effort to assign special laws to those processes (or parts of processes) that happen to be called measurements. Rather, I would say that the form of law is universal, and that even in what is called measurement, this universal law applies. This certainly was true in classical physics. But in quantum theory, many physicists began, either tacitly or explicitly, to talk as if special laws applied whenever something could be called a measurement or an observation (e.g., the "collapse" of the wave function, or the "reduction of the wave packet"). In my 1951 papers, as well as in our Rev. Mod. Phys. papers, it was pointed out tacitly and explicitly that measurements were to be regarded as special cases of universal law, for which there was no need to invoke additional laws, peculiar to measurement processes.

Now, part of the confusion went back to my Quantum Theory book and to my 1951 papers. In all this work, I realized intuitively that Bohr's formulation of quantum theory is basically consistent, and that v. Neumann's is basically inconsistent. However, in my effort to "clear up the confusion", I fell into the trap of talking in terms of v. Neumann's language forms, mainly in order to "answer" him or to show where his views were irrelevant. Thus, I failed to see that the term "quantum state" is totally incoherent both with Bohr's views and with the facts of the case (as regards the over all "quantum" situation). For this term implies that the wave function describes a separately existent "individual quantum system", which is, in some sense, "measured" or "observed", with the aid of an instrument. But because the universal law is taken to be unitary transformation of the wave function, one is forced to invoke special "laws" for the "measurement process". What form can these laws take? Evidently, this is almost entirely arbitrary. The difficulty is however to put these in a consistent form. Because this has never yet been done, we have what is called the "measurement problem", i.e., the problem of making a consistent description of the "measurement process". Moreover the very notion that measurement processes obey special laws of their own raises further "problems". For we shall then have to ask: "What about the 'metameasurement process', with which we observe, measure, and test our assumptions about the laws of the original measurement process? Does this have a yet further set of laws of its own? As I indicated in yesterday's letter, this leads inevitably to confused paradoxes about how the "mind of the observer" is supposed to be related to "the physical world".

Now in my view, all this is irrelevant and unnecessary. It could have been avoided, if one had understood that the term "quantum state" has no place in such discussions at all. That is to say (as Bohr insisted) quantum theory discusses only a statistical ensemble of observations and makes no attempt to discuss individual systems or events. In this respect it is like insurance statistics. But it is different from insurance statistics, in that there is a novel kind of wholeness, in the sense that the form of the experimental conditions and the content of the experimental results are a whole, in which analysis is not relevant. For this reason, the term "state of the system" is inconsistent with the entire "quantum" context, since it both tacitly and explicitly denies the wholeness, to which Bohr called attention.

My main objection to Bohr is his insistence that our language forms were such that in this context, no further discussion of individual contingent events was possible. This always seemed to me a rather arbitrary restriction.

The main source of confusion here is that most physicists wanted "to have their cake and eat it too". That is to say, they interpreted Bohr's remarks on the subject by saying that the "wave function is a complete description of the individual system" and yet "it is relevant only in a statistical ensemble." The need to say both of these things is what led people to talk in terms of the "collapse" of the wave function "in a measurement process". That is to say, "before the observation", the initial wave function is a "complete description of all that there is to be described", and "after the observation", the final wave function is once again "a complete description." This contradicted the "law of unitary transformation of the wave function" and necessitated a special law for the "measurement process."

As I said before, in my efforts to "criticize" these views, I adopted the term "quantum state" and thus became entangled in the very confusion that I wanted to "criticize". A better approach would have been as follows:

We begin with the wholeness of the form of the experimental conditions and the content of the experimental results. Each such whole situation is relevant only in a statistical ensemble of similar observations. One makes no attempt at all to discuss individual observations or events in terms of the laws of this context. Then, as Bohr has shown, we can give a consistent account of how the theory actually works, in each concrete situation.

What about the individual events in observations? Here I depart from Bohr by saying that they can be discussed. But to do this, we need a novel theoretical form, which has room for a description of individual events, that are mere contingencies in the "quantum context". So individual events can be discussed only by changing the theory in a fundamental way (by introducing what is novel and "not commensurable" with the quantum theory).

To a certain extent, I added to the confusion by failing to emphasise in my 1951 papers that the "hidden" variables were novel, and went outside the "quantum" context. Here, the situation was further confused by a widespread prejudice towards positivism, which implied that unless or until these "hidden" variables led to concrete new experimental predictions, they were "no different in essence" from the old theory. I should have insisted that a theoretical form may be radically new even when it is for the moment only being used to describe a familiar content. Thus, Hamilton-Jacobi wave theory of classical dynamics was a radically different form from Newton's particle dynamics, even though for a century or so it was applied to the same content as that of Newtonian dynamics.

My difficulty was that I tried to "answer" positivists, and thus, I inadvertently adopted their language forms, and fell into the very confusion that I wished to avoid. I should simply have said "Positivism is irrelevant here" and had nothing to do with any attempts to relate my views to positivism in any way at all, positive or negative.

In our Rev. Mod. Phys. papers we should have stressed that the theory is in certain ways, radically different both in form and in content from the quantum theory. Of course it can be criticised as still too similar in form to the quantum theory (e.g., it uses words like "hilbert space"). Indeed, I always said that it is merely a preliminary exploration of new forms, and not to be taken very seriously in the usual sense as a definitive theory. What the theory does is to explore how a consistent description of individual contingent events can be made in a context previously treated in terms of quantum theory. I really did not regard it as mainly an effort to solve what is called the measurement problem. As I indicated in yesterday's letter this formulation was probably mostly yours. I went along with it, partly because I was not clear on the irrelevance of the term "quantum state" and partly because I didn't want to delay your thesis too much.

Now to return to the v. Neumann formulation. This was, in my view, largely the outcome of the wish to say that existing theories provide a complete description of all that there is to be described. It is this assumption that prevents one from admitting that quantum theory is only relevant in a statistical context, and that something entirely different in form, either from quantum theory or from classical theory may be relevant for individual events. In other words physicists said: "Quantum theory is evidently a statistical theory. If there is a theory of individual events, it can be nothing other than a classical type theory. And if this doesn't work, then we have to say that the present quantum theory, with its classical interpretation, is already a complete description of all that there is to be described. Otherwise we would have to admit that the laws of the individual are unknown in form and in content. And this too would be disturbing to be contemplated seriously."

So von Neumann's theory (and related theories) are largely attempts to prove that no fundamental change is needed if we are to be able to describe individual events. But these attempts failed. For v. Neumann showed that fundamental change is still needed; i.e., one has to invoke special laws for what is called a measurement process. And as I have indicated, this kind of law inevitably involves spreading confusion, that entangles and mixes everything up. So one might as well give up the notion that quantum theory is a complete description of all that there is to be described, and go on to a radically different theory, for a new context, such as that of individual events.

Further confusion was introduced by the v. Neumann form of "proving" that a classical theory of individual events was not compatible with quantum theory. Now, from the outset, v. Neumann (and those who follow him) have introduced special new laws for the "measurement process" (e.g., reduction of the wave packet). Now, the essential point is that no "classical" type theory can have "special" laws that apply whenever something is called a measurement, while "ordinary" laws apply whenever it is not called a measurement. So, at the very first step, one introduces something radically outside the framework of classical notions. Why should one then even entertain the notion that "classical" type hidden variables could be introduced in a consistent fashion? It is like saying that a given material is in a gaseous state, and then "proving" by complicated arguments that it cannot be considered to be a solid.

Indeed, even in my 1951 papers, it was brought out (at least implicitly) that the "hidden variables" of the whole universe were relevant to each "system". So the classical notion of analysis of the world into disjoint systems was not relevant. This was a radically new form of discourse, and I should have emphasized this much more than I did.

Similarly, in our Rev. Mod. Phys. paper, the equations of motion of each system depend on the projection operators, P', and these depend on the total context of the universe. To call what happens a measurement is a source of confusion. Rather, we introduced a radically new form of universal law, in which the notion of special laws for processes called measurements was shown to be irrelevant. Even to use the word "measurement" in this context was misleading, and entangled us in confusion.

So all along the line, I failed to realize that it was precisely the common language form of physicists that is the source and sustainer of confusion. That is why it is urgent to realise the futility of "criticizing" and "answering" people like v. Neumann, or J. and P. To do this, you must at least implicitly admit the relevance of their language forms. And once you do this, you are trapped. So, as far as I am concerned, from now on, my work has no relationship at all to v. Neumann, J. and P., or others who base their ideas on the term "quantum state" or something equivalent.

In yesterday's letter, I said that the best thing to do with the confused "problem" of measurement is just to drop it. In your letter you write as if it were the only aspect of physics that ever interested you. If I may I would like to ask you whether this is actually the case. Once you wrote of seeing how art, science, and human life are aspects of one whole. Now, it seems that all this has reduced to asking what certain physicists mean when they say they are talking about measurements. Could this be a result of the pressures of the unhealthy environment in the USA?

Best regards

Dave

March 3, 1969

Dear Jeff,

I hope that by now you have heard something about a job offer. A few weeks ago, Yale University called me by phone from the USA to ask about you. They indicated a serious interest, and I gave you a strong recommendation. I hope that they have made a decision to give you a position.

I understand the difficulty of your situation, and why it is so hard for you to look into all these questions just now. Also, as I recall, your wife should soon be having a second child. I hope that all this goes well, and that things will soon clear up for you and for your wife.

Let me say that our correspondence has been very valuable for me. I think we are clearing up many points. It is important to consider v. Neumann's line carefully, because it has been woven implicitly into the structure of almost all that is done with quantum theory. Every one of us is v. Neumann, insofar our work is interpreted with his language forms and his way of thinking.

I have tried to indicate that v. Neumann's line is deeply confused (which means that all who work in quantum theory are also deeply confused). Part of the difficulty is that he does not say explicitly that he is radically changing the theory, when he is in fact doing so. Some of his changes are probably valuable, but they become destructive, when one treats them as "essentially not different" from the situation as discussed by Bohr and Heisenberg.

In recent correspondence with Popper, I saw further evidence of v. Neumann's deep confusion. In criticizing an article of Birkhoff and v. Neumann, Popper calls attention to a statement, in which these authors cite as evidence for the "non distributive quantum logic" the experiment of an electron beam going through two slits and showing "interference phenomenon".

Popper then asks: "Why don't these conclusions apply to elephants as well as to electrons?" From what v. Neumann and Birkhoff say explicitly, there is no way to justify the notion that the non-distributive logic is not universal.

My answer to Popper was to call attention to v. Neumann's "cut" between the "classical" and "quantum" parts of the world. V. Neumann would probably have said, "Elephants are large objects, and should therefore be treated in terms of 'classical observables'".

But then Popper could have asked: "How large should an object be in order to be treated in terms of classical observables?" v. Neumann would probably then have given his own private or particular version of Heisenberg's analysis of measurements, leading to the uncertainty principle. In this version, the apparatus would be treated in terms of "classical observables", while the electron would require "non distributive logic".

But then, Popper could have said: "The apparatus is constituted of electrons. How can you justify discussing it in terms of ordinary distributive logic?"

At this stage, v. Neumann could have moved his "cut", so as to call the apparatus "non distributive and quantum mechanical". But on the other side of the "cut" would be a "classical meta-apparatus". And so it would go, until one reached "the mind of the observer" (who is, of course, von Neumann, in the last analysis).

Such a regress involves endless confusion. Finally, when the original apparatus is "classical and distributive", it would be part of the conditions of the "meta apparatus." Thus there is no way to prevent the "apparatus from observing the meta apparatus", when it is being said that the "meta apparatus is observing the apparatus." It becomes impossible to decide which is observing which, and this is already confusion.

Even if we leave all this aside, we fall into hopeless confusion, when we reach the mind of the observer, i.e., von Neumann. The latter says: "I know that the contents of my mind are parallel to those of the physical world." (The principle of psycho-physical parallelism). But Popper can ask: "How do you know that these contents are parallel?" Von Neumann can answer: "I know the truth about the relationship between the contents of my mind and the physical world." Popper can say: "How do you know it is the truth?" And so, their confusion can go on indefinitely.

You may say: "All this is only von Neumann's bad philosophy, and it doesn't mean very much." But I hope you see that it is not merely philosophy. In order to answer relevant questions, such as those raised by Popper, von Neumann cannot do other than bring in his "cut" between "classical" and "quantum" parts of the world. This "cut" is a guaranteed recipe for maintaining confusion. For whenever one tries "to back von Neumann up against a wall" so that what he says can be "pinned down", he escapes, by moving the "cut". As a result, he can never see that he is lost in confusion. And any physicist who is influenced by von Neumann's language is implicitly and tacitly trapped in this kind of confusion (unless he admits that his work is just "pure mathematics", or else a set of "handbook formula").

Besides discussing von Neumann's confusion, I have tried to indicate how you and I have very different relevance judgements. I feel that while formalisms can sometimes be useful, they are in general a barrier and a block to understanding. You feel on the contrary that they are the very essence of understanding, in the sense that the main use of informal discussion is to help us produce better formalisms.

Moreover you feel that the "measurement problem" is very relevant, whereas, in my view it is not relevant. I feel that the main point is that the "individual quantum process" is unknown, and that attempts to discuss it in terms of "measurement theory" confuse us, because they give us the impression that we "know something" about such processes, when we do not. What is actually needed is to develop a radical new theory of such processes.

Finally, it would seem that your own work would be clarified, if you would emphasize (to yourself) that our views are radically different. So when you comment on my views, what you are really doing is to propose a rather different set of views, which you believe can meet the general situation to which my own views were originally directed.

Best regards

Dave

March 12, 1969

Dear Jeff

Saral and I were very happy to receive news of the birth of your child, Tania. We offer the two of you congratulations and best wishes on this occasion. We were also very happy to hear of the job offer at Yale. It sounds very good.

I want to repeat that I have learned a lot in our recent correspondence. I hope that eventually it can lead us both into something new, something beyond the limits of discussing things like the "measurement problem", "formalism", etc.

Best regards

Dave

March 13, 1969

Dear Jeff

Let me congratulate you once again on the birth of your child, and on your new job at Yale. It all sounds very good.

I received your long letter. On the whole, communication between us appears to be fairly good, right now at least (and we hope, for the future too). I shall answer your letter in detail later. For the present let me consider only a few "high points".

First, I agree that we need a new kind of language form for the context initially relevated in the quantum theory. I have some notions on this score, and will soon prepare a brief paper, making specific suggestions along this line.

Secondly, you are of course right to emphasise that formal and informal are a whole, in which disjunctive analysis leading to union (or synthesis) is not relevant. However, to do so at this time and in the way in which you do is misleading. I am reminded of a recent discussion with an artist friend of mine, centering on how his department head (and much of the staff) did not appreciate the fact that students were different from what they used to be, with the result that proposed "reforms" of curriculum were beside the point, confusing the issue, and not meeting widespread student dissatisfaction. He asked if I could say something relevant to students and staff. I proposed to discuss topic of "change". The response of the head of the dep't was "Change is no problem you just change. Is change a problem to Bohm?"

This was quite a fantastic reply. The fact was that this man was not able to meet change in a proper way. Of course, if change had not been a "problem" to him, he would not have created the mess that is going on now in his department.

So this man was, in his content, speaking as if he stood above all these difficulties, and saw change with perfect clarity. But in his whole form of communication and action, he was trying to avoid admitting the fact of change, as well as the need for further basic change. What a remarkable disharmony of form and content here! In content, the man speaks deep truths, and in form, he denies them altogether.

The fact is that our whole language, scientific and otherwise, now emphasizes a very sharp separation of formal and informal aspects. If I were now to say "formal and informal are really one whole", it would be appropriate for someone to watch my actions and see whether they did in fact cohere with this content. If they did not, my general effect would have been to add confusion, so that, on the whole, it would be "worse than useless".

Isn't it actually fantastically difficult today to speak and write with no disjunction of formal and informal aspects? That is why I suggest the terms "formal form" and "informal form". These relevate or call attention to the actual disharmony that pervades almost all work in physics today. Perhaps later, when we are a bit aware of this disharmony, as it operates from day to day, from moment to moment, we will find it appropriate to talk in terms of the wholeness of formal and informal aspects of language.

Secondly, when you speak of making the form and content of yesterday's communication the content of today's communication, I feel that what you say is not too clear. Can we actually do this? I feel that the essence, the "heart and soul" of any communication is in the actual energy, the feeling, the unspecifiable subtlety and complexity of mental movement. So if today I talk "about" yesterday's communication, my content is very likely to be chosen arbitrarily, fortuitously, and in a rather superficial and irrelevant way. Evidently, I cannot make the "heart and soul" of yesterday's communication into the content of today's communication. That is really impossible.

Is it therefore not misleading to say that we make the form and content of yesterday's communication into the content of today's communication? What actually happens is that today's communication is basically novel, but either in a relevant or in an irrelevant way.

Thus, yesterday's communication may "work in me" today, to lead to new perceptions and new actions (which include the communication of novel ideas, for example). Or else, they may have left a residue of unresolved "problems" in me. These "problems" will also "work in me", to lead to novel discourse. In some cases (e.g. functional or technical contexts), this discourse will remove the problems by "solving" them. But more generally, at deeper theoretical or psychological levels, the "problems" cannot be solved, because in fact, they are "confusions" and not "problems". So then, our discourse will consist in the creation of novel forms of confusion. But whatever we do, we are not actually able to talk "about" yesterday's communication. That is what we may say that we are doing. But if you observe, you will see that we never actually manage to do it. We are actually "working out" the implications of yesterday's discourse in today's actions, which include communication as a special form.

Finally, you are of course right to say that Popper is badly confused "about" the quantum theory. But Popper's question was still relevant "Why indeed does von Neumann apply non distributive logic to electrons and not to elephants?" Von Neumann cannot give a coherent answer to this question. Of course, Popper's whole discussion is also not coherent. But part of the reason for this is that he is trying to criticize von Neumann. Anyone who regards what von Neumann says "about" physics as relevant must inevitably be confused, just because von Neumann makes no coherent statements that are relevant to physics.

More later.

Best regards

Dave

March 14, 1969

Dear Jeff

This is a continuation of yesterday's letter, in answer to yours of March 6.

First, let me amplify what I said about the distinction, formal-informal. In present day physics, the formal language includes terms like Hilbert space, operator, commutator, eigenvalue, eigenfunction, analytic function, poles of the S-Matrix, (SU3), current algebra, etc., etc. In experimental physics, the language is much less formal. It includes terms like track in a cloud chamber, density of droplets, click of a Geiger counter, curvature of the track, viscosity, electrical resistance, electric current readings, pressure, temperature, etc. In every day life, we have terms like houses, people, windows, light and shade, clouds, air, wind, sky, stars, body and mind, etc. For more poetic language, we have beauty, immensity, harmony, love, etc. What I wanted to call attention to was the tremendous difference between terms in theoretical physics and those of experimental physics. But this difference is very small when we compare theoretical terms to those of everyday life and of poetry.

Now, we tend to take it for granted that this disjunction of language forms is natural and inevitable, even perhaps desirable and useful. But what I am doing is to question both the necessity for and the value of such a disjunction. I would propose that language is basically poetry – i.e., the act of "poesis", which is, the "making" of a new order of perception and communication. The first "sharpening up and fixing" of poetry is ordinary everyday prose. The next stage is the language of experimental physics, then that of theoretical physics and mathematics. But even these are basically poetry. Thus, the value of a piece of mathematical work is judged largely by its beauty or elegance. Isn't this much the same as to say that it is a good piece of "poetry"? And it is clear that Einstein was basically a poet, as was also Bohr (though to a lesser extent). Without poetry, there is no truth in science or mathematics.

I hope you see that I am going far beyond your suggestion that mathematical formalism is merely a short-hand notation for ordinary language symbolism. Of course, you can use the "ordinary language" term vector in Hilbert space, and you can write in "shorthand" notation, V_i. In my terminology, both of these are parts of the formal language.

The informal language includes terms like air, sea, lakes, mountains, elephants, etc. (That is why I emphasized Popper's question so much. The von Neumann school is very incoherent in its discussion of such informal terms.)

Next, let me come to the term "unambiguous communication". What can this mean? The word unambiguous means "not ambiguous", or "the opposite of ambiguous". But what does the word "ambiguous" mean? Surely this word is itself one of the most "ambiguous" words in the language. That is to say, "ambiguous" means that something cannot be defined precisely and without the possibility of confusion. How can one obtain a precise definition of indefinite and a clear definition of confusion? Since "ambiguous" is undefinable, so also is "unambiguous". Therefore, the very use of the phrase "unambiguous communication" is in itself the source of "ambiguous communication". And if Bohr begins with such an ambiguous communication, how will he be able to communicate to us what he means by phrases, such as "The language of everyday life, suitably refined where necessary to that of classical physics, is the only possible form of unambiguous communication". This phrase is "pure poetry", but I am not sure that it is "good poetry".

I would say instead that the words "unambiguous communication" can at best indicate a certain kind of situation in which people communicate, and then engage in further perception, communication and action. If in these further perceptions, communications and actions, they do not get lost in confused problems, this fact is taken to mean that the communication has been "unambiguous".

But if there are such "problems", then people are inclined to raise questions, such as "What do you mean?", "What can I say in response to that?" and "What ought I to do?" This response indicates, of course, that the communication was "ambiguous". But other than through an awareness of such responses, there is no way to say whether a communication was ambiguous or not. After all, you can tell me all about K. and S. I can *say* "It is all perfectly clear and unambiguous." But when you try to discuss their work further with me, you will see that I didn't understand, and that your communications were in fact "ambiguous", at least as far as I was concerned. So there is no verbal way to specify unambiguously whether or not a particular communication is ambiguous (and therefore much of Bohr's discussion of this subject is a serious disharmony of form and content).

In view of the above discussion, I would suggest that the division "ambiguous-unambiguous" is largely irrelevant, in the context that is now of interest to us. Therefore, I would not wish to pursue further your discussion of formalism and "unambiguous communication". Rather, I would suggest going along lines indicated in yesterday's letter to you.

What actually happens when we discuss yesterday's communications is that these communications, working in us, have led to something new today. Thus, they may lead to novel perceptions and novel kinds of action, as well as to yet newer forms of communication. Or they may lead to technical functional problems, requiring a solution. Or they may lead to confused problems. In this latter case, we are led to novel forms of confusion, unless we see the confused nature of our "problems", and thus drop them as irrelevant. One form of confusion is just the effort to solve our "problems" by generalizing the language form, without introducing fundamental or radical changes in it. Your discussion of "communication about yesterday's communication" is indeed a perfect recipe for the indefinite continuation of such confusion.

Did Bohr in fact "generalize the language of classical physics" and indeed, is it possible to do this in a coherent manner? His introduction of "wholeness" of the form of the experimental conditions and the content of the experimental results is a fundamental and radical violation of the language of classical physics. Indeed, nothing in this latter language is so "unambiguous" as the assumption, tacit and explicit, that the world can be treated as a union of disjoint objects (in interaction). This assumption is a far deeper and more fundamental part of the classical language, than is the mathematical formalism (differential operators, position, momentum, etc.)

It is only if we suppose that Bohr regarded the essence of the classical language as the mathematical formalism that we could even utter the phrase "quantum theory is a generalisation of classical physics." But even here, one of the first steps in Bohr's principle of complementarity is to imply the irrelevance of the notion that formal classical terms (such as position and momentum) can correspond to precisely defined numerical values. So it is only in a rather "peculiar" sense that quantum theory can correctly be described as a generalisation of classical concepts. This sense is that certain formal classical terms (position and momentum) are still used descriptively, but not inferentially. That is to say, valid inferences about position and momentum now have to be made in terms of the quantum algorithm, which is relevant only statistically (again a highly "non-classical" notion). If we once admit that the words "unambiguous communication" are highly ambiguous, then we are left with no reason at all for indefinitely continuing such a procedure. Rather, we can drop this phrase as irrelevant, and simply start to explore new forms of communication, without too much concerning ourselves with the question of whether they are "unambiguous" or not.

I think that deeply we are more or less in agreement on this point. Thus, the "hidden variable theories" of 1951 and 1966 imply going beyond "generalisations of quantum theory". But here, I would add that quantum theory similarly goes beyond being a "generalisation of classical physics". It is only because of Bohr's mode of verbalization that we tend so heavily to relevate the similarities between classical and quantum physics, and then to irrelevate the differences. If we were to take the informal form of language as what is most basic, for example, we would say that "quantum language" is radically "incommensurable" with classical language, but has certain rather superficial formal similarities with the latter (i.e., quantum operators and classical functions can, in some sense, coherently be given the same names, at least to a certain limited extent). But to put it this way makes it clear that we don't regard these formal similarities as being all that important. Thus, we can feel free to go further (as we did in our 1966 paper) and set up formal equations that depend crucially on the total context. This is very radically different from what is taken to be a relevant possibility in the classical language form. Of course, we have to go a lot further still on such lines. Indeed we have hardly begun to make the changes that are needed.

I shall discuss all these questions further in later letters. Meanwhile, I would like to ask you to consider the possibility that you will "stand in for me" at the Illinois Symposium. The talk itself will be delivered on a tape recorder. But they will need someone to answer questions. At present, Yakir Aharanov has agreed to do this.

But actually, you are much more familiar with this work than is Aharanov. I shall suggest to Aharanov and to Suppe (who is organising the Symposium) that if they are agreeable, they should approach you with a request to do this. So don't be surprised if you receive such a request. But don't take any action on your own initiative. Wait until they get in touch with you (they may decide not to do this after all).

I think it would help a great deal if you could do this. But don't do it if you think it would create a lot of difficulty for you.

Best regards

Dave

P.S. In terms of my paper for the Illinois Symposium, physics is perception-communication. But it is not "unambiguous perception-communication". Indeed, as you remark, the very "life" of a communication is in those features that are called ambiguous. Nobody would bother even to discuss physics unless something in what he thought of as physics "bothered" him. That is to say, he senses "problems", which are the residues of the working out of yesterday's communications, as these are operating in him today. Yesterday's communications are always dead and gone. We are always left only with their residues which are different. Thus, when one heard a given explanation yesterday, one was not "bothered". The mind, working all through the day and night, "made a problem" out of yesterday's communication. This "problem" arose because of some "ambiguities" in yesterday's communications. In many cases, these "problems" are confused, and therefore irrelevant (so that they should properly be dropped). Their function has in fact been to indicate the irrelevance of the notions to which yesterday's communication have led. In other cases the "removal of the ambiguity" is effected by means of new perceptions, new discoveries, new inventions, new means of practical action. Or else, one engages in "generalizations of yesterday's communications" which generally lead to novel forms of confusion.

You may ask whether I am not talking "about" communication, e.g., by saying how it goes from one day to the next. The answer is that my discussion is intended to be regarded as "poesis". That is to say, if you understand it, I suggest that your communication will come into harmony with my discussion (and in other ways). Thus, my discussion is intended to participate in the "making" of harmony in communication. It is not intended mainly to "give information about communication".

Actually, all discussion of communication functions as "poesis". But if it is in the form of the "about", its function is to "make" a disharmonious kind of communication.

More generally, all science functions as "poesis". For each theoretical language form leads to experimental instruments and modes of perception that are appropriate to that form. The question is then whether the over-all function is harmonious or not.

Thus, we see the ending of the distinction of the observer and the world. For if our discourse is "poesis", and we are aware that it is "poesis", then our discourse is seen to participate in the "making" of the fact, and indeed, of the "world" in which we live. It is no longer relevant to consider ourselves (along with our discourse) as disjoint from the content of perception-communication.

Why is generalisation of yesterday's communication usually a novel form of confusion? The reason is that to generalize it, something new has to be added. This does not cohere with the rest of "yesterday's communication", as it has something incommensurable with the latter. Indeed, generally speaking, we do not generalize a communication. We change it and say that it is essentially the same as before (except that it has been "generalized"). But the fact is that it is different and not coherent with what came before. Hence the confusion.

March 18, 1969

Dear Jeff

I can now say a bit more about the technical points raised in your recent letter to me.

Firstly, my question has always been this: I can understand that K. and S. are doing a piece of "pure mathematics". I can also understand that they could regard this mathematics as relevant to physics, in the sense that it provides a set of "handbook formulae". For example, the engineers may have a set of formulae relating the breaking strength of materials to their chemical composition. Let us write such a formula as

$$B = f(C)$$

where B is the "breaking strength" and "C" are the parameters describing the chemical composition. In such a formula the meaning of terms like "breaking strength" and "chemical composition" need never be considered by the mathematicians. They can safely assume that the engineers understand what is meant by these terms (e.g., the engineer can see why B does not, for example, represent the strength of an elephant and C the composition of the food that he ate). Consider then that another engineer wrote a very different formula

$$B = Q(W)$$

where B is the breaking strength of materials, and W is something of a different structure from C (e.g. pressure and temperature). Then a mathematician could meaningfully prove that the formula, $B = Q(W)$ is not in any way to be regarded as a special case of formula of the kind, $B = f(C)$. Similarly, if K. and S. regard the "quantum algebra" as one set of handbook formulae and the "classical algebra" is another such set, then it might make sense to prove that the former is not a special case of the latter. Do you agree that this is what K. and S. have accomplished?

What I feel to be lacking in your discussion of K. and S. is some indication of how and why the formal algebraic terms can be relevant to the situation that is under consideration by physicists. I am serious here, as I feel that [the] only consistent formulation of quantum theory is that of Bohr, and that if one accepts his formulation, it is immediately evident on informal grounds that there can be no "hidden variables" of the kind considered by K. and S. To "prove" that there are no hidden variables would in terms of Bohr's notions be a source of confusion, similar to the "proof" that there are no "quadrangular triangles." (The very effort to "prove" this – i.e. to accept the relevance of the "problem of showing the impossibility of quadrangular triangles" is a sign of deep confusion in one's language forms and related modes of thought.)

Of course, I take it that when you discuss K. and S., you are not starting from Bohr's point of view, so that you are not "guilty" of the kind of confusion described above. But then, what informal description are you using? K. and S. are implicitly using von Neumann's informal description. You have stripped the proof of K. and S. of all of von Neumann's informal terms, such as "observables", "measurement", etc. (which imply his famous "cut" or disjunction of the world into quantum and classical parts.) But, as far as I can see, this leaves you with no informal terms at all. Thus, either you are regarding the work of K. and S. as "pure mathematics" or else, you are regarding it as a discussion of "handbook formulae". In either case, your own discussion would not be very relevant to physics.

The above indicates why I urgently suggested to you the need to make clear what is the informal language that you would use in connection with the formalism of K. and S. I do not feel that your paper actually does this.

Best regards

Dave

P.S. Elsewhere, you asked whether the distinction of formal form from informal form is formal or informal. I hope you can see now that this distinction is basically informal. Indeed, from this, it follows that the informal is the dominant side of any language form, even that which is called formal. Thus, when v. Neumann sets up his formal terms, it is always the informal form (which is one whole with von Neumann's character structure) that determines what will be the specific form of the formalism. Thus it is von Neumann's basically informal character structure that determines his predilection to give so much attention to logical forms, while he almost ignores the informal language that is actually dominating him, and making him what he is. He does this, in effect, by treating the informal language in a sloppy (or as you say "dishonest") way. As a result, the informal language (e.g., the "cut") is always leading to insoluble formal problems. So all that is done with the formalism is in the long run determined by the neglected and accidentally determined informal form. Therefore, von Neumann is the arbitrary, the fortuitous, the contradictory, and the confused. The irony is that he becomes this, just because he wants to reject it, and instead to be the purely logical, necessary, rational and clear. This contradiction is generally characteristic of scientists of our age. That is why v. Neumann's work has had such a wide "resonance".

Let us now consider another point. As I indicated in a previous letter, nobody ever "generalizes" what was said yesterday. Actually, he changes it in ways that are "incommensurable" or incompatible with what was said before. Therefore, even to say "We can't go back to classical physics" is likely to lead to confusion. Of course, in one sense, this is true. We can't go back to anything that existed yesterday, because today it is different, in an "irreversible" way. But what we can do is to consider the new relevance of classical physics, in the context of all has been learned over the past 100 years or so.

For example, in an article in Nuovo Cimento (about 1954 or 1955)[13] Mario Schonberg showed that, in some sense, quantum theory is a special case of classical physics, provided that we change what is meant by "classical physics" in a certain way. And, of course, you should have no objection to such changes, since you already accept the vast change implicit in replacing classical physics by a commuting algebra of "propositions" (whose fantastic implications have not even being considered by those who work along these lines). Let me say, therefore, that in a certain sense, Schonberg shows that quantum algebras are sub algebras of classical algebras. However he gives good reason for assuming that "classical algebras" are not commutative. So is he going back to "classical physics" or not? (He does use "classical phase space".)

I shall go into a further development of such views in later letters. For the present, I mention them only to point out how arbitrary is the K. and S. notion that "classical physics" can only be described in terms of commuting algebras.

March 20, 1969

Dear Jeff

I am glad that you accepted professor Suppe's invitation to "stand in" for me. Of course, you must present your own views and not mine. (In any case, as we both agree, to present "my views" is impossible, even for me, as they are different today from what they were yesterday.)

One important point. I was not refused a visa. The fact is that I did not apply for one, because I was, for various reasons, unable to get away at the time. Please tell this to Professor Suppe, and ask him to make it clear that the question of my getting a visa was never raised.

Professor Cohen's suggestion of a lecture tour sounds good to me. Please let him know that I am interested, but ask him if he can defer action on this, until certain questions have been clarified, with regard to my own personal affairs.

Now to come to the main content of your letter, let me say that, in my view, we are not basically in disagreement at all. However there has been some confusion in the effectiveness of the language with which we have been communicating.

Firstly, it was very unfortunate if my paper gave the impression of "coming down on Kuhn's side" with regard to paradigms. All I meant to say is that in a highly implicit sense, paradigms are special cases of the notion that science is basically perception-communication. Actually, I meant to suggest that paradigms are totally inadequate as a description of scientific research, except in so far as they contain a "germ" of the notion that science is perception-communication.

What I meant to say is that there never was paradigm, classical or quantum theoretical. Science was always in a movement of ceaseless and fundamental change,

[13]The article was published in 1957, i.e Schönberg (1957). Presumably Bohm remembered Schönberg working on the article in 1954/5 when he was in Brazil. See Talbot (2017) for more on Schönberg (a.k.a. Schenberg) – CT.

from moment to moment. The notion of a "quiet period of normal science" was never more than a "comforting illusion".

For example, consider the development of Hamilton-Jacobi theory. I say that informally and in germ, this theory contains notions (e.g. of waves) which were radically different from those suggested initially by Newton. Please recall, however, that in my view, the informal dominates the formal, even in the work of those people who relevate the formalism to appear to be basic (though evidently such people do not realize they are dominated by the informal notions that they largely ignore). Thus, the H.J. theory contained in essence many of the notions that later had to be forced on physicists by the basic experiments leading to the quantum theory. The proof of equivalence of H.J. theory and Newtonian Dynamics is, in my view, a side issue, which is not very relevant. It merely shows equivalence in some very limited formally defined context. The whole point of research is that a new idea relevates a new context. Thus, formalistically inclined physicists were, in a sense, perverting the scientific spirit, by trying to show that the H.J. theory was "in essence the same as Newton's theory". In other words, it is just the effort to construe Newton's theory and H.J. theory as contributions to the "same paradigm" that I want to characterize as confused.

Perhaps I could put it this way: My main point is that to assert that two different notions are "essentially the same" has generally been a source of confusion in scientific research (though it has on certain occasions been relevant to do this). So to "construe" the H.J. theory so that it is part of the "same paradigm" as Newton's theory is also confusion.

I hope you see from this that in my view, there is no paradigm at all. There is only a ceaseless flux of different theories.

Perhaps you will try to make this clear at the Symposium, if you have the opportunity.

Of course, it follows from what you say that "theories about theories" are also in ceaseless flux. Indeed, even what I say at this moment is in this movement of ceaseless flux. I do not wish to suggest that I am stating "eternal verities". I am only putting forth something in which, for the present, I see no contradiction, and which is, in the present context, both relevant and valid, in my view. I only ask what is your present response to what I say. The future must be left to take care of itself.

It follows then, that you and I are basically in agreement with regard to "history". This is always being "constructed anew", in the light of new relevance judgements.

Now, let me say something about the theories of malaria. I feel that you are not taking seriously enough the "third theory" that diseases are basically due to lack of resistance, induced by social and psychological causes. In my view, this notion will prevail in the long run, so that in 50 years or less, germs and virus theories will be as "dead as the Dodo".

What I mean to say is that when human beings are generally resistant to disease, it will no longer be very relevant to distinguish between say, malaria, typhoid fever, and the common cold. In other words it is only the word "malaria" that causes us to regard this as an "entity", distinct from "typhoid fever". Consider, for example, what is called "rheumatism". This word leads us to regard as an "entity" what is actually the result of a vast conglomeration of causes, mainly psychological (i.e., psychologically induced strains in muscle tissue). Similarly, the "common cold" has a vast conglomeration of causes, mainly those factors that weaken us and make us susceptible to it. To relevate certain germs as "the cause" of the disease is just a verbal device, that tends to confuse us.

So your description of how scientists would look at malaria (page 3 of your letter) is an example of what I mean to say. First, we begin with the notion that damp night air is causally relevant in producing malaria. But then we will discover that damp night air does not always produce malaria, and that malaria can exist even without damp night air (e.g., when the pools are so small that the air is not made humid by them). This leads us to question this notion of causal relevance. Then we suggest "microorganisms carried by mosquitoes that generally breed in pools of water are causally relevant". But then we find out that many people exposed to all sorts of microorganisms (including malaria) do not fall ill. This leads us to suspect that microorganisms are a side issue, and that lack of resistance is the basically relevant "causal factor".

A similar situation would prevail if someone were in the habit of stepping out of windows and hurting himself. He would say: "Gravity is causally relevant in breaking my bones. I therefore need special devices that strengthen my bones (e.g. metal braces and shock absorbers), so that will not be hurt when I step out of windows." But later, he might realize that what is basically relevant is his habit of stepping out of windows. By not going on with this habit, he ceases to break his bones, and still needs no special devices to "protect him from gravity". Similarly, we now say that it is the "malaria germs" that are making us ill. But it may be that it is because we live the way we do that these germs are able to hurt us. If we cease "stepping out of the window", we won't need pesticides, drugs and antibiotics.

I hope it is clear from the above that I am emphasizing that one's informal notions as to what is causally relevant dominate one's thinking about illness (or anything else). The empiricist scientists imagines that it is his experiments that dominate his conclusions. But the fact is that his experiments are strictly limited and controlled by what he is informally and tacitly regarding as causally relevant. Indeed, even to relevate a disease called malaria is tacitly a judgement of causal relevance. If men were generally resistant to disease, they would give little relevance to observed differences in forms of bacteria. Instead of saying "This man became ill because malaria germs entered his blood", they would say "This man became ill, because he was weakened by worry, stress, conflict, bad food, and other factors, so that he could no longer meet the normal hazards of life." It would be quite irrelevant that it happened to be a malaria germ that "got him". It could equally well have been an "accident" due to a "mistake in driving a motor car", caused by the fact that his mental alertness (along with his physical activity) were destroyed by his "way of life".

So I am saying that scientists quite naturally slip into radically different and "incommensurable" ways of thinking of what is causally relevant in a certain context. However, because of certain verbal habits of using language, built up over thousands of years, they tend to say "It is essentially the same 'entity' that we are talking about." In my view, there is no entity that could be called malaria. The word "malaria" relevates certain notions as to what is causally effective. When we go from the "damp night air theory" to the "germ theory", we have radically changed the meaning of the word "malaria". And when we go to the "lack of resistance" theory, we have even more radically changed this meaning. (E.g., the distinction between malaria and other diseases ceases to be basically relevant.)

So, instead of saying, as you do that the paradigm does not exist until we transcend it, I say that the notion of paradigm is basically irrelevant. What is relevant is the ceaseless change and flux of ideas, in their context of perception and action. Therefore, while it is true that scientists have "made" Newton's theory and H.J. theory "commensurable" by saying that this is the case, I emphasize that this mode of discourse is confusion and therefore not relevant.

Finally, with regard to "non-quadrangular triangle", what you say about computers may well be true. But it is my view that to compare human reason to a computer is, generally speaking, a way of getting lost in trivial irrelevancies. Can a computer make the kind of relevance perceptions that are needed, before we can see whether a theory is worth considering or not. Von Neumann's computers can direct our ballistic missiles. They can also direct us, in our little "games" of "escalation of conflict". But can they show us the irrelevant and confused notions that have caused us to resort to such computations in the first place? What I meant to indicate by the term "non-quadrangular triangle" is just that relevance perception the dominant question, and that formal logical "proofs", such as those given by the computer, are a rather minor side issue.

Best regards

Dave

P.S.

I would add that the heritage from ancient Greece tends to cause us to tremendously over-relevate the importance of "proofs". Actually, "proof" is a rather minor issue. The deepest thing in mathematics is to perceive the relevance of new questions, and the most elegant forms of mathematics are those that are self-evident, not needing "proofs" at all.

March 21, 1969

Dear Jeff,

I am supplementing yesterday's letter to you.

First, let me sum up by saying: In my view, it is not just "new paradigms" that are "incommensurable". Rather, each new idea contains something "incommensurable" with what comes before. Even a "flash of understanding" is in some ways "incommensurable" with "what the person who wrote the article had in mind". So, each moment is, in some ways "incommensurable" even with the moment that came before. However, it may be that what is "incommensurable" will reveal its full potentialities only in time. But if a scientists says: "It is all essentially the same", these potentialities will be revealed in the form of confusion.

Now, I want to discuss the "malaria example" a bit more. Is there an "entity" called malaria? If so, what could it be? What is its "essence"?

For example, do certain symptoms always indicate malaria? Evidently not. Most of the symptoms of malaria can be induced by drugs. And certainly, all of them can be induced psychosomatically. Vice versa, not everybody who "really has malaria" displays all the symptoms. So this does not answer our question, of defining the "essence" of malaria.

Evidently, malaria is a whole in which analysis is not relevant. Thus, if you have the "damp air theory", you say: "A sign of genuine malaria (and not psychosomatically induced symptoms, for example) is that the person was exposed to damp night air["]. If you have the germ theory, you say that a person really has malaria only when he has malarial microbes in his blood. If you hold to the theory that disease is mainly due to our "way of life" you would say that the difference between "psychosomatically induced malaria" and "bacterially induced malaria" is not very relevant, in the long run. Both such individuals are ill, because of the "way of life", as is the man who is prone to a motor car accidents, quarrelling with his wife, etc.

Of course, once a man has "fallen ill" you may give him treatment for "temporary alleviation of symptoms of disease". Quinine will "cure" him of "malarial symptoms" while tranquilizers will "cure" him of "quarrelling symptoms". But in this view, the man is still ill, after he has been "relieved" of certain dangerous symptoms. (Quarreling with people can, for example, be even more dangerous than malarial fever.) The man who has been "cured of malaria" still follows his destructive "way of life", so that he is still prone to get ill from other causes (common colds, quarrels leading to heart ailments, motor car accidents, etc.). Indeed, he is still very ill, even when his malarial symptoms have vanished.

I thus wanted to indicate that we have three notions of "illness" which are not completely "commensurable". Indeed, what is common to all three is little more than a word, i.e., "illness". In the first notion, there are many kinds of illness. One of these, called malaria, is caused by breathing in damp night air. The second notion is similar to the first, in that there are still many kinds of illness. But it is different in that the illness called malaria is caused by germs, injected into the blood stream by mosquitoes. In the third notion there is only one illness, which is our "way of life". Thermonuclear bombs and malaria are symptoms of the illness.

The question at issue is then: "What does it mean to say that a man suffers from a disease?" In case A, it means that he has breathed in something poisonous. Thus, to be exposed to poison gases and to malaria are regarded as basically similar. One could equally well say: "I have been poisoned by the noxious vapours that hang about this place in the night." In case B, we say: "I have been infected by living organisms." So even here, the two theories really imply something "incommensurable". For "poisoning" and "infection" are not entirely reducible to some "common measure". And in case C, we say: "I have been weakened by my 'way of life' so that I cannot meet normal hazards, such as bacteria, motor car driving, and the difficulties of establishing harmonious human relationships."

Now to come to another point. Of course, it is a fact that physicists have relevated certain paradigms, and that this is the result of a kind of choice. Actually, it is not a "decision", since at no stage does anybody ever consciously decide to do this. Rather, it is a kind of automatic, unconscious choice. But my whole point was to say that this kind of approach leads inevitably to confusion.

This is an important way in which I differ from Kuhn. He takes the attitude that "the way scientists work" is an "objective fact" that he can "look at and describe", but that is not influenced or influenceable by what is said about it. I take the view that "philosophy of science" is poetry (i.e. "poesis"). If what I say moves people to do science differently, then it helps "make" a new kind of science. Actually, whoever does "philosophy of science" is doing "poetry". But if he treats it as "prose" (i.e. "objective scientific fact"), it will just be confused poetry, which will move people toward confusion.

So I am suggesting that if you will "listen to my poetry", you will perhaps see its truth, and thus be "moved" by it. If it doesn't "move" anyone, it has no real value. But it makes no sense to regard it as "propaganda". I don't want to manipulate people's minds. Rather, I want them to "listen" to what I say, and if it is not true, to tell me so. If it is true, and people see it to be true, then they will be "moved" whether they like it or not, (as a sudden perception of a beautiful lake in the mountains could "move" you).

To come back to the "malaria example" for a moment, what I had in mind is that the theory that disease is due to our "way of life" is analogous to a "quantum theory of disease" (in the sense that disease is described in its wholeness). The bacterial theory is analogous to "classical theory of disease" (i.e., little "particles" called germs, moving in "well defined orbits" are the "carriers of disease"). The "bad air" theory is analogous to Medieval or Greek (or alchemical) physics. (Which used notions of "vapors", "spirits", "noble metals" etc.). So you are wrong to say that my views apply only to the quantum-classical development.

Finally, I must say that I could not understand the "computer proof" that the angles at the base of an isosceles triangle are equal. Perhaps it is (as you suggest) that my understanding of mathematics is "peculiar". (I would say it is peculiar in the sense of not having been so completely "computerized" as is the general thinking of modern mathematicians and scientists.)

When I look over this "proof", I cannot help but get the feeling that it is overlooking something very deep. Probably, the difficulty lies in articulating the full meaning of the word "congruence". For example, in non-Euclidean geometry, the proof that you cite does not necessarily follow (E.g., AB and BC could both be geodesics of equal "length", but of different "shape".)* Nowhere in this "proof" does one state the Euclidean character of the space. On the other hand the "standard proof" involves the idea of rotation, perpendicularity, etc., and thus takes full account of the Euclidean character of the space.

---

*E.g., in Riemannian Geometry of finite lines, all this follows. In non-Riemannian geometry the conclusions are even stronger.

If you look carefully at the "computer proof" you see it has all sorts of tacit assumptions in it: Some of these are:

1. All lines of equal length are congruent. This is *proved* in Euclidean geometry by considering the operation of translating the lines so that they have a common vertex, and rotating one into the other.
2. The line BC is congruent to the line CB. This is proved in Euclidean geometry by a 180° rotation about the midpoint of the line. (It is not true in non-Euclidean geometry, i.e. not in general.)

If you add up these two steps, it is equivalent to the "schoolboy proof" of bisecting the third (unequal) angle. So where is the great advantage of the "computer proof"?

I think this illustrates a basic feature of the "computer approach" to things. It is that the computer operator deludes himself into believing that he can "tell" the computer something, and that the computer can then "tell" him something new. Actually, the computer operator is the one who inserts what is new, but he does so unconsciously and subliminally (otherwise, he would see the triviality of what he is doing). As a result, such novelty is arbitrary and fortuitous, hence generally confused. It is very similar to what the followers of von Neumann have to do about the experimental conditions, in the way described in my talk. Therefore, the disease of "von Neumannitis" has now "mutated" to become the disease of "computeritis". But ultimately, of course, this is all just a symptom of our basic disease, i.e., our "way of life". So malaria and the "computer proof about the isosceles triangle" are all symptoms of one illness.

Now, I would like you to listen to some "poetry" on the subject of computers and mathematicians. Of course, it is a fact that mathematicians have come to regard mathematics as syntactical, formal, etc. It is a fact that mathematicians believe that they have thus "stripped" mathematics of all that is "informal" and "vague". But as I have indicated, what they actually have done is to make themselves completely unaware of the informal notions that dominate every formal step, in a way that is confused (just because the informal is now entirely fortuitous, as a result of it being sub-liminal). Therefore, in my view, modern mathematicians are destroying mathematics. Insofar as mathematics contributes to man's life, they are destroying man. For example, they are lending their authority to the delusion that it is possible to engage in formal, syntactical activities, that have no memory beyond themselves. Thus, the man who makes hydrogen bombs can say: "I am only following the rules of society. I will do whatever is asked of me, according to the rules. I am like a computer in this regard." But of course, he is dominated by vague informal notions, such as fear of being called a non-conformist, fear of not being patriotic, fear of attack by another country, fear of not having a job, etc.

Similarly, when a mathematician formalizes "my informal ideas" about "non quadrangular triangle", I wonder if he too is not being dominated by vague informal notions, such as: "If I formalise things, I will be nice and certain that they are right." This is what I called the heritage of ancient Greece. They not only gave us the idea of proof, but also the idea that proof yields "absolute certainties" and "eternal verities". But if proofs depend on vague informal notions of all kinds, they can be dangerous delusions, leading us to believe in the "certain truth" of what is actually false or confused.

Best regards

Dave

P.S. Perhaps I could add that in my talk, it is implicit that there never was a quantum paradigm. Thus, I cite the Heisenberg-Bohr view of quantum theory – then that of Schrödinger, Pauli, Jordan, Dirac, von Neumann, von Neumann and Birkhoff, Field Theory, Renormalized field theory, S matrix theory, SU3 and similar group theories, current algebras. These are all radically different, both formally and informally. It is only by saying that they were all "quantum theory", thus implying that they were all "essentially the same" that physicists could intensely relevate their superficial similarities and thus be blinded to their deep and fundamental differences.

Likewise, there never was a classical paradigm. Newton's original ideas were elaborated and changed fundamentally by some Jesuit priest (Boscovitch), in the direction of thoroughgoing atomism (see L. Whyte for a discussion of this point).[10] Laplace, D'Alembert, etc., changed things again in a radical way (principle of virtual work) Lagrange, Hamilton et al brought in wave theory. Other physicists brought in ether theories, "gyroscopic ether molecules", vortex models of the atom. Then came Rutherford's planetary model of the atom. Think how radically different all these theories are. Did Kelvin's vortex model treat of the "same entity" as Newton, Laplace, Lagrange, Hamilton, etc. I say it was very different. Indeed, if Bohr had not come along, hydrodynamicists would soon have found in such models a notion of "stable limit cycles" which could have at least qualitatively explained spectral lines and the stability of atoms, and perhaps even quantitatively. The success of such a theory would have taken most of the impetus out of developments of Bohr's emphasis on language and wholeness (perhaps out of Einstein's point of view also). So it is in large measure fortuitous that we now take relativity and quantum theory, as if they were paradigms. What is actually happening here is just unceasing change, in which in each step, novel features emerge that are "incommensurable" with what came before, while certain features remain similar to what came before (though not necessarily "commensurable" with the latter).

[There follows a type-written document, "Some Comments on Maynard-Smith's Contribution"[11] – CT.]

---

[10]See Whyte (1961) – CT.

[11]Published in Waddington (1969), pp. 98–105. See Introduction, p. 3., n 10 – CT.

References

Bastin, T. (Ed.). (1971). *Quantum theory and beyond.* Cambridge: Cambridge University Press.
Bohm, D. (1952a). A suggested interpretation of the quantum theory in terms of "hidden" variables i. *Physical Review 85*(2), 166–179.
Bohm, D. (1952b). A suggested interpretation of the quantum theory in terms of "hidden" variables ii. *Physical Review*, *85*(2), 180–193.
Bohm, D., & Bub, J. (1966a). A Proposed Solution of the Measurement Problem in Quantum Mechanics by a Hidden Variable Theory. *Reviews of Modern Physics*, *38*(3), 453–469.
Bohm, D., & Bub, J. (1966b). A refutation of proof by Jauch and Piron that hidden variables can be excluded in quantum mechanics. *Reviews of Modern Physics*, *38*(3), 470–475.
Schönberg, M. (1957). Quantum kinematics and geometry. *Nuovo Cimento, Supp. VI*, 356–380.
Suppe, F. (Ed.). (1974). *The structure of scientific theories*. Illinois: University of Illinois Press.
Talbot, C. (Ed.). (2017). *David Bohm: Causality and chance, letters to three women*. Berlin: Springer.
Waddington, C. H. (Ed.). (1969). *Towards a theoretical biology 2*. Edinburgh: Edinburgh University Press.
Whyte, L. L. (Ed.). (1961). *Roger Joseph Boscovich SJ FRS, 1711–1787: Studies of his life and work on the 250th anniversary of his birth.* London: George Allen and Unwin.
Wigner, E. P. (1969). Are we machines? *Proceedings of the American Philosophical Society*, *113*, 95–101.

Chapter 9
Folder C137. Kuhn and Incommensurability. Summary on the "Quantum". April–August 1969

April 19, 1969

Dear Jeff

I just received your letter. First, let me say that you did a very good job at the Symposium,[1] considering the conditions under which you had to work. This is not merely my own opinion. Let me quote from a letter of Suppe. "Your session went quite well . . . the discussion of your paper was among the best of any session. Much of the credit is due to Jeffrey Bub, who did a very impressive job of standing in for you. Professor Causey's comment . . . was quite unsatisfactory. As Jeffrey pointed out, he completely missed the point of the paper . . . Fortunately, Jeffrey also saw this and so presented a short formal reply. This focussed the subsequent discussion, and the irrelevancies of Causey's commentary were subsequently ignored."

I would say that the situation you describe at the Symposium was not compatible with communication. So don't blame yourself because you didn't reach perfection. More generally, we can learn from this experience that new ideas have to be communicated in the presence of only a few people at a time, under very informal circumstances. My real "mistake" was to present such a paper at such a meeting. But, after all, a "mistake" is actually "mistaken", only if one repeats it. Confusion is confusion only when repeated. We should change the word to "reconfusion" and let the word "confuse" go back to its original meaning, i.e. to melt or flow together, in amalgamation.

[1] See C136, p. 263, n 6. The taped talk by Bohm, "Science as Perception-Communication" is given in Suppe (1974), pp. 374–391. A criticism by Professor R.L. Causey is given on pp. 392–401, a reply to Causey by Bub on pp. 402–408, and a discussion including Bub, Kuhn, Causey, Achinstein, Putnam, van Fraassen and (Patrick) Suppes on pp. 409–419. A "Reply to Discussion" by Bohm was added later, pp. 420–423—CT.

C. Talbot (ed.), *David Bohm's Critique of Modern Physics*,
https://doi.org/10.1007/978-3-030-45537-8_9

Now, for a few general comments on your letter.

Firstly, let me say that I would reformulate my basic thesis somewhat, to make it more clear. Let us drop the word "incommensurable" as "reconfused". I shall instead use words like incompatible, incoherent or disharmonious . I now introduce the notion that each theory has what are regarded as centrally relevant or pivotal aspects, while it has others that are regarded as of peripheral or secondary relevance. Thus, in the "bad air" theory of malaria, the centrally relevant notion is that of "poisoning" due to "evil vapours, miasmic principles, etc." which are exuded into the damp night air. In the microorganic theory of malaria, the centrally relevant feature is that the disease is carried by small particles, which are injected into the blood by mosquitoes. These particles (microorganisms) multiply and produce substances that are poisonous. But in this theory, such poisoning is of peripheral relevance. In the psycho-social theory, it is admitted that there are particles carried by mosquitoes, that can multiply and poison the blood. But these are regarded as having peripheral relevance. The centrally relevant feature is that man has for thousands of years lived disharmoniously. This has produced a general lack of resistance to bacteria, along with other "illnesses", such as the tendency to be accident prone, the tendency to get heart disease from ambition and quarrelsomeness, the tendency to suffer from wars, concentration camps, mass holocausts, etc. So the centrally relevant feature is man's disharmony, from which he suffers in a vast number of ways. Malaria is surely one of these ways. In this view, it doesn't matter much whether you die of malaria, of traffic accidents, concentration camps, or thermonuclear bombs.

What happens here is that the centrally relevant features of different theories are not coherent or compatible with each other. Rather, the centrally relevant feature of one theory (bacterial) is coherent only with peripherally relevant features of the next theory (psycho-social).

How can it be known which feature of a theory is centrally relevant? This is an *artistic*, *poetic* question not capable of being answered by "looking at nature only". Thus 19th century physicists implicitly assumed that the centrally relevant features of classical physics were in the *formalisms* of differential equations. So they concluded that the Hamilton–Jacobi theory was "essentially the same" as Newton's theory. I am suggesting that this is "mediocre poetry". I feel that the informal terms, such as "wave" or "particle" are centrally relevant, as the "germs" of an unlimited number of possible theories. So the "proof" that H.J. and Newton are equivalent *formally* under certain limited conditions is only of peripheral relevance. In my view, formalisms change with kaleidoscopic ease. Classical physics went through a long series of radically different formalisms, and so did quantum theory. Indeed, I would say that formalisms are a "dime-a-dozen". They are therefore generally of peripheral relevance. (E.g. like the hundreds of sketches by an architect, all of which play a momentary roll in the realisation of the poetic and creative insight of a really great architect, like Michelangelo.)

Incidentally, you say that Kuhn objected to "historical inadequacies in my account of quantum theory". Could you tell me what these are? If he doesn't accept my statement that there has been a kaleidoscopic succession of formalisms in quantum theory, this has nothing to do with "historical fact". Rather, most physicists are "poetically" saying that the differences between these formalisms are only of peripheral relevance. I am "poetically" saying that they are of central relevance. (This shows that "history is basically poetry").

To come back to the "malaria example", let me remind you that we (at least implicitly) distinguish between a disease and it's symptoms. Indeed, the function of a doctor is to "heal" a person, or to restore him to "health". I looked up the word "heal" in the dictionary. It has the same root as the Anglo-Saxon word "hale", which means "whole". So it turns out that Bohr and malaria are concerned with the same question, i.e. wholeness! When a man is ill, he is "not whole". The doctor who "heals" him is "making him whole". Health means wholeness.

This implies that the content of medical science is extremely ambiguous indeed. Different epochs, in different societies, with different notions of wholeness, can have very different kinds of medical science. These kinds will generally be incompatible or incoherent in their centrally relevant features, but what is centrally relevant in one may be accepted as peripherally relevant in another.

Our own epoch is characterized by the notion that medical science is concerned mainly with establishing wholeness or harmony at the physico-chemical level of man's functioning. Other sciences, (psychology, sociology, etc.) aim to establish wholeness or harmony at other levels of human function. All these sciences can interact (as indicated by terms like "psycho-somatic", "psycho-social" etc., etc.) But broadly it is supposed that one can relevantly begin by treating them as distinct and separate. Thus, we may say that modern science is characterised by the notion of "fragmentary wholeness". In physics, this was manifested in the classical atomic theory, and in medicine (or biology) it is the theory that disease, heredity, etc., are all determined by various particles (viruses, bacteria, DNA molecules, etc.).

Bohr came out with the notion that "fragmentary wholeness" is not relevant in the context of the "quantum". There is emerging slowly a new view of medicine, in which also the notion of "fragmentary wholeness" is dropped as irrelevant.

Now to come to your more detailed comments. Most of these involve "reconfusion" about the question of how different aspects of a theory are to be "poetically emphasized". Thus, if one uses the "mediocre poetry" of modern science, one can say (with Putnam) that not every theoretical change involves a change of methodology. But the fact is that each different theory calls for some change in the method of experimentation. We usually tend to say that these changes are "small", "unimportant" etc. But this is a relevance judgement, therefore poetry, and not "observation of the facts of nature". Who can tell whether a new instrument of experimentation (e.g. the laser) is not the "germ" of a vast new kind of methodology? Is it not better to keep away from such relevance judgements? What function do they really have? I suggest that the only function is "to make us feel that we are in secure and familiar territory." Therefore they are irrelevant, and perhaps worse than useless (since they may lead us to overlook "germs" of fundamental change.

With regard to Kuhn's comments I was not trying to say "What is implies what ought to be". If Kuhn says "That is not how it was", I say that this reply is possible only because of certain "poetic" relevance judgements of Kuhn's. In otherwords, Kuhn cannot tell us "what was" except in terms of his "poetry". If you talk in terms of my "poetry", you have a different notion of "what was". So the real argument is whether you will listen to my "poems" or to Kuhn's "poems". Which of these kinds of "poetry" actually "moves" you? (i.e. history is poetry).

Implicitly I was not using history as evidence. Rather I was using it "poetically" to try to illustrate my point. Thus, one can tell children legendary stories of "great historical figures". Like tribal mythologies these cannot relevantly be taken as literal statements of "what actually happened." The trouble is that "historians of science" believe that they are telling us "what actually happened". They too can do nothing but write "poetry and mythology", but they "reconfuse" the issue by calling it "historical fact". Perhaps in future articles I should try to make this clearer.

I hope you see that the malaria example is not merely a "trivial" example of "wholeness", "meaning change", etc.

You say that "something new is called for". It is not enough to say that "poetry is necessary". One has actually to produce the "new poetry". All this is correct, of course But I feel that one has to begin with the "germs of new poetry". I think that the malaria example was such a "germ". By bringing in the fact that "health" means "wholeness", the "germ" is made to begin to develop. All this was implicit in what I said originally. But as you say, most scientists have ceased to be able to be "poets". (More accurately, their "poetry" is routinised and mediocre & therefore equivalent to a set of "cliches"). So one has to write something that awakens the creative energy of those who listen or read. How is this to be done?

I think you see that my intention in using historical examples was just the same as yours – i.e. to relevate certain things. However, this relevation is neither epistemological nor ontological in function. Rather it is to provide a kind of "imagery" for communicating the centrally relevant content.

Finally, it is amusing to read of the "mystique" around my name, as an ancient Central European sage.

Best regards

Dave

April 21, 1969

Dear Jeff

This is a brief addition to yesterday's letter, where I discussed the notion that health means wholeness.

It occurred to me that every city has a Department of Health (or Public Health). This really means a "Department of Wholeness". In this phrase, one sees very vividly what is wrong with our civilisation. The words "Department of Wholeness" are a contradiction, a disharmony of form and content, which in themselves constitute a source of "disharmonious wholeness" in society.

What is really meant by "Health Dep't" is "Dep't of Physical and Chemical Wholeness". The psychologist is in the "Dep't of Mental Wholeness" (though now he may be in the "Dep't of Psycho-somatic Wholeness"). The sociologist is in the "Dep't of Social Wholeness". The priest is in the "Dep't of Spiritual Wholeness."

But this very approach that splits wholeness into "Departments" is man's illness. Wholeness is relevant only in the whole of life, and not in the form of "departments". So it has no meaning to "heal a man in part". And since each man is social as well as individual, any man who lives in a "sick" society participates in this "sickness". The word corruption means to co-rupt. "Rupt" is to break up (i.e., disrupt, erupt, etc.) Co-rupt means to break up the whole into fragments. This is the nature of our "Departmentalised" society, which produces the Departmentalised (i.e., Co-rupt) kind of human being. In other words, to say "Department of Wholeness" is to utter the formula for universal corruption.

From this, it follows that it is without meaning to try to develop a special subject, called Wholeness in Physics. Unless we begin with notions as broad as "Health in General" (i.e., Wholeness as a Whole), we shall fall into co-ruption once again. In a way, we cannot relevantly separate "Wholeness in Physics" from "Health into Physics". Physics is now (like almost everything else) fragmented and therefore co-rupt.

With regard to poetry, you can't begin with an epic. Begin with one or two lines of it, and go on from there.

Best regards

Dave

May 3 1969

Dear Jeff

I hope that by now you have begun to digest the results of the symposium. I can see that it is not likely to be worth while to address such a gathering again, as communication is very restricted in the situation that you described. You probably did as well as could be hoped for in such an unfavorable situation.

I thought I would try to wind up our discussion of the von Neumann approach to quantum theory (if this can now be done). Firstly, von Neumann's theory is not properly an object of discourse. Actually what we are doing is to continue, articulate and extend the content implicitly called up in our minds as we read von Neumann's papers. We are also analyzing and criticizing this extended content, revealing contradictions, and perhaps some aspects that seem to be worth preserving, in some modified form, in new theories.

Now, I can only tell you what happens to me as I do this sort of thing, in the hope that to do this communicates something relevant to you.

When I consider von Neumann's way of combining formal and informal terms, I get a sense of wild spinning confusion. The informal terms completely disorientate whatever I try to do with the formalism. His use of words like "observable", "observer", "instrument", "classical observables", etc., bring in unknowable and uncontrollable implications, which mix in and fuse together (i.e., con fuse) with whatever I try to do with the formalism.

When I try to do what you advocate, and divest von Neumann (or Kochen and Specker) of all their informal terms, then I am left with pure mathematics (i.e., it isn't even capable of being called a "handbook formula"). I have terms like algebra, commutator, embedding, mapping, etc.. I trust K. and S. to prove that von Neumann's algebra, (inspired by his consideration of what physicists did in the activity called "quantum theorising") cannot be embedded into a commutative algebra. But if there are no words like "observable", "instrument", etc., one does not see how this is to be related to words like "hidden variables". Somehow, at least implicitly, you have to suppose that algebraic terms correspond to "observables", or else you can't even begin to connect the algebra with physics. If such an implication is present in your work, even in germ, then this is the "thin edge of the wedge" for bringing in the wild spinning confusion that I feel to be characteristic of a proper response to von Neuman's articles.

In my Illinois talk, I showed, in an informal discussion that the wave function is part of a description, in which are also involved the description of the experimental conditions and the experimental results. This description is a whole, in which analysis into disjoint components is not relevant. It is thus already clear that the notion of a distribution that is independent of these conditions cannot be significant, in this context. I do not see why we need to "chew the question over", by going on with K. and S. It is already clear that the whole quantum formalism is only a very special mode of dealing with such a description, and that entirely new approaches are available (which I am now looking into). To go on with K. and S. seems to be inhibiting, rather than helpful, (at least to me). For it must bring in the confusion of von Neumann, as soon as it goes even as much as ϵ beyond "pure mathematics".

Best regards

Dave

[The following document and the letter following it are not dated. They are presented in the order given in the archives—CT.]

SUPPLEMENTARY NOTE

This is a supplementary note, aimed (hopefully) at making the situation around the "quantum" clearer.

Firstly, with regard to the discussion of what I call the Bohr–Heisenberg view (pages [space in original]), let me say that in a purely informal way, we can see that a classical type explanation in terms of hidden variables is not possible. For, as pointed out in my talk,[2] the description of the observed object is amalgamated with a description of the form of the experimental conditions and the content of the experimental results, in such a way that analysis into disjoint components is not relevant. So it has no meaning to say that the observed object exists in a state that can be described apart from the experimental conditions. But the essence of a classical type theory is that the observed object is describable, in a way that has no essential relationship to the experimental conditions (i.e. these drop out as mere intermediary links of inference between the experimental results and the state of the observed object). Therefore, it is already clear that no classical type theory can be relevant here. And in my 1951 papers,[3] this came out in the fact that the "hidden variables" involved the apparatus, hence the experimental conditions (a similar involvement was implicit in our Rev. Mod. Phys. papers).[4]

Since it is not relevant to assume that experiments or "measurement processes" obey special laws of their own, this implies that a "hidden variable" explanation means that the whole universe is in principle involved in each aspect or part. In this way, Bohr's notion of wholeness of the form of the experimental conditions and the content of the experimental results finds a parallel, in that each parameter depends for its meaning and relevance, in the totality of parameters.

Of course, neither my 1951 papers nor our Rev. Mod. Phys. papers constitute fully adequate treatments of this situation. I regard them both as preliminary exploratory steps, useful perhaps in that they may prepare our minds for the realization that something radically new and different is relevant, in this context.

Now, it seems to me that the above informal discussion makes the situation concerning hidden variables clear. Indeed, in some intuitive sense, it is more or less what I had in mind since 1951 and before, though in my explicit language I became entangled in confusion because I did not see the irrelevance of terms like "quantum state" used by v. Neumann and by most other physicists.

However, I now see also that most physicists are trained and conditioned to believe that they really understand something only when they have expressed it in terms of formal equations. For example, with Hiley, I noticed this quite often. Time and time again, I used to make informal statements that were perfectly clear to me. But Hiley seemed puzzled, and this puzzlement was removed, only when I added four or five equations, related by formal reasoning. (Since then, I think he has begun to understand the irrelevance of these equations). It is similar to what happened when I wrote my book. The conclusions were often self evident informally, but physicists were conditioned not to believe them, unless they were preceded by a canonical number of equations. These latter somehow seem to make physicists "feel better" about the matter, because they "know" that "what happens in logically connected equations must be true", while informal statements are, by comparison, "unreliable".

[2] See p. 301, n 1 above—CT.

[3] i.e. Bohm (1952a, b)—CT.

[4] Bohm and Bub (1966a, b)—CT.

Now, if people like v. Neumann feel that "they really understand" only after they have placed, let us say, fifteen or twenty equations between the premises and the conclusions, I would have no fundamental objection to them doing this. The real difficulty is, however, that while they are doing this, their attention (and ours) becomes so focused on the formal logical reasoning that one forgets about the informal premises almost altogether. Thus, it appears that these equations themselves constitute "truth" and "understanding". But because one ceases to attend to the informal premises, these are left "freely floating", so that they can wave about in an arbitrary and fortuitous manner, leading to wild confusion in what is being done with the formal logical chains of reasoning.

I think this has been brought out in my talk[5] (pages [space in original]). Von Neumann pays no attention to the experimental conditions, and indeed dismisses them tacitly as irrelevant, when he talks in terms of a "cut" between classical and quantum parts of the world. Nevertheless, these conditions are still relevant, and they come in "by the back door", whenever an actual experimental situation has to be discussed.

Perhaps I could put it this way. As Bohr has shown, the wave function can consistently be regarded as having only statistical significance, and as not having any significance at all, for individual phenomena. But, of course, these phenomena do take place, and one can then ask "How do we deal with these phenomena?" Bohr is satisfied to say that our language is such as to make this question irrelevant and meaningless. Other physicists, believing that they have understood Bohr, actually change his views by saying: "The wave function is the most complete description of all that there is to be described. Therefore, insofar as the individual system has describable features, these must be described in terms of the wave function".

For example, a uranium nucleus "has one wave function" before it decays. After it is observed to decay, it "has another wave function." Thus, the wave function has been said to undergo a non-unitary transformation. But the "laws" of physics say that the transformation has to be unitary. Therefore, one is led implicitly to treat the situation as if whatever is called a measurement has "special laws of its own".

Now to do this is to become entangled in a fantastic quagmire of confused questions. I indicated some of these in certain letters (e.g. the need for "metalaws" of "metameasurements" etc., going on to the mind of the observer). Here, I shall draw attention to yet further aspects of this confusion.

What happens, in effect, with a von Neumann type theory, is that in an implicit and tacit way, the experimental conditions become involved in the "laws of the measurement process". For in the last analysis, the "non-unitary process of collapse" depends on these conditions (i.e. whether it "collapses" to a "state in which x is diagonal" or "a state in which p is diagonal" is determined by these conditions). But because these conditions are not mentioned explicitly, they are taken into account only tacitly and implicitly, in a sub-liminal way. Which features are taken into account depend on fortuitous choices by the individual physicist, which generally do not agree with what is done by other physicists. Thus, in a certain sense, the

[5] See p. 301, n 1 above—CT.

"laws of the measurement process" do come to depend on the "mind of the physicist who is talking about them." For it is the "state of this mind" that, in each concrete case, determines the actual content of the "laws of collapse".

Of course, all this is sheer confusion and irrationality. But because of the feeling that "equations connected by valid formal logic are truths" one can fail to notice that the informal premises are just as mixed up as are the premises of a corporation that uses the most modern computers to project a better "image", and to persuade people to buy more of products that are basically of no use to them.

I say therefore that the formalism is not merely a harmless "comfort" to those physicists who feel lost, unless every step is mapped out in a formal logical way. Rather, it is acting as a genuinely destructive factor. For it is self-evident from the very first step, that the "quantum" situation introduces informal notions that are not compatible with classical type theories. One can see this directly and informally by considering the over all experimental situation, as Bohr did. Or if one uses the v. Neumann type of language, one can see that the laws ascribed to the "measurement process" involve the experimental conditions in an inseparable way. No classical type theory can have the experimental conditions involved inextricably in the statement of its "laws of motion". So the impossibility of classical type hidden variables is contained directly and self evidently in the very first step, in the informal premises of the theory. Any treatment which makes this seem to emerge only after a long chain of subtle formal reasoning is in itself misleading and confusing. It consists of what might be called "hiding the obvious behind the smokescreen of learned verbiage". But even worse, it causes us to lose sight of these informal premises, which now "float freely" in an arbitrary and fortuitous manner, to introduce all sorts of unforeseeable new kinds of confusion.

One may ask why there is such a widespread wish to believe that quantum theory is in principle a complete description of all that there is to be described. This is, in my view, just a manifestation of the "conservative tendency" described in my talk.[6] There is evidence that the individual measurement has something new in it, not contained in current quantum theory. But our whole heritage leads us to say "It is not different in essence from what we already know". But since it is different, we become involved in confusion, in trying to act as if it were not different.

Basically, we seem to find it hard to say "Even the foundation of our own thinking rests on the unknown." Rather, we prefer to say "We know everything in principle already; and it is only a problem to find how this knowledge can apply in each specific situation. So we talk of the "many body problem", the "measurement problem", etc. Could we not rather talk of "the unknown that is implicit in the many body context, the measurement context, etc."? Or would this make us too uncomfortable?

I hope it is clear to you that it is not my intention to try to solve the "measurement problem". Indeed, I do not see that is relevant to treat measurement as a problem. Rather I feel that the individual process in the quantum context is unknown. The unknown cannot be a problem. One can inquire into the unknown, to discover what

[6]See p. 301, n 1 above—CT.

is relevant in new contexts, how it is to be described, etc. But it is only in the context of what is known that a problem can be formulated.

The situation is just that we have no adequate description of the individual phenomenon in a "quantum" context, and what we wish to do is to see if such a description can be discovered. This involves "poesis", rather than "architecture".

With regard to those who continue the v. Neumann school (J. and P., K. and S., Gudder, etc.) my feeling is that unless and until someone goes carefully into the informal premises, and especially the experimental conditions, it is best to keep away from any attempt to relate my views to theirs, positively or negatively. For as I explained in a previous letter, the only relevant action with regard to persistent and entangling confusion is to perceive its irrelevance and keep away from it. Perhaps you would find it interesting to discuss the informal premises behind the equations of K. and S.?

I am serious in the question of asking you to discuss the informal premises behind K. and S., as I believe that if this were done properly, it would remove a large part of the confusion in which modern quantum theory has been entangled. The following steps would be involved:

Firstly, one has to justify the replacement of classical type probability functions in phase space by commuting algebras of "Yes-No" questions. These algebras have eigenfunctions ψ_+ corresponding to "yes" and ψ_- corresponding to "no". In principle, they allow linear combinations of eigenfunctions

(1) $\psi = a\psi_+ + b\psi_-$

However as long as the "Hamiltonian" commutes with all the "Yes-No" operators, a consistent subsidiary condition will be that no such linear combinations occur.

The danger is that as soon as you bring in a quantum theory with non-commuting algebras, the "equations of motion" will lead to linear combinations of the "classical wave functions" as described above. Perhaps, if your "cut" between "quantum" and "classical" parts of the world were permanently fixed, this could be avoided. But it is essential to such a theory that the "cut" be freely movable, so that we can say, of large scale systems, that either quantum theory or classical theory can be applied to them. However, this means that when a "system" was treated "quantum mechanically" it would have wave functions of type (1), but yet, as soon as we treated it "classically" these will be replaced by $\psi = \psi_-$; or $\psi = \psi_+$.

Of course, you can use the argument of DLP to say that in typical cases, this would have no practical significance. But the question of showing the impossibility of a hidden variable is not a "practical" question. If there is something illogical in your chain of reasoning, you simply haven't proved anything at all.

So I am willing to accept that K. and S. have proved that you can't embed a quantum algebra into a commuting algebra. But I don't see that this has any bearing as yet on the question of whether the quantum theory admits hidden variables, of a classical type. The reason I say this is because, as all accept, there would have to be some relationship between the "quantum state of a system" and the "classical state of an observing instrument". And there is good reason to believe that the above described illogicality would be relevant in discussing this relationship. Thus, it is an evident fact that the relationship in question would have to depend on the form of the experimental conditions. How do K. and S. propose to discuss this form? Will they use the "quantum algebra", or will they use the "classical algebra"? If they use the "quantum algebra", wave functions of type (1) will appear, and we will have just the illogical features described above. If they use the "classical algebra", they will be implying a permanently fixed "cut", such that the experimental conditions cannot be described "quantum mechanically". This, of course, is inadmissible. So perhaps they will postulate a set of "meta-experimental conditions", which are treated by the classical algebra, while the "original experimental conditions" are treated by the quantum algebra. But this leads us back to von Neumann's regress to the "mind of the observer" with his "principle of psycho physical parallelism." Do K. and S. have any proposals for getting out of these difficulties? Or what is more to the point, do you?

I am not saying that this cannot be done. I really don't know. But if you do make such a consistent discussion, it will be of great value, whatever the outcome.

It seems to me that if one doesn't carry out such a discussion of the experimental conditions, than the work of K. and S. can have little significance. At most, we could say perhaps that they regard quantum theory as a set of "handbook formulae" or rules for computing the average results of ensembles of experiments. These rules involve the use of a non commuting algebra. They then regard classical theory as another set of "handbook formulae", expressed in terms of a commuting algebra. What they "prove" is, at most, that the "quantum" formulae cannot be regarded as special cases of any kinds of "classical" formulae (I am not even sure that they are able to do this). But without a discussion of the experimental conditions, they cannot regard their theory as a description either of an experiment or of some classical "system" in movement. Rather, it is like a formula in the handbook, which depends on whatever empirical description the "user" is able to provide. But then, as I have shown, the "user" is always forced to take the experimental conditions into account, at least subliminally. And if even unconsciously he is regarding the theory as anything more than a set of "handbook formulae", he is caught in irrationality and inconsistency.

My impression is that you, along with most of those who work in the field, are not satisfied to regard the quantum theory as a set of "handbook formulae", and nothing more. Therefore, you are caught in such inconsistencies.

All this confusion seems so pointless to me, because, as I have shown, the impossibility of classical type "hidden variables" is evident informally from the very outset. In addition, one can see the general direction in which one has to go. (Parameters involving the whole universe in the description of each aspect.) So I can't see why you are so interested in K. and S. Far from "clearing the ground", it seems to me that until can discuss the experimental conditions consistently, there are "cluttering it up" with irrelevant debris.

Dear Jeff

Enclosed is a talk that should help clear up things between us. There are some supplementary notes at the back.

I have written to Boston, and talked to people at Yale. I hope you get an offer soon.

Best regards to you and to your wife.

Dave

June 20, 1969

Dear Jeff

Received your letter, and your manuscript commenting on my talk.

With regard to Wittgenstein, what you say is interesting. But it is also significant that Wittgenstein changed his views many times, and probably never resolved the questions he raised in a coherent way. (E.g. Near the end of his life, he is reported to have commented on his work "What a fool I've been!").

With regard to Bohr, it is just his insistence on using 'common sense language' that I have always felt to be the source of his incoherence. He is right to say "We are suspended in language, . . . etc." But his further implication that we cannot basically alter the "language of common sense" was a prejudice (i.e., a pre-judgement). The same holds for the general point of view which tries to insert the rival "quantum" content, either into "common sense" or into "classical language in physics." Thus, my use of "cells in phase space" was valid only as a means of showing the incoherence of classical language in this context. It calls for the dropping of classical mode of description in informal discourse, not merely in physics, but in everyday life as well.

I am preparing some papers, in which I start to do this (which I shall send you when they are ready). But beyond this, I can only say that I feel your comments tend toward re-establishing a conservative approach.

Best regards

Dave

July 2, 1969

Dear Jeff,

This is a brief letter, which I hope will help to clear up the questions in which we have been a bit "hung up" in our correspondence.

Firstly, I can see now, in looking over the whole situation, that I have gone too far in asserting that Bohr's view of quantum theory was the only consistent one. In fact, Bohr's views were verbally consistent, i.e. in terms of informal language. But in many ways, he wasn't clear with regard to deeper issues that are "hard to put in words".

Bohr asserted the wholeness of the form of the experimental conditions and the content of the experimental results. But he also said that the "informal language and description" had to be that of "common sense" referred, where necessary, to that of "classical physics". This latter is, of course, just the language of "analysis into parts". And he also implied that the mathematical language (i.e. Hilbert spaces, operators, etc.) had nothing to do with the description of the experimental arrangement. Thus, implicitly, he fragmented our description in several ways.

(1) A sharp division between the language of the theory and the language for describing the experiment.

(2) A sharp division between the "analytical" domain of "common sense" and a special "wholeness domain" which included the quantum theory and possibly other contexts, (e.g. social, psychological) in which he thought "complementarity" might apply.

But he introduced a disharmony between the "content" of "wholeness", characteristic of the "quantum" context, and the "form" of fragmentation, characteristic of his whole way of talking about "life in its totality". That is, it is not coherent to introduce a division between special spheres in which "wholeness" applies, and "life as a whole" (in which "fragmentation" generally applies). (In this regard see the enclosed typed draft). Rather "wholeness" implies that our way of talking is unbroken and undivided, so that there is no sharp break in language from one context to another.

Einstein understood this implicitly, when he insisted that the mathematical language of a unified field theory could describe the experimental arrangement, and would supersede the "common sense" description. But then Einstein and Bohr did not communicate, because their language forms were so different.

In a way, Dirac understood the point, when he said that a "matrix is an observable". This was a very deep perception, indicating that the experiment had to be described in a new way. But it got lost in mathematical formalism, and little further progress was made along this line.

Von Neumann had a similar perception, when he described the experiment in a new mathematical language (projection operators, etc.), but this too got lost in formalism, and a confused version of Cartesian philosophy, which is itself not coherent.

To sum up, Bohr, Heisenberg, Dirac, Einstein and von Neumann all failed to communicate, except in what could be done with the formalism. Thus, implicitly, the formalism became the essential point, because it was all that was capable of being coherently communicated.

I am currently writing a paper, starting from a different informal language, which articulates in an unbroken way to "become the mathematics". I should be able to send you a copy in a few months. What is essential is that the language of "common sense" is given up, at the very outset, so that even in "everyday life" this will cease to be our basic mode of communication.

With regard to your proposals (in your last letter) that I start from certain formal theories, and try to give them an informal extension, I must say that I am not capable of operating in this way. When I start from a formalism, I feel that I am "lost in a wild chaos of fragmentations" in which "nothing makes sense", and in which "I am trying to impose a coherent order on meaningless formal manipulations." If you can start with a formalism and come out with a good informal language, I can only say: "More power to you!" But you might as well ask me to swim the English Channel, as to ask me to do what you want done.

Finally, about Wittgenstein, it is interesting but dangerous to quote him. He did change his views many times, and he was never satisfied with them, to the very end. Most of the things you can quote are notes taken by other people, in fragments of what he said. Anybody can read anything he likes into such quotations.

With best regards

Dave

[The following undated typed document on fragmentation[7] appears here in the archives—CT.]

Society is in a condition of fragmentation. That is to say, nation is arrayed against nation, race against race, group against group, and individual against individual. Living in this way, each human being learns a pattern of behaviour, in which he acts in an incoherent manner, proceeding from contrary urges, contrary assumptions and contrary, aims. Each man begins to feel that this mode of living is "natural" and, indeed, merely an expression of "his very self". Over the centuries, children raised by fragmented adults living in a fragmented society have learned this pattern, which has, in turn, been passed on to later generations.

It may be said that, in a certain sense, mankind is "ill" with fragmentation, and that from earliest times, man has sought to "become healthy". Of course, since the pattern of fragmentation is generally regarded as "natural" it would appear that the way to "health" would require us to undergo a very fundamental and radical transformation, in which we would "become whole". It is significant that the root of the word "health" is the Anglo-Saxon word "hale", meaning "whole". Similarly, the word "holy" has the same root (e.g. as in the German "heilig") and also means "whole". But over the centuries, man has added to fragmentation, in his very search for "wholeness". Thus organized religions were set up, which promised mankind ultimate one-ness and harmony but which, at the very outset, introduced fragmentation, first by dividing the "sacred" from other aspects of life, and then by dividing people from each other, according to their beliefs. Later, men inquired scientifically into how to overcome various kinds of fragmentation (physical, mental, social), and in answer to this question set up "Departments of Health" by which was really meant, of course, "Departments of Wholeness". But the very word "Department" means "part" and this does not cohere with the aim of "becoming whole" that is implicit in man's

[7] Bohm's ideas on fragmentation were extended later in Watson (1971), Chap. 3 and Bohm (1980), Chap. 1—CT.

search for "Health". (For example, different "Departments of Wholeness" could follow contrary policies and thus come into conflict, as to how to "make men whole"). Therefore, in the long run, science brought in a kind of fragmentation similar to that produced by organized religion.

It is clear, indeed, that any means of defining "wholeness" in some particular way is already a separation between what is called wholeness and other aspects which make up "life as a whole". This separation is in essence a form of fragmentation, which adds to all other forms and intensifies the feeling that "we are broken up and therefore need to integrate or unify all the fragmentary aspects". Thus, it is just in our unending quest for unity or one-ness that we have introduced the most subtle and persistent source of fragmentation, i.e. the verbal division between "the fragmented beings that we are" and "the wholeness that we wish to achieve".

Understanding that the "effort to become whole" is one of the most important means of maintaining our very pervasive tendency to fragmentation in every aspect of life, we may ask what it means to consider "wholeness" as an end or aim of any specified form of action (e.g. "changing society" or "changing the child by better education"). To ask this is not to frame a rhetorical question but, rather, to indicate the need for an inquiry, that is actually of very serious concern to each human being. For no one really wants to go on with the disharmony, conflict and general destruction that are the inevitable outcome of fragmentation. Yet, unless we are whole, from the very beginning, whatever we do to "overcome fragmentation" is just a continuation of fragmentation into more subtle forms. Can we simply "stay with" this question, without giving any ready or pat answers, and discover how fragmentation actually originates in each one of us, and propagates through every aspect of our lives?

Aug 14, 1969[8]

Dear Jeff

I just received your report for the NSF. It is very interesting. Of course, the "forest" of strange mathematical symbols passes over my head, like a wave, leaving little or no impression. But in some very vague and elliptical sense, I can see that you are communicating to the formal logical mathematical person something that I believe has to be said in simple and informal language, which is that quantum theory raises a new kind of question in physics. I could never understand your formal chains of reasoning, unless perhaps you gave me a three hour lecture in which I could ask you questions freely. Nor can I see any reason why I should understand it, as I am working in a very different line.

Where we differ is about Bohr. I believe that Bohr's work did confuse the issue in a certain way, thus raising "impossible questions" in which v Neumann and others got caught, trying to give "answers". In other words the confusion in the "orthodox interpretation" is in a certain sense the evitable reaction of physicists, who are trying to "clear up Bohr's confusion" rather than "just dropping it and doing something

[8]First letter with this date—CT.

new and fresh." Bohr's confusion was the insistence on using "ordinary common sense language" to describe an experiment and a very different formal mathematical language for theoretical inferences. This division in languages leads to disharmony. V. Neumann was only trying to "resolve the disharmony" by suggesting a single formal language that would cover both theoretical inferences and descriptions of the results of experiments. In my view, his mistake was his failure to notice that it is the informal everyday language that has also to be changed, thus leading to a radically different notion of what physics is and what scientific research is.

I am now developing a new grammar of informal discourse, which is different from the ordinary grammar and syntax of our everyday language. I have talked about this at several universities, and encountered a lively interest. But it cannot be communicated in a letter. It will take several articles at least.

Best regards

Dave

Aug 14, 1969[9]

Dear Jeff

PLEASE READ THE ENCLOSED AIR LETTER FIRST

Continuation of contents of air letter.

Pursuant to the discussion of your comments on Bohr & Aage Petersen's conversation (p.18 of your report),[10] you quote Wittgenstein's question: "What is it that we cannot say?" I wonder whether this question makes sense. If we say "what we cannot say," then, in some sense, we have already "said it". One has a vague intuition of what Wittgenstein may probably mean by his question. I would put it this way: The attempt to represent the whole of reality generally leads to contradictions and incoherence. Indeed, any attempt to grasp "wholeness" (including Bohr's) is pervaded by incoherence of a generally very subtle nature. For example, to try to separate logical form from "the whole of reality" is incoherent because logical form is an aspect of "the whole of reality". Logical form is a key feature of human reality and human reality cannot be separated coherently from "the whole of reality".

Propositions are reality. That is, they are real as marks on paper or as noises. And they are real as significances, i.e., movements on many levels in our brain, nervous system, and muscles. Similarly, language is reality.

When you say that propositions cannot represent logical forms, you are using a proposition to represent some feature of logical form, i.e. its unrepresentability. So while we feel that there is something important that you want to say, one is also confused by your failure to say it coherently. This is not accidental. Rather, as I indicated in the paper I sent you a month ago, this difficulty is inherent in any "idea of wholeness". That is basically why Bohr's work leads to incoherence, in the long run.

---

[9]Second letter with this date—CT.

[10]Suppe (1974), pp. 404–405—CT.

One can also remark here that Bohr's distinction between the experimental conditions and the results tends to separate these results from the total experimental situation, so as to make these results an object of further theoretical discourse, leading to abstract inferences divorced from the informal description of the experimental conditions. In the new language forms (grammar) that I am suggesting, there is no separation between the description of the experimental situation and the "results". In other words, the term "results of an experiment" is not grammatically allowed in the new language form. I would thus modify your statement in the middle of p. 20 to read: Bohr's statement is so obviously absurd that it must be untrue.

I would like to repeat that while Wittgenstein made important discoveries, it is dangerous to take quotations from him. Many of these are very incoherent, unless they are put in the full context of the continued development of his views, in which later statements frequently contradicted earlier ones, and in which one knows that he was unsure of his views, even at the very end.

Best regards

Dave

P.P.S. I just received your letter, written from New Haven. You are right to say that our points of view are very different (and have probably been so, from the very beginning). I would say that to treat "wholeness" as a form, logical or otherwise, is basically incoherent. For "wholeness" means "undivided", if it means anything at all. You are establishing what I would call a "Department of Wholeness", (i.e. Wholeness as Logical Form). Thus, you are splitting "logic" off from "reality as a whole". Therefore, such an approach is fragmentation – a denial of wholeness in action, while wholeness is asserted in verbalisation.

I can only say, for the present at least, that I wish you luck in your effort to "reach the horizon".

Dave

Aug 16, 1969

Dear Jeff,

In continuation of my letter of a few days ago, I now see with great clarity that we have always been poles apart in our basic views, while we have communicated largely on superficial details of the forms used in physics. Rather than go on with endless but fruitless arguments about these details, I would like to give form to the basic questions on which we do not meet.

(Your interest in Plato confirms the difference between us. I have always felt that Plato's views are very harmful and indeed even destructive, when applied in physics, or in any other field or area of life.)

Perhaps we could go back to a question that you told me you raised when you were very young: "How is it possible for anything to exist at all?" This was part of a yet more extensive question "How is it possible for me to exist?"

My view is that to raise the question of possibility is to bring in a certain logical form, i.e. the formal distinction between the possible and the impossible. Your particular way of raising the question implies that you take the meaningfulness of this distinction for granted, even when it is applied, not to some limited or particular content, but to the totality of all that exists, including you, me, society, nature, and the movements and processes whereby we verbalise formal distinctions, such as that between the possible and the impossible.

Now, I would say that if you are going this far in your questions it is necessary also to question the form of your question. In particular, is the distinction between the possible and the impossible a coherent form of discourse, when applied in such a broad context? I myself am inclined to answer that it is not coherent. But if your answer is otherwise, how would you justify it? Wouldn't it finally come down to a purely personal feeling that this is your question, and that you cannot communicate with someone else, to whom this is not a coherent question at all? Thus, the question presupposes that you exist, and therefore, you are not carrying your question to the very end. Surely, it is necessary to begin without any presupposition that you exist, or else your question is not a serious one. Rather, you take it for granted that you exist (and therefore that the world exists), and your only question is to find some "logically coherent" way of describing how you exist without contradiction. But I am quite serious in saying this is only a "game of questioning". For example, my own view is that "you and I" are nothing but contradiction, and that as long as "you and I" are in the discourse, logical coherence is not possible. Therefore your question has no meaning. It is equivalent to the question: "How can a contradictory entity exist in a logically coherent way?"

Perhaps if we adopted Plato's view of an ideal world of pure coherent forms that are "whole", then we could give meaning to the question "How can I exist?" But as far as I can see, these forms are pure verbalism. I do not observe such forms anywhere. In mathematics, for example, I see no more "truth" than in the forms of other aspects of life. In my view, logical form is a contingent result of our particular human evolution and human history, up to the moment. This form is changing all the time. However, it is not mathematicians who are changing it, with new kinds of "logic". For such mathematicians generally still use the old "logic" in the "metalanguage" with which they talk about their "new logical forms". A genuine development of logical form has to penetrate into the whole of language, especially into the "everyday common sense language" (that Bohr treats as eternally fixed almost as sacred.)

Your question is "How is it possible for me to exist?" and my question is "What does it mean to extend logical form into the context of the totality of all that exists?" You are your question. I am my question. Because our questions are so different, we are two basically unrelated and isolated individuals. Is it possible for us to go beyond the limited forms of these questions? If not, then each of us can only hope to find another isolated individual who fortuitously shares his own basic question, so that communication may be possible. But I am inclined to believe that at bottom, every individual has a different question. Thus, communication between them ultimately has no meaning, until people are able to see all these different questions and to cease to give so much importance to them.

What is needed is to question the very form of questioning in general. And to question the form of questioning as such is to question logical form as well.

I am not denying the value of logical form all together. Rather, I suggest that each particular form (e.g., "logic as it is now recognised today") is of limited value, and that such form is always changing and evolving, as we use it. It is therefore necessary to question each particular notion of logical form. And it is necessary to be aware that each question may presuppose a further logical form in an unspecifiable and highly implicit way. We will then always be alert to observe the new logical forms that are implicit in our new questions.

Let me also add that I am questioning the forms of my own questions, as well as those of your questions. We can in this way help each other to question the forms of all the questions that we have accepted fortuitously in the past, and when we are both beyond these particular accidentally raised questions, we can come to a new kind of question that has meaning the both of us. That would be communication between us.

When I read the proceedings of the Illinois Conference, I saw that everybody there did indeed have a different question, and that there was in fact little communication except perhaps on superficial details. My picture of the philosophy of science is now one of a crowd of individuals, each speaking a different language, but imagining that all speak the same language (like the Tower of Babel).

Finally, let me say that I am exploring a new grammar and syntax of everyday language, in which confused questions of the kind to which our present language leads us do not arise, because they are seen, at the very outset, to have no meaning.

Best regards

Dave

Aug 21, 1969

Dear Jeff,

To finish up the discussion of your paper (i.e. the report) I would like to emphasize that in my view, the quantum theory is not unique, in the way in which you suggest that it is.

My example of malaria would have been better, if I have noted that "Health" means "Wholeness". Thus, physics and medicine (as well as other sciences) have ultimately been shaped by some notion of Wholeness.

Take pre-technical societies, for example. Subjects like witchcraft and astrology are perfectly natural, within their notions of Wholeness. I talked recently with a man who still believes in both. He pointed out to me that they are based on the notion of analogy; i.e., things that are similar belong to one Whole. So if Mars is "warlike" in color, this will take part in some "whole quality of aggression – universally present – in man and in nature." Witchcraft is based on a similar notion of Wholeness. Recently, an African witch-doctor appeared on television, and we could see that in the context of his society, his methods of establishing psychological harmony were more effective than ours in our society.

Our technological society is inseparably intertwined with a mechanistic notion of Wholeness. The world is constituted of parts, and the whole is either the sum of its parts (when there is no interaction) or more than the sum of the parts (when there is interaction, leading for example to "collective properties"). But what is crucial is the acceptance that analysis into parts is relevant. In physics, the parts are atoms or elementary particles. In medicine, they are cells, viruses, germs, DNA particles, etc.

One can now discern a new society taking shape, in which the notion of Wholeness as a synthesis, following analysis into parts, will cease to be relevant. In medicine, one sees that a disharmonious society leads to death and illness, whether in the form of wars, riots, slums, racial battles, etc, or in the form of cancer and susceptibility to infections, or to various forms of insanity and psychological disturbances. In physics, one is beginning to see that analysis into particles is not very relevant. Einstein suggested this in connection with unified field theory and Bohr in connection with the attempt to give a verbally consistent description of the experimental situation in a "quantum" context.

Each notion of wholeness implies its own criteria of what constitutes an adequate test, not compatible with that of the others. Thus, modern medicine is quite unable to confront coherently the question of testing the adequacy of the procedures of the African witch doctor, in his own context. Quantum test are not compatible with classical tests, and these in turn are not compatible with tests used in pre-Newtonian physics.

I hope this helps clear things up a bit.

Best regards

Dave

Postmark Aug 29, 1969

[date added—CT.]

Dear Jeff

Enclosed is my answer to the comments on my paper.

Best regards

Dave

[There follows a typewritten version of "Reply to Discussion" published in Suppe (1974). See p. 301, n 1 above—CT.]

References

Bohm, D. (1952a). A suggested interpretation of the quantum theory in terms of hidden variables i. *Physical Review*, *85*(2), 166–179.
Bohm, D. (1952b). A suggested interpretation of the quantum theory in terms of hidden variables ii. *Physical Review*, *85*(2), 180–193.
Bohm, D. (1980). *Wholeness and the implicate order*. Abingdon: Routledge and Kegan Paul.
Bohm, D., & Bub, J. (1966a). A proposed solution of the measurement problem in quantum mechanics by a hidden variable theory. *Reviews of Modern Physics*, *38*(3), 453–469.
Bohm, D., & Bub, J. (1966b). A refutation of proof by Jauch and Piron that hidden variables can be excluded in quantum mechanics. *Reviews of Modern Physics*, *38*(3), 470–475.
Suppe, F. (Ed.). (1974). *The structure of scientific theories*. Illinois: University of Illinois Press.
Watson, F. (Ed.). (1971). *The social impact of modern biology*. Abingdon: Routledge.

Appendix A

Correspondence between Jeffrey Bub and Angelo Loinger[1]

Universita Degli Studi Di Pavia
Instituto Di Fisica Teorica

Dr. J. Bub
Birkbeck College
London University
London (England)

4/10/1966

Dear Dr. Bub,

I have seen your papers on the hidden variables, published in "Revs. Mod. Phys." (July 1966)[2]

I wish to stress the following points: (i) quantum theory is a structure which is not independent of classical physics – this has been properly emphasized by Bohr, but many young physicists seem to have not realized it; (ii) even from the Einsteinian viewpoint, according to which quantum theory is, in a sense, incomplete, the quantal measurement problem is a perfectly legitimate one, and I feel that it has been rather definitively settled in our papers in "Nucl. Phys."[3] and "Il Nuovo Cimento"[4]; (iii) in principle, a completely deterministic theory is quite conceivable, but it is difficult for me to believe that such a theory can be constructed in a cheap way, e.g. by a

[1]Jeffrey Bub found this correpondence among his papers and kindly made it available. For more on Loinger et al. see C130, p. 17, n 4. For Bohm's correspondence with Loinger see C130 pp. 17–29—CT.

[2]I.e. Bohm and Bub (1966a, b)—CT.

[3]Daneri et al. (1962) or Wheeler and Zurek (1983), pp. 657–679—CT.

[4]Daneri et al. (1966)—CT.

C. Talbot (ed.), *David Bohm's Critique of Modern Physics*,
https://doi.org/10.1007/978-3-030-45537-8

simple refinement (with the aid of suitable parameters) of the current formulation of q.m.

I believe with Einstein that if we shall abandon Bohr's philosophy, we shall have only one alternative: a general, unified field theory in the Einsteinian sense of the term.

With my best regards, I am

Yours sincerely

A. Loinger

November 12th 1966

Professor A. Loinger,
Universita Degli Studi di Pavia,
Istituto di Fisica Teorica,
Via Taramelli 4,
Pavia,
Italy.

Dear Professor Loinger,

Thank you very much for your letter and the reprint you sent me. I would have answered you before, but there was a little delay in forwarding your letter from London to Minnesota (where I am working at present). In the meantime, I received a reprint of an article by K.S. Tausk ("Relation of Measurement with Ergodicity, Macroscopic Systems, Information and Conservation Laws" – International Centre for Theoretical Physics, Trieste, Internal Report 1471966)[5] and spent some time re-thinking your articles from his point of view.

Unfortunately Tausk uses a different notation from that used in your articles so that at first sight it seems as if he has wilfully ignored or misunderstood your theory. However, if one translates his argument back into your notation (as used in Nuclear Physics 33 (1962) 297–319) there appears to be, at the very least, some confusion in your paper.

In section 9 you consider a Hamiltonian H of the macro-system and a basis $\Omega_{k\nu i}$ such that the following ergodicity conditions are satisfied:

$$\mathbf{M}\{\sum_{i=1}^{s_{k\nu}} |(\Omega_{k\nu i}, U(t)\Omega_{k\mu j})|^2\} \approx \frac{s_{k\nu}}{S_\mu}$$

$$|\mathbf{M}\{\sum_{i=1}^{s_{k\nu}} (\Omega_{k\nu i}, U(t)\Omega_{k\mu j})^* (\Omega_{k\nu_i}, U(t)\Omega_{k\mu' j'})\}| \ll \frac{s_{k\nu}}{S_\mu}$$

[5]Tausk's paper is not available. See Freire (2015), Chapter 5 and Pessoa et al. (2008) (also https://www.researchgate.net/publication/225121330_The_Tausk_Controversy_on_the_Foundations_of_Quantum_Mechanics_Physics_Philosophy_and_Politics) for more on Tausk. See C130, 25 for Bohm's favourable comment on Tausk. The extraordinary denigration of Tausk by Daneri, Loinger and Prosperi is revealed in these letters—CT.

$$(\mu \neq \mu') \tag{9.1}$$

M denotes the operation of time averaging:

$$\lim_{T\to\infty} \int_0^T dt \ldots$$

$U(t) = \exp(-iHt/\hbar)$

Now you suppose that in every channel $\mathbf{C}_k$ there exists a cell $\mathbf{C}_{ke_k}$ with a number of dimensions enormously larger than that of the other cells $\mathbf{C}_{k\nu}$, in such a way that:

$$1 - \frac{s_{ke_k}}{S_k} \ll 1, \qquad \frac{s_{k\nu}}{S_k} \ll 1 \qquad (\nu \neq e_k) \tag{9.6}$$

Thus, the expressions (9.1) are evidently used in the sense:

$$\mathbf{M}\{\sum_{i=1}^{s_{k\nu}} |(\Omega_{k\nu i}, U(t)\Omega_{k\mu j})|^2\} \approx \frac{s_{k\nu}}{S_\mu} \approx \delta_{\nu e_k} \tag{A}$$

$$|\mathbf{M}\{\sum_{i=1}^{s_{k\nu}} (\Omega_{k\nu i}, U(t)\Omega_{k\mu j})^* (\Omega_{k\nu i}, U(t)\Omega_{k\mu' j'})\}| \ll \frac{s_{k\nu}}{S_\mu}$$

$$\text{i.e.} \ll \delta_{\nu e_k}$$

$$\text{i.e.} \approx 0 \quad (\mu \neq \mu') \tag{B}$$

Consider a (micro-)state vector of the macroscopic system which is initially entirely within channel $\mathbf{C}_k$, i.e. a general vector of the form:

$$\Psi_{(k)}(0) = \sum_{\mu j} \omega_{k\mu j} \Omega_{k\mu j} \qquad \sum_{\mu j} |\omega_{k\mu j}|^2 = 1$$

where:

$\omega_{k\mu j}$ are coefficients

$\Omega_{k\mu j}$ are basis vectors of the cell $\mathbf{C}_{k\mu}$

(0) refers to time $t = 0$

The norm of the projection of $\Psi_{(k)}(t) = U(t)\Psi_{(k)}(0)$ onto the cell $\mathbf{C}_{k\nu}$ is:

$$\begin{aligned}\mathrm{Norm}(P_{C_{k\nu}}\Psi_{(k)}(t)) &= \sum_{i=1}^{s_{k\nu}} |(\Omega_{k\nu i}, \Psi_{(k)}(t))|^2 \\ &= \sum_{i=1}^{s_{k\nu}} |\sum_{\mu j} \omega_{k\mu j}(\Omega_{k\nu i}, U(t)\Omega_{k\mu j})|^2 \\ &= \sum_{i=1}^{s_{k\nu}} \sum_{\mu j} |\omega_{k\mu j}|^2 |(\Omega_{k\nu i}, U(t)\Omega_{k\mu j})|^2 \\ &+ \sum_{i=1}^{s_{k\nu}} \sum_{\mu j \neq \mu' j'} \omega^*_{k\mu j}\omega_{k\mu' j'}(\Omega_{k\nu i}, U(t)\Omega_{k\mu j})^*(\Omega_{k\nu i}, U(t)\Omega_{k\mu' j'})\end{aligned}$$

Now take the time average of this quantity:

$$\begin{aligned}\mathbf{M}\{\mathrm{Norm}(P_{C_{k\nu}}\Psi_{(k)}(t))\} &\approx \sum_{\mu j} |\omega_{k\mu j}|^2 \delta_{\nu e_k} + 0 \\ &\approx \delta_{\nu e_k} \\ \text{since } &\sum_{\mu j} |\omega_{k\mu j}|^2 = 1\end{aligned}$$

Hence, the ergodicity condition (A) means that any state vector initially within the cell $\mathbf{C}_{k\mu}$ in channel $\mathbf{C}_k$ will spend most of the time within the cell $\mathbf{C}_{ke_k}$ in channel $\mathbf{C}_k$ during any sufficiently long time interval. Condition (B) essentially allows the above statement to be extended to include an initial state vector anywhere within the channel $\mathbf{C}_k$ (i.e. not confined initially to a particular cell).

The point I wish to make here is that the ergodicity conditions describe an ergodic behaviour within a particular channel $\mathbf{C}_k$; indeed, k denotes the set of values of the macroscopic constants of the motion (besides the energy) which characterize a particular macro-state, so these quantities do not change during the free evolution of the system defined by $U(t) = \exp(-iHt/\hbar)$.

Now in Section 9 you consider a macro-system which is initially in the state

$$\begin{aligned}\Psi_0 &= \sum_{k\mu j} (\Omega_{k\mu j}, \Psi_0)\Omega_{k\mu j} \qquad (9.2) \\ &= \sum_{k\mu j} \omega_{k\mu j}\Omega_{k\mu j} \quad , \quad \sum_{k\mu j} |\omega_{k\mu j}|^2 = 1 \quad \text{(for brevity)}\end{aligned}$$

i.e. a linear combination of state vectors from different channels.

In this case:

$$\mathrm{Norm}(P_{\mathbf{C}_{k'\nu}}\Psi(t)) = \sum_{i=1}^{s_{k'\nu}} |\sum_{\mu j} \omega_{k'\mu j}(\Omega_{k'\nu i}, U(t)\Omega_{k'\mu j})|^2$$

[corresponds to (9.3)]

i.e. the contributions from $k \neq k'$ drop out (i) because the channels have been chosen orthogonal and (ii) because $U(t)$ describes a time-evolution which does not carry a given vector out of the channel to which it belongs at $t = 0$ (so that $(\Omega_{k'\nu i}, U(t)\Omega_{k\mu j}) = 0$ for $k \neq k'$).

Hence:

$$\mathbf{M}\{\text{Norm}(P_{\mathbf{C}_{k'\nu}}\Psi(t))\} \approx \sum_{\mu j} |\omega_{k'\mu j}|^2 \delta_{\nu e_k}$$

[corresponds to (9.4)]

Here

$$\sum_{\mu j} |\omega_{k'\mu j}|^2 = p_{k'} \neq 1 \qquad \text{[corresponds to (9.5)]}$$

but

$$\sum_{k'} p_{k'} = \sum_{k'\mu j} |\omega_{k'\mu j}|^2 = 1$$

There is a difference between this case and the previous case which is not clearly brought out in your article. The projection of a vector Ψ onto a manifold V spanned by some orthonormal set $\varphi_1, \varphi_2, \ldots$, is, by definition:

$$P_V \Psi = \sum_n (\varphi_n, \Psi)\varphi_n$$

Hence, if the projection of a normalized vector Ψ onto a certain manifold has the norm 1 (i.e. $\sum_n |(\varphi_n, \Psi)|^2 = 1$), Ψ is entirely within this manifold. However, if the norms of the projections of the vector Ψ onto the orthogonal manifolds $V_1, V_2, \ldots$ are respectively $p_1, p_2, \ldots$ then this means only that Ψ spans all the manifolds $V_1, V_2, \ldots$ on which its projections has a non-zero norm – not that the probability is p_1 that Ψ is actually in V_1, p_2 that Ψ is actually in V_2, etc. Trivial as this may seem, I think it is worth emphasizing because the projection postulate states that the probability is p_1 that the vector Ψ will aotually be in V_1 after a certain measurement, p_2 that Ψ will be in V_2, etc. Because you speak loosely of "the probability of finding a system in a certain state or manifold at time t", you obscure the distinction between the "probability of finding ..." in the sense of the probability interpretation of quantum mechanics (which involves a certain measurement) and "the probability of finding ..." in the sense of ergodic theory (which involves a certain dynamical motion of the free macro-system). In order to make this distinction absolutely clear, I suggest that, when discussing the ergodic motion of the macroscopic system, you say something like: "The system will spend a fraction $p_k \delta \nu e_k$ of any sufficiently long time interval within the cell $\mathbf{C}_{k\nu}$" instead of: "The probability of finding the system in the macro-state $\mathbf{C}_{k\nu}$ is $p_k \delta \nu e_k$". (If you object to this because this is not what you mean, my point is that precisely what you mean has not been made explicit.)

Now it should be clear from the above remarks that the results (9.4), (9.5) do not mean that the state vector, initially of the form (9.2), will spend a fraction p_k of any sufficiently long time interval in the cell $\mathbf{C}_{ke_k}$. You claim, however, that it is legitimate to interpret this result in the sense that, after a sufficiently long time interval, the system is macro-scopically in one of the cells $\mathbf{C}_{ke_k}$ with probability p_k (see your comments before equation (9.9′)).

But this inference, that after a sufficiently long time interval the system is macroscopically in one of the cells $\mathbf{C}_{ke_k}$ with probability p_k, does not follow from any assumptions about the ergodicity of the system. It follows, rather, from the assumption that a macro-system can only be in one particular macro-state at any particular time (although this need not be the equilibrium macro-state), which is surely what you have set out to prove by your ergodic theory.

Thus, you state (page 298):

"In order that an objective meaning may be attributed to the macro-states of the large bodies, it is of course necessary that – by virtue of the laws of quantum mechanics and of the structure of the macroscopic bodies – states incompatible with the macroscopic observations be actually impossible." (Let me call this your thesis.)

By "objective" you presumably mean that a macro-state is a state of the system which exists independently of any observation. A macro-observation reveals to the observer that macro-state which exists independently of our observations. But the existence of such objective macro-states has not been shown by an analysis appealing only to the laws of quantum mechanics and the structure of macroscopic bodies.

I do not think that you can even claim to have proved the weaker thesis, that "by virtue of the laws of quantum mechanics and of the decomposition of Hilbert space into cells by the assumption of macro-observations, states incompatible with the macro-observations are actually macroscopically unobservable". (I take it that you consider that your analysis directly proves this weak thesis, because the argument on page 315 is apparently designed to show the equivalence of the weak thesis with the strong thesis – see from "One could think that the following objection can be raised against our point of view. ... " until end of section.)

The ergodic motion of the apparatus carries the state vector of the apparatus into a particular equilibrium cell within a certain channel. But you first assume that, after the interaction, the state vector of the apparatus is actually in a definite channel, although before an observation has been made the observer can predict only the probability of the final equilibrium macro-state. In other words, you assume that, although microscopically the initial state of the system is represented by a linear superposition of state vectors from different channels, because the system is macroscopic, in the sense that only macroscopic observations are made on it, the state vector of the system will actually be in the equilibrium cell of one particular channel after a sufficiently long time, although this channel is unknown and is not predicted by quantum mechanics. This is an assumption (which, of course, everybody believes is true); it has certainly not been proved on the basis of your ergodic theory, and it is clearly incompatible with the linear equation of motion of quantum mechanics.

As far as I can see, your theory of measurement is merely a formal account of the amplification process which takes place in such measuring instruments as cloud chambers, bubble chambers, photographic emulsions, spark chambers, scintillation counters, photomultipliers, Cerenkov counters, etc., after these instruments have already "absorbed" a particle in a certain eigenstate of the observable measured. Thus, the ergodicity conditions describe the transition of the instrument from an initial state in a non-equilibrium cell in a particular channel to a final state in the equilibrium cell of that channel. Different channels are associated with different eigenstates of the observable measured. If the state of the object system is represented initially by a linear superposition of eigenstates, then it is assumed (not proved) that the state of the apparatus is, after the interaction, in a certain channel (why? merely because you have asserted that states incompatible with macro-observations are impossible?), although the actual channel is unknown and only predictable with a certain probability, determined by the initial superposition of eigenstates of the object system. This, of course, is the measurement problem which you set out to solve. The whole point of a theory of measurement is to explain what happens before the amplification process. viz. (a) why or how a given instrument reacts only to one eigenstate in a linear superposition and (b) what determines the eigenstate in a given experiment. The point of a hidden variable theory of measurement is surely to show that since quantum mechanics answers (b) by saying that nothing determines the resulting eigenstate, (a) cannot be answered at all within the framework of formal quantum mechanics.

So, if this criticism of your theory based on Tausk's paper is invalid and has merely revealed my lack of understanding, I still think that Professor Bohm's reply to your letter provides a conclusive argument against all theories of measurement within the framework of quantum mechanics which purport to solve the measurement problem by an analysis of the measuring instrument as a macro-system (i.e. following Jordan and Ludwig's "philosophy").

In your letter to me you also mention that a hidden variable theory of the kind suggested in our papers would be a "cheap way" of constructing a deterministic rival to quantum mechanics. I think everyone would agree with you on this point. Nevertheless, even a very naive theory, such as that proposed in our papers, could provide a first step towards a deterministic unified field theory of the kind you are willing to accept in principle. Firstly, a hidden variable theory can suggest experimental situations where the predictions of quantum mechanics might break down. Even though the the predictions of a naive hidden variable theory would almost certainly be false as well under such circumstances, this would not alter the fact that the hidden variable theory had performed the very important function of suggesting the experiment. Secondly, (and here I am sure you will disagree with me), a hidden variable theory provides a model which shows clearly the insolubility of the measurement problem within the framework of formal quantum mechanics, and it suggests the direction a future deterministic theory should take if it is to incorporate a solution to this problem.

As to your other point, that quantum theory is not independent of classical physics, I am not quite sure why the hidden variable theory proposed in our papers ignores this fact.

I hope you will let me know what you think of these remarks.

Sincerely,

Jeffrey Bub.

Dr. J. Bub
Institute of Technology
Dept. of Chemistry
University of Minnesota
Minneapolis, Minnesota
55455 U.S.A.

Pavia, November 23, 1966

Dear Doctor Bub,

you have misunderstood our paper. I shall limit my answer to few comments. My friend Prosperi will reply to your objections in a more systematic way.

(1) The quantal measurement problem of quantum mechanics is essentially a consistency problem: loosely speaking, it consists in proving the compatibility of the results respectively obtained by treating: (a) the micro-object (I) by means of q.m. and the macro-apparatus (II) as an "external" object, and (b) the global system I+II by means of q.m.

This is a perfectly legitimate problem; as a matter of fact, nobody discusses this legitimacy, with the only exceptions of the rare prophets of the hidden variables.

(2) Bell's result in Physics 1, 195 (1964) - which is actually posterior to the paper of the same author in Revs. Mod. Phys., July 1966 – kills the interest (and just from a realistic point of view) of any hidden variable theory. In any case, I shall consider seriously the theories of this kind only when they will give some new result. Incidentally, I remark that the task of inventing a deterministic theory starting from a theory constructed by means of an injection of suitable hidden parameters into q.m. is analogous to the task of discovering the Newtonian mechanics starting from statistical mechanics in the Gibbsian formulation: you should be very very fortunate for accomplishing a task of this kind!

(3) Our $u_{k\nu}(t)$ is always a quantal probability. Accordingly, the "probability of finding ..." is always (if not stated otherwise) a probability in the sense of q.m.. I think that our language is perfectly clear to any unprejudiced reader. In particular, we have never written the statement of the second sentence of the second paragraph at p. 4 of your letter ("You claim, however, ...").

(4) Our concept of macroscopic observable has been directly derived from the well-known considerations of von Neumann, van Kampen, and Fierz.

(5) The second and the third paragraph at p. 5 of your letter are deprived of sense. Consider e.g. formulae (10.1′), (10.2′) and (10.3′) at p. 313 of our paper. We do not claim that "the state vector of the system will actually be in the equilibrium cell of one particular (unknown) channel after a sufficiently long time", but only that, so far as the macrovariables are concerned, the time evolved of W' and the operator (10.3′) give the same results.

This is not an assuption, but it has been proved in our paper, and it is valid – I repeat – only for the macroscopic variables of the apparatus, but not for any mathematically conceivable dynamical variable.
(6) Of course, the "weaker thesis" has not been proved in the strict sense of the term, but it has been made very plausible by means of qualitative reasonings.

We do not claim that superpositions of vectors corresponding to different macrostates cannot be realized. Indeed, this possibility is inherent to the mathematical structure of q.m. and cannot be eliminated. In our paper we have simply pointed out that, as a consequence of a qualitative reasoning, it is impossible to construct an apparatus, satisfying the specified requirements, which allows to measure a quantity, whose eigenstates are superpositions of macroscopically distinguishable states. From here we then conclude that, practically, a superposition of macroscopically distinguishable states is equivalent to a mixture.
(7) We have never claimed to have suppressed the postulate of the reduction of a pure state to another given pure state, not even to have explained the above reduction by means of the time evolution of the undisturbed system. We have only proved that in the expression of the quantal probability that the apparatus is found in one or in another of the possible macrostates at the end of the measurement, no interference term are present, and consequently the reduction (in the above sense) of the wave packet of I+II can be interpreted as a simple increase of information (transition from a mixture to one of its composing elements). In this way, we do not need the decisive intervention of a second apparatus (III) as in von Neumann's theory.
(8) Prof. Bohm has written to me a long letter, but I must say that his criticism of our approach is not justified. But I shall write directly to him.

Yours Sincerely

A. Loinger

P.S.– Tausk is so presumptuously idiot, that he does not deserve a detailed refutation. Thus I shall emphasize only two points: (i) Our formula (5.3) – which T. does not like – is due to von Neumann and is universally accepted. – (ii) So far as formulae (3.5), (3.6), (3.7) of T.'s pseudopaper – or, better, Eqs. (7.1), (7.2), (10.3) of our paper – are concerned, I remark, first of all, that T. does not consider the fundamental role of the time evolution of I+II after the interaction, and secondly, that T. has not understood that when $|d_0\rangle$ varies in V_0, the $|d_r\rangle$'s of his (3.2) vary in the respective V_r and consequently they cannot be orthogonal to all, with the exception of one, the vectors of the fixed basis $\{|X_{rk}\rangle\}$ (last paragraph of p. 13), and not even, in general, to all, with the exception of one, the subspaces Ω_{rk} (second paragraph of p. 14). Accordingly, the interference terms are only apparently absent from his (3.7). In the second line of our (10.3), interference terms among different macrostates of the apparatus are surely still present.

Universita Di Milano
Instituto Di Scienze Fisiche "Aldo Pontremoli"

April 17, 1968

Dr. Jeffrey Bub
University of Minnesota
Institute of Technology
Department of Chemistry
Minneapolis, Minnesota 55455
U.S.A.

Dear Doctor Bub,

Thank you very much for the preprint of the paper[6] on D-L-P theory.

In my opinion, your ideas on the problem are not correct. Further, it seems to me that in several points (e.g. when you speak of "maneuver") your manuscript is not very kind to us.

Have you sent a copy also to Prof. Rosenfeld?

With the best regards, I am

Yours Sincerely

A. Loinger

P.S. – Why do you quote the pseudopaper of Tausk, which is a monument of stupidity?

Universita Di Milano
Instituto Di Scienze Fisiche "Aldo Pontremoli"

May 13th, 1968

Dear Doctor Bub

Thank you very much for your kind letter of May 5th.

In my opinion, in your treatment of the relationship between Bohr's ideas and our approach there is an absence of nuances, i.e. you take too many things too au pied de la lettre. Rosenfeld and I feel that if Bohr had considered the problem of the consistency of the two procedures:

(i) I with Schröd. eq., II "external";

(ii) I+II with S. eq.,

(which is an inescapable problem, since no axiom in q.m. gives an upper bound for the number of the degrees of freedom), he would have arrived at a point of view not substantially different from ours. Notice that in our approach classical physics is still present: it is hidden, in a sense, behind the various "plausibility considerations". On the other hand, the emergence of a mixture assures a sort of reconciliation with Einstein's ideas.

[6]Published as Bub (1968), a development of the above letter of November 12, 1966—CT.

I hope that we shall be able to discuss more properly this question at the Cambridge meeting,[7] in July.

Yours Sincerely

A. Loinger

References

Bastin, T., (Ed.). (1971). *Quantum theory and beyond*, Cambridge: Cambridge University Press.

Bohm, D., & Bub, J. (1966a). A proposed solution of the measurement problem in quantum mechanics by a hidden variable theory. *Reviews of Modern Physics*, *38*(3), 453–469.

Bohm, D., & Bub, J. (1966b). A refutation of proof by jauch and piron that hidden variables can be excluded in quantum mechanics. *Reviews of Modern Physics*, *38*(3), 470–475.

Bub, J. (1968). The Daneri-Loinger-prosperi quantum theory of measurement. *Nuovo Cimento B*, *LVII*(2), 503–520.

Daneri, A., Loinger, A., & Prosperi, G. (1962). Quantum theory of measurement and ergodicity conditions. *Nuclear Physics*, *33*, 297–319.

Daneri, A., Loinger, A., & Prosperi, G. (1966). Further remarks on relations between statistical mechanics and the quantum theory of measurement. *Nuovo Cimento B*, *44*(1), 119–128.

Freire Jr., O. (2015). *The quantum dissidents: Rebuilding the foundations of quantum mechanics (1950–1990)*. Berlin: Springer.

Pessoa Jr., O., Freire Jr., O., & De Greiff, A. (2008). The tausk controversy on the foundations of quantum mechanics: physics, philosophy, and politics. *Physics in perspective*, *10*, 138–162.

Wheeler, J. A., & Zurek, W. H. (1983). *Quantum theory and measurement*. Princeton: Princeton University Press.

[7]See C133, p. 184, n 6 and C136, p. 250, n 1. A paper was given at Cambridge by G.M. Prosperi, "Macroscopic Physics, Quantum Mechanics and Quantum Theory of Measurement", Bastin (1971), pp. 55–64 with a reply by Bub, "Comment on the Daneri-Loinger-Prosperi Quantum Theory of Measurement", pp. 65–70—CT.

Appendix B

A Letter from Jeffrey Bub to David Bohm on an Article by Rosenfeld

[Rosenfeld was a supporter of the Danieri, Loinger and Prosperi approach to "measurement in quantum mechanics" and presented his own paper based on their approach Rosenfeld (1965). For more details see Freire (2015). Jeffrey Bub has provided a copy of this six page letter to David Bohm on Rosenfeld's article. Rather than transcribing it – a difficult task – I have presented an image of the original believing it demonstrates Bub's considerable forensic mathematical skills. There were no publications arising from this work, presumably because dealing with DLP (from both Bohm's philosophical and Bub's technical analysis) was considered sufficient—CT.]

C. Talbot (ed.), *David Bohm's Critique of Modern Physics*,
https://doi.org/10.1007/978-3-030-45537-8

1.

November 12th, 1966.

Dear Professor Bohm,

In my last letter to you I mentioned that Rosenfeld's article on the Daneri-Loinger-Prosperi paper actually obscures the essential argument. The following should explain what I meant. Unfortunately, it is again necessary to stick closely to the actual words and notation used by Rosenfeld and to compare this with the D-L-P notation, because Rosenfeld plays both on the meaning of words like "probability" and also on the D-L-P notation. This time I'll spare you my handwriting, though.

After dividing the state space of the macro-system into energy shells, Rosenfeld subdivides these shells into <u>cells</u> (page 225, bottom). "Each cell thus corresponds to a manifold of s_k orthogonal states Ω_{ki} (1 =1, 2, ... s_k) which will not in general be eigenstates of the energy. There are a great many cells in each energy shell, nearly all of which contain numbers of states s_k very much smaller than the total number S of states of theshell; but <u>some</u> exceptional <u>cells</u> [note the plural], corresponding to states of equilibrium of the body under specified external conditions, contain numbers of states of the same order of magnitude as S. Whatever the initial state of the body, its natural evolution will eventually carry it, after a time of the order of the relaxation time, into the <u>cell</u> [singular] corresponding to the equilibrium under the given conditions. It will then, at least within any reasonable time, remain in <u>this cell</u>, except for occasional excursions to smaller cells, corresponding to the usual fluctuations."

Rosenfeld then states (page 226) that if the initial state vector of the macro-system is $\Omega_{\ell j}$, the time-evolved state $\mathcal{E}(t)\Omega_{\ell j}$, where $\mathcal{E}(t) = \exp(-iHt/\hbar)$, will at time t be within the cell $\{\Omega_{ki}\}$ (i.e. the cell spanned by the basis vectors Ω_{ki}, 1 = 1, 2, ... s_k) with probability: $\sum_i |(\Omega_{ki}, \mathcal{E}(t)\Omega_{\ell j})|^2$. Now, as I have pointed out in my previous letter to you, this is just not true. This expression is the norm of the projection of the vector $\mathcal{E}(t)\Omega_{\ell j}$ onto the cell $\{\Omega_{ki}\}$, i.e. the overlap integral of $\mathcal{E}(t)\Omega_{\ell j}$ and $\{\Omega_{ki}\}$. If this quantity is unity at time t and $\Omega_{\ell j}$ is normalized, then it is true that $\mathcal{E}(t)\Omega_{\ell j}$ is entirely within the cell $\{\Omega_{ki}\}$. If it is not unity, then $\mathcal{E}(t)\Omega_{\ell j}$ is certainly not in the cell $\{\Omega_{ki}\}$ at time t, because it also spans other cells. Of course, if a ~~measurement~~ suitable measurement is performed at time t, then <u>the probability of obtaining a result corresponding to the eigenmanifold $\{\Omega_{ki}\}$</u> is just this expression, and the projection postulate further states that if this result is obtained, the state vector will be within this cell after the measurement.

Anyway, this is a small point which only confuses his argument a little.

($\neq \mathcal{E}(t)\Omega_{\ell j}$, because a measurement is performed at time t)

2.

Consider now Rosenfeld's ergodicity conditions:

$$M\{\sum_t |(\Omega_{ki}, \mathcal{E}(t)\Omega_{lj})|^2\} \approx s_k/s$$

$$|M\{\sum_t (\Omega_{ki}, \mathcal{E}(t)\Omega_{lj})(\Omega_{ki}, \mathcal{E}(t)\Omega_{l'j'})\}| \ll s_k/s \quad (l \neq l')$$

(1) (Rosenfeld's numbering)

The point to notice here is that, because Rosenfeld has so far not mentioned "channels" (see D-L-P), these conditions seemsto refer to any cells within a particular energy shell, whereas th D-L-P conditions are explicitly formulated within a particular channel of a particular energy shell. (They have an extra index. D-L-P's index ν corresponds to Rosenfeld's k.)

To see how this really confuses the whole argument, follow Rosenfeld further. On page 227 he says:

"If we start from any state

$$\psi_0 = \sum_i c_i \varphi_i$$

of the atomic system, the state of the total system after the interaction is terminated will be approximately expressed as

$$\sum_r c_r \varphi_r \Phi_r$$

and at any later time t, as

$$\sum_r c_r \mathcal{E}_0(t)\varphi_r \mathcal{E}(t)\Phi_r$$

Thus, once the correspondence between the states φ_r and Φ_r has been established during the short time τ by the interaction between the two systems, it subsists after they have again separated. We may now leave the atomic system to its fate and all we have to do is to study the evolution of the measuring body and its very essential consequences for the nature of the information which the measuring process can give us.

This last stage, which may be described as the formation of a permanent record of the measurement, is represented in our model by the natural evolution of the macroscopic apparatus from the unstable state Φ_r in which it has been brought by the interaction with the atomic system to the state of equilibrium which corresponds to the external conditions resulting from the measurement. In ord r to obtain a precise formulation of the situation, we must be a little more specific about the choice of the cell subdivision of the energy shell. Each state Φ_r will in general be a superposition of the states Ω_{ki} defining the cells, but we may always choose the latter in such a way that a definite group of cells $\{\Omega^{(r)}_{ki}\}$ corresponds to each state Φ_r, without overlap between these groups."

These "groups of cells $\Omega^{(r)}_{ki}$" evidently correspond to the D-L-P channels, i.e. the superscript (r) corresponds to the D- L-P subscript k. Rosenfeld seems to imply that, immediately after the interaction, the apparatus is in a certain definite

3.

state Ψ_r (specific r) within a certain "channel" (group of cells), although it may not yet be in the equilibrium cell within that "channel". Note that this has not yet been proved, and is presumably to be derived in some way from the magical ergodicity conditions.

Rosenfeld then computes the "correlation coefficient:

$$C_{k,rr'} = \sum_i (\Omega_{ki}, \mathcal{E}(t)\Psi_r)(\Omega_{ki}, \mathcal{E}(t)\Psi_{r'})^* \quad \ldots\ldots\ldots (5)$$

between the probability amplitudes for any two states Ψ_r, $\Psi_{r'}$, and any cell $\{\Omega_{ki}\}$; in particular, $C_{k,rr'}$ represents the probability of finding at time t the state evolving from Ψ_r in the cell $\{\Omega_{ki}\}$." (I have already pointed out that this statement is only true if a measurement is performed at time t)

Substituting:

$$\Psi_r = \sum_k \sum_{i=1}^{s_k} (\Omega_{ki}^{(r)}, \Psi_r)\, \Omega_{ki}^{(r)} \quad \ldots\ldots (3)$$

with

$$\sum_{ki} |(\Omega_{ki}^{(r)}, \Psi_r)|^2 = 1$$

~~xxxgxkx~~ into (5) we get:

$$C_{k,rr'} = \sum_{\ell j} \sum_{\ell' j'} (\Omega_{\ell j}^{(r)}, \Psi_r)(\Omega_{\ell' j'}^{(r')}, \Psi_{r'})^* \sum_i (\Omega_{ki}, \mathcal{E}(t)\Omega_{\ell j}^{(r)})(\Omega_{ki}, \mathcal{E}(t)\Omega_{\ell' j'}^{(r')})^*$$

Consider now:

$$C_{k,rr} = \sum_{\ell j} |(\Omega_{\ell j}^{(r)}, \Psi_r)|^2 \sum_i |(\Omega_{ki}, \mathcal{E}(t)\Omega_{\ell j}^{(r)})|^2 + \sum_{\ell j \neq \ell' j'} (\Omega_{\ell j}^{(r)}, \Psi_r)(\Omega_{\ell' j'}^{(r)}, \Psi_r)^* \sum_i (\Omega_{ki}, \mathcal{E}(t)\Omega_{\ell j}^{(r)})(\Omega_{ki}, \mathcal{E}(t)\Omega_{\ell' j'}^{(r)})^*$$

The time average:

$$M C_{k,rr} \approx s_k/s + 0, \text{ by ergodicity conditions} \quad \ldots\ldots (7)$$

Rosenfeld states that this is ~~xxxyxxxxxx~~ "negligibly small, except for the cell $\{\Omega_{e_r i}\}$, for which it practically amounts to certainty; in other words, we may write instead of Eq. (7)

$$M C_{k,rr} \approx \delta_{ke_r} \quad \ldots\ldots\ldots\ldots (8)"$$

Now this is very misleading, because it seems as if there is one <u>unique</u> equilibrium cell $\{\Omega_{e_r i}^{(r)}\}$ associated with each Ψ_r. Of course it has been implied that there is only one equilibrium cell within each group of cells (channel) specified by a fixed r. (These equilibrium cells have weight s_{e_r}/s. Clearly s_{e_r}/s cannot be ≈ 1 for all r. There seems to be a question of normalization here which Rosenfeld hasn't faced. In the D-L-P theory, the weights of the equilibrium cells $C_{\kappa e_\kappa}$ are $s_{\kappa e_\kappa}/s_\kappa$, where s_κ is the dimensionality <u>of the channel C_κ</u>. In Rosenfeld's notation this would be $s_{e_r}^{(r)}/s^{(r)}$, where $s^{(r)}$ is $\sum s_k$ (all cells in group $\{\Omega_{ki}^{(r)}\}$). So Rosenfeld's ergodicity conditions seem to be a little faulty.) Nevertheless, (using an index t instead of r to make the point clear) <u>each</u> equilibrium cell $\{\Omega_{e_t i}^{(t)}\}$ -- <u>for any t</u> -- is mathematically as good as any other, i.e. they all have $s_k/s \approx 1$ for $k = e_t$ (any and all t). So, strictly speaking:

$$\lim_{T\to\infty} \frac{1}{T}\int_{t_1}^{T} C_{k,rr}\,dt \approx s_k/s,$$

This means summation over all k from the group $\{\Omega_{ki}^{(r)}\}$ — fixed r.

4.

and:

$$\frac{s_K}{s} = \frac{\text{dimensionality of cell spanned by } \Omega_{Ki}\ (i=1,2,\ldots s_K)}{\text{dimensionality of energy shell}}$$

$$\approx 0 \text{ if } \{\Omega_{Ki}\} \neq \text{ some equilibrium cell } \{\Omega_{e_r i}\}$$

$$\approx 1 \text{ if } \{\Omega_{Ki}\} = \text{ " " "}$$

(the comment about normalization applies here)

Thus, we may say (instead of Rosenfeld's comments before (8):

$$M\, C_{K,rr} \approx s_K/s$$

which is negligibly small, except for the cells $\{\Omega_{e_r i}\}$, for which it practically amounts to certainty; in other words we may write instead of Eq. (7):

$$M\, C_{K,rr} \approx \delta_{Ke_r} \quad (\text{for any } r)$$

Now this result, which follows strictly from Rosenfeld's mathematics (as distinct from his implicit assumptions) is completely unsatisfactory. For if the ergodic motion of the free macro-system after a measurement interaction can carry an initial state Ψ_r into an equilibrium cell $\{\Omega_{e_t i}\}$ with t ≠ r, then the equilibrium cells certainly cannot be correlated with states of the object system, i.e. they can give no information about the state of the object system. Clearly, therefore, Rosenfeld must be making the additional assumption (which is made explicitly by D-L-P and embodied in their ergodicity conditions) that the time evolution operator $\mathcal{E}(t)$ of the free macro-system never carries the state vector Ψ_r out of the group of cells $\{\Omega_{e_r i}\}$ into which it has been brought by the interaction with the object system. Only then can he claim that:

$$M\, C_{K,rr} \approx s_K/s \qquad \ldots\ldots\ldots\ldots (7)$$

$$\approx \delta_{Ke_r} \quad \text{(same r both sides of the equation)} \ldots.(8)$$

But now, consider again the "correlation coefficient" $C_{K,rr'}$. Applying Rosenfeld's ergodicity conditions (the D-L-P conditions do not apply to a term of the form $\sum(\Omega_{Ki}, \mathcal{E}(t)\Omega^{r}_{e_r i})(\Omega_{Ki}, \mathcal{E}(t)\Omega^{r'}_{e_{r'} i})$ for r ≠ r') we find for the time average:

$$M\, C_{K,rr'} \approx \delta_{rr'}\, s_K/s \qquad \ldots\ldots\ldots\ldots(6)$$

Rosenfeld now claims (comments after (6)): "This leads us to the conclusion -- indicated by the factor $\delta_{rr'}$ in Eq. (6) -- that the correlations between the different states Ψ_r, $\Psi_{r'}$ disappear when the state of equilibrium is reached; they are wiped out, so to speak, in the process of recording the result of the measurement."

At first sight this statement, which is the crux of Rosenfeld's conclusion, seems incontestable. However, if we remember that an additional assumption, not embodied in the ergodicity conditions as formulated here, viz. (is necessary) that $\Psi_r(t) = \mathcal{E}(t)\Psi_r(0)$

5.

(which is orthogonal to the group of cells $\{\Omega^{(r')}\}$)

always remains within the group of cells (channel) $\{\Omega^{(r)}_{\kappa i}\}$, i.e. $\Psi_r(t) = \Psi_r(t)$, ~~is necessary,~~ we see that the term $\delta_{rr'}$ in equation (6) in fact has nothing ~~kaxda~~ at all to do with the operation of time-averaging. It depends only on the orthogonality of $\Psi_r(t)$, $\Psi_{r'}(t)$, which follows from the orthogonality of $\Psi_r(0)$, $\Psi_{r'}(0)$ and the above assumption on $\mathcal{E}(t)$. Thus:

$$C_{\kappa,rr'} = \sum_{\ell j} \sum_{\ell' j'} (\Omega^{(r)}_{\ell j}, \Psi_r)(\Omega^{(r')}_{\ell' j'}, \Psi_{r'})^* \sum_i (\Omega_{\kappa i}, \mathcal{E}(t)\Omega^{(r)}_{\ell j})(\Omega_{\kappa i}, \mathcal{E}(t)\Omega^{(r')}_{\ell' j'})^*.$$

Ψ_r is a linear superpostion of states $\Omega^{(r)}_{\kappa i}$ from a definite group of cells $\{\Omega^{(r)}_{\kappa i}\}$ containing the states $\Omega^{(r)}_{\kappa i}$ in the linear superposition Ψ_r (bottom page 227). Since $\Omega_{\kappa i}$ is fixed, it must belong either to $\{\Omega^{(r)}_{\kappa i}\}$, or to $\{\Omega^{(r')}_{\kappa i}\}$, or to neither group. Since these groups of cells are orthogonal, by the assumption on $\mathcal{E}(t)$ either $(\Omega_{\kappa i}, \mathcal{E}(t)\Omega^{(r)}_{\ell j}) = 0$, or $(\Omega_{\kappa i}, \mathcal{E}(t)\Omega^{(r')}_{\ell' j'}) = 0$ (or both). Hence:

$$C_{\kappa,rr'} = \delta_{rr'} C_{\kappa,rr}.$$

Again, the last expression on page 229, representing the interference term in a measurement is:

$$\sum_{r \neq r'} c_r c_{r'}^* (\chi_s, \mathcal{E}_a(t)\varphi_r)(\chi_s, \mathcal{E}_a(t)\varphi_{r'})^* C_{\kappa,rr'}.$$

Here $C_{\kappa,rr'}$ is zero if r ≠ r' simply because Ψ_r, $\Psi_{r'}$ are assumed to belong to different orthogonal manifolds at all times, not because of a destruction of interference as a result of the ergodic motion of the macro-system.

Note that D-L-P are not guilty of this confusion. For example, look at D- L-P page 312, equation (10.3) ("the probability of finding ~~t~~he system I in the state χ_s and the system II in the macrostate $C_{\kappa\nu}$ at the time t"):

$$\sum_{i=1}^{S_{\kappa\nu}} |(\chi_s \chi_{\kappa\nu i}, \exp[-\tfrac{i}{\hbar}(H_I + H_{II})t] \sum_r c_r \varphi_r \Phi_r)|^2$$

$$= |c_\kappa|^2 |(\chi_s, \exp[-\tfrac{i}{\hbar}H_I t]\varphi_\kappa)|^2 \sum_{i=1}^{S_{\kappa\nu}} |(\Omega_{\kappa\nu i}, \exp[-\tfrac{i}{\hbar}H_{II} t]\Phi_\kappa)|^2$$

$$\longrightarrow |c_\kappa|^2 |(\chi_s, \exp[-\tfrac{i}{\hbar}H_I t]\varphi_\kappa)|^2 \delta_{\nu\kappa}$$

Note that all terms with $r \neq \kappa$ are dropped because of orthogonality before time averaging (represented by arrow).

This corresponds to Rosenfeld's expressions at the bottom of page 229 ("the probability of "finding the value b_s -- associated with the eigenstate χ_s -- for the quantity B" of the object system and of having the apparatus in one of the states of the cell $\{\Omega_{\kappa i}\}$):

$$\sum_i |\sum_r c_r (\chi_s \Omega_{\kappa i}, \mathcal{E}_a(t)\varphi_r \mathcal{E}(t)\Psi_r)|^2$$

$$= \sum_{r,r'} c_r c_{r'}^* (\chi_s, \mathcal{E}_a(t)\varphi_r)(\chi_s, \mathcal{E}_a(t)\varphi_{r'})^* C_{\kappa,rr'}$$

$$\longrightarrow \sum_{r,r'} \cdots\cdots \delta_{rr'}\delta_{\kappa r} = \sum_r |c_r|^2 |(\chi_s, \mathcal{E}_a(t)\varphi_r)|^2 \delta_{\kappa r}$$

Apparently Rosenfeld does not realise that the summation over r,r' can be dropped, because $\Omega_{\kappa i}$ must be orthogonal to all except one Ψ_r, i.e. --

(see next page)

6.

$$\sum_i \left|\sum_r c_r (\chi_s \Omega_{ki}, E_a(t)\phi_r E(t)\Phi_r)\right|^2$$

$$= \sum_i |c_r (\chi_s, E_a(t)\phi_r)(\Omega_{ki}, E(t)\Phi_r)|^2$$

(r refers to the unique group of cells containing the cell $\{\Omega_{ki}\}$)

$$= |c_r|^2 |(\chi_s, E_a(t)\phi_r)|^2 \sum_i |(\Omega_{ki}, E(t)\Phi_r)|^2$$

The time average of <u>this expression</u> is, of course;

$$|c_r|^2 |(\chi_s, E_a(t)\phi_r)|^2 \, \delta_{k \epsilon r}$$

which means that after a sufficiently long time the state of the apparatus is in the particular equilibrium cell $\{\Omega^{(r)}_{ki}\}$, that the object system state is ϕ_r, and that the probability of this event is $|c_r|^2$ (where the initial object system state is $\sum_r c_r \phi_r$). If we do not bother to look at the macro-instrument at this time, we can therefore predict that the probability (of "finding the value b_s for B") is:

$$\sum_r |c_r|^2 |(\chi_s, E_a(t)\phi_r)|^2 \, \delta_{k \epsilon r}$$

i.e. a summation over the probabilities of all the possible events (equilibrium states of the macro-system).

Rosenfeld derives this last expression by taking:

$$M C_{k,rr'} = \delta_{rr'} \delta_{k \epsilon r}$$

which follows from <u>his</u> ergodicity conditions. This confuses the whole issue, because it makes it look as if the result depends essentially on the averaging. (Actually this result is, strictly speaking, incorrect within the context of Rosenfeld's derivation, because it does not represent the probability of "finding the value b_s for B" <u>and</u> of "having the apparatus ... in one of the states of the cell $\{\Omega_{ki}\}$", if by this is meant a <u>particular</u> cell (within a particular group of cells $\{\Omega^{(r)}_{ki}\}$). <u>If this is</u> what is meant, then clearly there should be no summation over r. Only if Rosenfeld really intends "having the apparatus eventually in any equilibrium cell", does this expression follow.)

So it seems as if, like Jauch and Piron, D-L-P and especially Rosenfeld have missed the point, begged the question, or just proved nothing essentially new at all, unless I have completely misunderstood their papers. This is always a possibility and I'd be glad to be re-assured by you on this point.

Sincerely,

References

Freire, Jr., O. (2015). *The quantum dissidents: Rebuilding the foundations of quantum mechanics (1950–1990)*. Berlin: Springer.

Rosenfeld, L. (1965). The measuring process in quantum mechanics. *Progress of Theoretical Physics Supplement, E65*, 222–231.

Appendix C

The Numbers/Maths "Game"[8]

Information from Jeffrey Bub:

Suppose you observe a system over time and notice that it transitions between two states that you label 0 and 1 once every second, with an equal probability of a transition to the same state, or to the alternative state. You might conclude that the system behaves randomly, like a coin that has an equal probability of landing heads or tails, according to the probabilistic rule:

$\text{prob}(0 \to 0) = 1/2$
$\text{prob}(0 \to 1) = 1/2$
$\text{prob}(1 \to 1) = 1/2$
$\text{prob}(1 \to 0) = 1/2$

But then you look at sequences of states: 00, 01, 10, 11. You notice that these transitions are deterministic:

$00 \to 11$
$01 \to 10$
$10 \to 01$
$11 \to 00$

So a better description would be in terms of four states, not two: the four distinct ordered pairs of 0's and 1's.
In fact, the system transitions as follows:
001100110011 ...
and the pairs 00, 01, ... carry full information, not 0 and 1.

As I recall, the point I wanted to make was that if you define the states as 0 or 1, then everything looks random (of course, only an automaton could see it this way; a human

[8]C131, p. 63 and p. 65—CT.

C. Talbot (ed.), *David Bohm's Critique of Modern Physics*,
https://doi.org/10.1007/978-3-030-45537-8

would immediately see what's going on). If you define the states as pairs, then you see that the system is deterministic. That is depending on what features you abstract as specifying the 'state' of the system (in the sense of a 'complete' specification of the system), you get very different descriptions of the behavior. I related this somehow to Bohm's talk of distinctions, relations, orders, patterns, and structures. I was thinking, broadly, of hidden variables and quantum mechanics, where we were trying to relate the ('incomplete') probabilistic description of quantum mechanics to a deterministic description in terms of new ('complete') states.

Appendix D

On the Failure of Communication between Bohr and Einstein[9]

D. Bohm and D. L. Schumacher,

Birkbeck College, University of London

The most relevant point concerning the discussions between Bohr and Einstein[10] is that they did not communicate, in spite of serious effort to do so. It is even now difficult to face the simple fact of their failure to communicate in a similarly serious fashion, without prejudgment concerning the truth of either position. Although in some superficial sense Einstein may have admitted the consistency of Bohr's arguments, it is clear that he did not understand the novel implications of what Bohr was trying to say. Likewise, it seems evident that for this reason Bohr did not fully understand what made it impossible for Einstein to see the full meaning of his novel contribution to scientific discourse. It would therefore not be relevant to suppose, for example, that Bohr was clear and Einstein confused (or vice versa) and that the confusion of one or the other of these men brought about his failure to grasp the content of what the other man said. Rather, the essential point in this situation was a mutual kind of confusion in which each failed to see what was relevant to the other: thus both contributed to the failure of communication which was not a property of either person. The fact of a failure of communication in this situation was much more significant than the content of what was to be communicated. It led physics to split into mutually irrelevant fragmentary parts which tended to develop fixed forms, rather than to engage in a genuine dialogue in which each would change, permitting something new to emerge.

The first point of confusion in the argument was Einstein's use of the word "disturbance" in connection with the quantum context. Einstein's whole argument depended on the statement that two distant atoms do not "interact" (so that by "observing the

[9]Birkbeck archives document B44, undated—CT.

[10]No references are given but see references given in Appendix E—CT.

C. Talbot (ed.), *David Bohm's Critique of Modern Physics*,
https://doi.org/10.1007/978-3-030-45537-8

spin of one of them", one knows that of the other, without any "disturbance" at all). But the word "interaction" has meaning only in a classical context, in which one atom exerts a "force" on another, this effect being described in terms of Newton's laws applied to well-defined orbits. On the other hand, if one looks at the many-body wave equation, one sees that to bring in words like "potential energy of interaction" to the description is even from a purely formal point of view a rather inappropriate metaphor. Formally, what we have in the quantum context is an abstract "potential function" which enters the equation for a 3N dimensional "wave", in which description there is no place relevantly to discuss localized entities (such as particles), or localized events (such as measurements), or anything else to which the word "interaction" could meaningfully be applied.

Now what Bohr did was to say all this only implicitly, in terms of his notion of the wholeness of the experimental conditions and the experimental results. But Einstein could not see the relevance of Bohr's arguments, because in them, tacitly and informally, Bohr gave a central place to the term "observing apparatus". Of course, Bohr did not mean by this to accept the positivist notion that measurements or observations are the most essential aspects of physics. Nevertheless, informally he could not avoid communicating to Einstein the implication that he regarded observation to have a fundamental relevance in the content of physics. The key point was, of course, that the terms "observing apparatus" and "object" were being tacitly taken by Bohr to be informal forms of discourse, whereas Einstein supposed that he was taking them to be the aspect of essential content of discourse. However, by so heavily emphasizing these forms in his answer to Einstein, Bohr was informally strengthening Einstein's feeling that Bohr was taking the observing instrument as part of the basic content of physics. Thus, Bohr's response to Einstein was not relevant, in that it did not properly meet Einstein's tacit, informal and unstated objections and reservations, and instead, tended to strengthen these rather than answer them. Further, the notion of wholeness became confused as it tended to become a content of discourse referring to the impossibility of separating the "apparatus" from the "object". Rather, it is the form of the experimental conditions and the content of the experimental outcome which are a whole in the sense of Bohr's quantum description.

In order to reach Einstein, Bohr should perhaps have begun immediately, and without complicated arguments, to question the relevance of the term "interaction" in the quantum context. Since the word "interaction" has no meaning in this context (even though most physicists use it continually in their work with quantum theory), it follows that the notion of "no interaction" also has no meaning. Thus, Einstein's conclusion that the properties of an atom could be measured without "disturbing" that atom, or some other atom, depended on the use of irrelevant descriptive language. Without this conclusion, Einstein's inference that there are separately existent "elements of reality" corresponding to each atom is untenable.

To demonstrate the direction in which Einstein's descriptive language was taking him, Bohr could have called Einstein's attention to the kind of non-linear field theories that Einstein felt to be capable of providing a universal mode of description. In such theories, there can be no question of a separately existent particle. Rather, the word "particle" is only a convenient name for a pattern or structure of fields that pervades all space, but has a relatively localized pulse-like region of high intensity (or perhaps even of singularity). In such a description, each "atom" actually interpenetrates all the others (including those comprising any "measuring apparatus"). So, at least on the face of the matter, it would appear that there was no reason for Einstein not to consider the long range coordination of responses implied in the experiment of Einstein, Podolsky and Rosen as a whole (similar, in certain ways, to what Bohr suggested). Certainly in the "non-relativistic" context to which they had restricted themselves ($c \to \infty$), it would not appear at all strange to have a description of this kind, because in this case all "parts" of the universe would have to be in immediate and direct contact.

Of course, Einstein would have objected at this point by saying that to him, the relevant context for field theory was also that for other relativistic notions, especially that of a signal, with a finite universal limit of velocity of propagation. Since the EPR experiment, regarded in this way, would imply the instantaneous transfer of a certain communicable content ("information"), Einstein would never have regarded such a point of view to be worthy of serious consideration, not even for a moment.

In this connection it is essential to recognize that for Einstein, the notion of "signal" was so basic that he was probably not aware of how deeply it pervaded his thinking in its role as a universally relevant form which is in the background of all his discourse. It is rather like water, of which fishes may be unaware. Therefore, Bohr's principal task in this situation would have been to call Einstein's attention to his tacit relevance judgments which made it impossible for him seriously to consider other informal language forms, e.g. Bohr's, as at least potentially reasonable points of view which could usefully be explored.

In doing this, Bohr would have been able to indicate that Einstein was using "signal" as a form of discourse, rather than as part of the basic content of physics. In this form there is implicit a distinction between a "signal" and its "significance". That is, Einstein was making a separation of form and content in his communication. Evidently, these are an unanalysable whole, (e.g., it is the "significance" that justifies calling a set of events a "signal" rather than "noise"). So Einstein hinself was using a purely formal distinction of "signal" and "significance" as basic to his terms of description in the context of physics. Bohr could then have pointed out that if "signal" had been mistakenly regarded as part of the basic content of physical description, then one could have been led to a view which would also have made "significance" (or "meaning") an essential part of this content. This would have put the conscious observer who thinks about the meaning of his observations into a very central role indeed. Thus Einstein's theory of relativity is just as open to such a misinterpretation as is Bohr's, in which one could similarly regard the distinction between observing apparatus and observed system to be part of the basic content of physics.

In this way, Bohr could probably have made it clear to Einstein that every theory tacitly assumes the relevance of certain formal distinctions as basic to its terms of discourse. Assuming the relevance of these distinctions does not commit us to the assumption that they are a necessary part of the content of physics (or that they are of any "ontological" significence). In this way, Einstein's feeling that Bohr was making the whole universe depend on the observer (or on his measuring instruments) could have been shown to be unfounded. Einstein then could have seriously entertained the notions that Bohr was putting forth for consideration.

But then, Einstein would almost certainly have been led to question certain features of Bohr's presentation. For Bohr tacitly suggested that his informal forms of discussion of the quantum context should have universal relevance. To be sure, he emphasized that physical theory must develop in novel ways. But it was generally implied that these new developments could be assimilated as extensions of the "quantum" mode of description, in the sense that the latter would remain relevant in such novel contexts. Specifically, the paper on field measurements by Bohr and Rosenfeld indicated that under certain assumptions the "quantum" mode of description can be extended to the context of field theory, so that in principle one could go on to discuss relativistic theories in such terms.

It is not likely that Einstein would have been satisfied with this view of the harmonizing of relativity and quantum theory. Firstly, in the Bohr-Rosenfeld paper there appeared informal notions (such as particles of effectively infinite charge and mass), which do not seem to be relevant to the atomic constitution of matter as accepted by the authors. Surely it is significant here that as yet there is not even a consistent formal theory of elementary particles that is "relativistic". In view of the persistent difficulty of producing such a theory over a period of forty years or so, one may reasonably wonder whether it is in fact possible to do this at all. And without such a theory, Einstein (and probably a large number of other physicists as well) would feel that the formal treatment of the measurability of fields is too abstract to be very relevant to the serious situation in physics. To this, one could add that, just as Bohr emphasizes, the consistency of the whole mode of description depends critically on the smallness of $e^2/(ch)$. This might tend to suggest that it is only when "relativistic" considerations are unimportant that Bohr's mode of description can be considered relevant. And it need hardly be added in the present context that Bohr's mode of description was also his mode of communication.

In addition to the more or less technical questions, Einstein would almost certainly have criticized Bohr's way of describing the experimental context solely in terms of the language of classical physics with words like "position" and "momentum"). Einstein's point of view implied that ultimately one would cease to use words like "the position and momentum of a particle" or "the value of a field quantity at a given point" as basic terms for description of the experimental conditions and the experimental results. Rather, in terms of a non-linear field theory, one can only talk about various aspects of the whole field. The words "momentum" and "particle", etc., and the phrase "field at a given point" can best be regarded as abstractions, relevant in the general context of this whole, just as words like "window" and "doorway" are abstractions relevant in the context of a house. So, as we would not feel obliged to describe a house as an interacting aggregate of rooms, windows, doorways, etc., each with its position and momentum, we would similarly not feel it necessary to discuss cloud chamber photographs and geiger counter clicks in terms of "paths of particles". Rather, we would say that these furnish indications that are relevant to the condition of the whole field, so that descriptions like "the value of a field quantity at a given point" would also have little or no relevance.

Without necessarily adopting Einstein's particular way of approaching this question, one can say that he also had certain insights which are relevant for physics. For implicitly at least, he was pointing out that new kinds of theories implied a new *informal* language of description of the experimental situations and the experimental results that is in harmony with the formal language. On the other hand, by adopting "classical" terms for the informal language of physics, and an entirely different set of "quantum" terms for its formal language, Bohr was introducing a certain further kind of disharmony into the whole situation of the mode of description and communication in physics. In this sense one can say that implicitly at least, Bohr was subscribing to the current usage in which the formal terms of the quantum theory, such as "operator", "commutator", "Hilbert space", etc. are sharply contrasted with informal terms, such as "the curvature of a track", "the click of a Geiger counter", etc.

When one first learns quantum theory, one is forcefully struck by the almost fantastic difference between these two kinds of terms. Of course, Bohr emphasized that they are related in the sense that the "noncommutation of two operators" corresponds to the mutual exclusiveness of the experimental conditions needed to measure them. But just emphasizing such a point in the language adopted merely continues this sort of disjunction of the language forms with a relationship of correspondence between them; this is not the same as wholeness, which in this case would be a harmony of the formal with the informal.

Even if in conversation Bohr called attention to the wholeness of formal and informal languages, the fact is that in his writing he gives the distinct impression, tacitly and informally, of accepting the current disjunction of these languages in physics to be natural and unavoidable. Surely Bohr would be the first to admit that the informal implications of what he writes after careful and due deliberation are at least as significant as what he says informally, and in many ways, even more so, otherwise he would not have taken the trouble to revise his manuscripts many times. If there is lack of harmony between the implications of the form of the writings and the content of what is expressed informally, this would suggest that Bohr did not manage to reach clarity on this point. And this is very probably one of the ressons why communication between Bohr and Einstein broke down.

One could sey that Bohr first admits classical language as a basic informal mode of description, and then in some way denies its universal relevance by saying that the content of a description which makes relevant such experimental conditions which would allow measuring the momentum is not compatible with the content of a description which makes relevant those experimental conditions which allow measuring the position. (For Bohr would not speak of the experimental conditions as *content* of the description.) This is a form of discourse that is not in complete harmony with Bohr's intention of wholeness. Such disharmony is similar to that which arises in the use of words like "inseparable". This word depends for its meaning on first asserting the notion of "separation" and then denying it. However, such a procedure still makes the notion of separation *relevant*. So to assert "inseparable wholeness" is a form of confusion, because "wholeness" implies the *irrelevance* of "separation": putting the two words together in this way implies their *mutual relevance* (e.g., one could relevantly imagine the possibility of separating the whole into parts, but that as a matter of contingent fact it is not actually possible to carry out such a separation). Similarly, Bohr appears to assert the concepts of classical dynamics and then to deny their universal relevance, so that there must apparently be such a denial depending on the contingent fact of the applicability of the quantum algorithm. A harmonious description would require that one does not begin with a language form implying the relevance of "disjunction into dynamical components" and then try to deny the relevance of this disjunction in the same context in which such relevance is necessarily implied in a tacit and informal way.

It is here that Einstein's insights are significant although they are confused by much that is irrelevant. They indicated especially that formal and informal terms of description need not be sharply contrasted and implicitly contradictory.

If Bohr could have understood the deeper meanings in Einstein's views, at least to the extent outlined above, then he could also have admitted the need to take seriously the possibility that the "quantum" mode of description may have limited relevance. For reasons given here already (as well as for other reasons) he could have seen that the quantum mode of description may not be relevant in any context which makes relevant the informal "relativity" language of Einstein, e.g. the context of the "non-relativistic" descriptions. Similarly, Einstein could have seen that the notion of a "signal" cannot be relevant in the "quantum" context, because the "significance" of such a signal implies a pattern or order of events, which can be present only in the limit where many "quanta" are involved. It has been said, for example, that in a very sensitive condition, the human eye can respond to two "quanta". But clearly, this response could not have enough "significance" to make it relevant to call these "quanta" by the name "signal".

The difficulty in the exchange between Bohr and Einstein was, insofar as it can be summarized, that each attached to his statements a very implicit universal relevance which is essentially a commitment to a notion of scientific truth. Bohr's implicit position was that eventually, all that is relevant in relativity theory could be harmonized with the language of the quantum description. Thus it was just Einstein's notion of scientific truth which he ruled irrelevant, and thus ruled out full communication with Einstein. For one's notion of scientimic truth is also one's notion of harmony. Likewise, Einstein implied that all that is relevant in quantum theory, which for him had very little to do with complementarity, could be assimilated in terms of a dynamical description for a complex nonlinear set of fields. Thus Einstein's notion of scientific truth was associated with dynamics and geometry as forms, which harmonized with the notion of "signal".

It is therefore extremely easy to regard "signal" as a form of disjunction and to regard "wholeness" as a form of union, in the context of these remarks. Rather, "signal" stands for a verbal <u>form</u> which implies in this context the analytic <u>content</u> of disjunction and union (in the formal languages of dynamics, geometry, and logical inferences), whereas form and content <u>are</u> a whole, e.g. the <u>form</u> of the experimental conditions and the <u>content</u> of the experimental results, (which is not the <u>object</u> of relevant discourse). It might be added that "individual effects" and "(quantum) ensembles" are another expression of this whole. Of course if these remarks are interpreted in a way which assumes a distinction between form and content, a certain kind of irrelevant discourse could accrue, as does, for example, when there is an implicit separation between a "theory" and "what it asserts". The "separation of subject and object" is a separation of form and content, and thus even a phrase like "expression of wholeness" could lead to a continuation of such discourse.

Communication between Einstein and Bohr could have been opened up if each had become aware of his implicit judgments of relevance, and if both had, thus, gone out to explore new contexts in which neither relativity nor quantum theory would be considered to be basically relevant. Such communication would have been creative, rather than merely a means of conveying each point of view to the other. In such communication one is not talking "about" quantum theory or "about" relativity. Such talking "about" has a very limited relevance; when it is carried beyond a limited context it can for example present essentially unresolvable "problems", or it can create opposing "sides" which cannot meet.

Similarly, just in writing these remarks it is not the main intention to talk "about" Bohr's view or "about" Einstein's view as if these were objects of discourse. (E.g., it is not the intention "to explain Bohr more clearly".) Rather, the "intention" is to say something essentially new (which is not really an "intention" at all, since this implies "interpretation", or a separation of form and content). This has so far not been said because all those who discussed these matters tended to continue the breakdown of communication between Bohr and Einstein.

So in the present work it is not an intention to "talk about our own views", or to "criticize others". The main relevant point which can be summarized briefly is that physicists have not been in full communication: if each of those concerned can be aware of implicit judgments, his own and those of others, then full communication can begin.

Appendix E

On the Role of Language Forms in Theoretical and Experimental Physics[11]

D. Bohm and D. L. Schumacher, Department of Physics,
Birkbeck College, University of London

One of the most remarkable characteristics common to the developments in modern physics is revolutionary change in fundamental notions which has been brought about by recognition of new forms of language in physics, rather than by the introduction of new physical content into previous language forms. For example, Einstein resolved the confused situation concerning the relationship of mechanics and electrodynamics which prevailed toward the end of the nineteenth century by ruling the Newtonian form of physical description in terms of absolute space and time to be irrelevant. His insight was based on the perception that it was no longer fruitful to continue with efforts to change the content of physics, as described in terms of the older language forms (e.g. by making new assumptions about the properties of the ether). Thus room was made for thinking in terms of a different language form, one that allowed a new kind of description of space and time as mutually related and dependent on general physical conditions, such as velocity, acceleration, gravitational fields, etc.

Similarly, Bohr saw that the general structure of discoveries concerning black body radiation, atomic spectra, etc. as well as the mere fact of the stability of atoms, implied the irrelevance of the whole general framework of the laws of classical dynamics, (descriptions in terms of continuous motion in well-defined orbits). That is to say, he perceived the uselessness of efforts to change only the content of these classical laws (e.g. by looking for mechanical explanations of the "quantum"). And, as happened with Einstein in a different context, room was then made for a new kind of language form that allowed for novel modes of description (which for example did not continue the traditional description in terms of separation of the observer from the observed).

---

[11] Birkbeck archives document B88, undated. References as given in the original—CT.

C. Talbot (ed.), *David Bohm's Critique of Modern Physics*,
https://doi.org/10.1007/978-3-030-45537-8

The full significance of these revolutionary changes in language form tends to be missed, because of a certain implicit but crucial distinction between formal language forms and informal language forms. The formal language form is, of course, a set of explicit and well-defined rules, such as algorithms, geometries, and structures of logical inferences. The informal language form is, however, just that of ordinary discourse (which is needed for the general description that is a subject for scientific inquiry). As is well known, such discourse also has a vast set of rules, mostly tacit and implicit. These rules are always changing as the language is used, but in ways that cannot be fully stated or formalized. Nor is it possible by means of formal and explicit rules to make a sharp distinction between formal and informal forms. Rather, this distinction is also largely contained tacitly in the informal language.

It is necessary to emphasize that the main contributions of both Einstein and Bohr were in the introduction of novel modes of informal discourse rather than in the development of new formalisms. Thus, Einstein introduced a basically new way of using informal language through the essentially linguistic requirement that absolute velocity shall have no relevance (in special relativity) and that absolute acceleration shall have no relevance (in general relativity). To satisfy these requirements, he introduced new informal terms of description, such as "signal" and "field", and he was thus enabled to develop quite new kinds of formal laws, which eventually gave a broad and yet specific content to his informal notions.

Similarly, Bohr also introduced a new kind of informal description into physics. This "quantum" mode of physical description had its origin in the perception of the irrelevance of physical descriptions in terms of well-defined orbits and also of the irrelevance of the disjunction between "observing instrument" and "observed system". The "quantum description" is a form of unanalyzable wholeness. This means that the description of the experimental results is not separable from the general experimental conditions, which provide a kind of general context, (rather as the description of modulation of a wave is not separable from a context given by the description of the wave that is modulated.) And as happened in Einstein's case, this new informal description led to new kinds of formal laws and to a great deal of new content.

In certain crucial ways, the development of physics subsequent to the basic work of Einstein and Bohr has been such as to hide the significance of such change in the informal language. Indeed, what has happened is that the formal language has been undergoing continual modification and extension, while very little change has taken place in the informal language. One can thus conclude that there has been a growing tendency to regard the informal language as not very relevant. Or if the informal language is admitted to be relevant, the usual view may perhaps be that Einstein and Bohr have already pretty well settled such questions, so that the main task of physics today is to get on with the development of new formalisms which will predict wider ranges of experimental results within existing forms of description.

This tendency to emphasize the formal language is especially noticeable in the work of von Neumann and those who follow the general course that he initiated. It has perhaps not been sufficiently noticed that von Neumann started from a certain kind of informal description which was very different from that of Bohr. A basic term in von Neumann's language was the "quantum state". This quantum state was described formally as a vector in Hilbert space (defined in a purely mathematical sense). But informally it was treated as a kind of separately existent "physical" counterpart of the wave function. In this way, the basic informal terms were in effect identified with the formal terms, so that the overall language was dominated by the formalism. This language dominated by the formalism was in fact an extension of the informal language which had already tacitly been established in classical physics, where mathematically defined terms (such as points described through coordinates or field intensities that were functions of coordinates) were taken to represent directly or "stand for" corresponding physical entities. Thus, in essence, von Neumann was trying to continue the general mode of informal description of classical dynamics, in contrast to Bohr, who radically altered this informal mode of language and description.[12]

There was, however, a serious problem in von Neumann's extension of the informal language form of classical dynamics to the "quantum" context. For if the "quantum state" is thus described as separately existent "physical reality", it is not at all clear how we can obtain knowledge (or information) about this "state". To try to solve this problem, von Neumann introduced a further set of "classical observables", which he assumed could be "known" in essentially the same way that one is supposed to be able to "know" the values of quantities that appear in ordinary classical description. Thus, he discussed in terms of two disjoint dynamical descriptions, the "quantum" and the "classical". Whether a given "physical system" was described "classically" or "quantum mechanically" in this language, was to a certain extent arbitrary. But what was not arbitrary was that somewhere there had to be a disjunction (or what von Neumann called a "cut") between the contexts in which these two descriptions applied.

It was never clear whether the use of these two disjoint dynamical descriptions is actually consistent. Indeed, because of some kind of (generally tacit) dissatisfaction with von Neumann's approach, this field became the subject of many efforts to solve what was eventually called the "measurement problem". A number of different schools have arisen in this field, and what is most striking in the literature on this subject is the general inability of different schools to agree on what the problem is and on what can be done about it.

This situation need not be regarded as surprising, for in the "measurement problem", we actually have a formal elaboration of what might be termed the "central problem of philosophy", at least as far as Western thinking is concerned. This problem was tacitly formulated by Descartes when he adopted two disjoint descriptions of the world, one in terms of "matter" or "extended substance" and the other in terms of "mind" or "thinking substance". Descartes supposed that these two disjoint "substances" were somehow related by God. But in a later historical period, when an appeal to God ceased to be accepted (at least academically) as a relevant solution of philosophical difficulties, what was left was the problem of how these two disjoint descriptions could be consistently related. This problem was carried forward by Kant and many of those who followed him, in the form of questions roughly equivalent

[12]Bohr (1963).

to "What am I?", "What are things in themselves?", "What are the limits of my knowledge about them?", and "How do I know what I know?".

Clearly, von Neumann is, in a way which is intimately related to this entire academic tradition, treating the "quantum state" as a separately existing "thing in itself" which is somehow "observed" through the mediation of a disjoint "classical observable" that can be "known" by a human being (i.e. it can be part of the content of his "mind").[13] So the disjunction of "knowledge" and "what is known" is inherent in the form of the language of von Neumann's approach. And as a more careful study shows, this disjunction leads to essentially the same kind of unresolvable difficulties that arise in the "central problem of philosophy". In informal discussion aimed at establishing the consistency of his treatment, von Neumann is led to talk in terms of a "cut" between the "mind of an observer" and the "physical world" and of a "psycho-physical parallelism" of these two disjoint domains, similar to a corresponding parallelism of "mind" and "matter that Descartes supposed to have been established by God.

In more modern philosophy, it has been noted specially by Wittgenstein that "problems" of the kind discussed previously are not relevant to the content of our knowledge, scientific or otherwise. Rather, they arise as an elaboration of a certain kind of language, in which the form of the description and the content are sharply disjoined. Form and content are however an inseparable whole which is not itself an object of relevant discourse (not further analyzable).

Nevertheless, in everyday life, there are a great many objects (such as tables, chairs, trees, etc.) which can be taken as a kind of content that is in some sense separate from or independent of the form of the description. That is to say, we can consistently talk "about" such objects and not get into difficulty, since we can get relevant "non verbal" observations by coming into contact with them. Thus, our descriptions can in turn add to our knowledge "about" such an object. On the other hand, when we come to talk about what is more abstract, such a form is not actually relevant. Consider, for example, the quantum theory. Is this a relevant object of discourse? At first sight, one might think that he is "looking at the quantum theory" as if it were present in the form of a "mental object". But a little reflection shows that there is no such "mental object". Whatever appears in the imagination is in essence a kind of "display" of the content of what is words (or in thought). The description of quantum theory is quantum theory. Or, in other words, quantum theory is the discourse rather than the object of discourse. There is no relevant way to talk "about" quantum theory, as if we could "gaze" at the quantum theory with the "eye of the mind". Similarly, while we can consistently talk "about" objects of triangular shape, there is no relevance to talking "about" a geometrical triangle. For the "geometrical triangle" that we imagine is merely a kind of "display" of the content of our discourse concerning the triangle. The "triangle" of geometry is not distinct from the description of it. So, rather than talk "about" geometrical "objects", what we can actually do is "talk geometry", i.e. make geometry the content and form of our discourse. Likewise, we

[13] von Neumann (1955), Chap. VI.

cannot relevantly talk "about" quantum theory, but instead, we can "talk quantum theory".

Of course, we all learn as children to imagine that we are looking at "mental objects" and gathering information about them which is independent of what we say and think about them. Many philosophers and mathematicians (especially influenced by Plato) have used a form of language in which it is implied that "mental objects" actually exist, and that one can accumulate knowledge "about" them by looking at them "inwardly". In this way, one is led to believe that through the "mental object", one "knows" the "real object", which is however also "a thing in itself, outside the observing subject". For example, one may thus feel that he "sees" the "quantum state". But since the "quantum state" one "sees" is actually only a "display" of the content of a description, it follows that there cannot be a "problem" of obtaining "knowledge about it". That is to say, the "quantum state" is already a description, and von Neumann's "classical observables" are merely another description disjoint from the first one. The use of two descriptions to cover what is basically a single context is then, in effect, rather like an attempt to describe the first description in terms of the second. Such efforts are bound to lead to insoluble problems, similar to those arising from the question, "How do I know what I know?"

Questions of this kind imply a sharp distinction, which is actually generally irrelevant, between form and content. Thus, the form in which quantum theory is described is quantum theory; the form in which knowledge is communicated is knowledge.

Only in a limited context is it relevant to talk in terms of content as if it were separable from the form of description (e.g. in connection with potentially or actually perceivable objects). But more generally, to change the form of an abstract kind of description is to change its content as well. This is indeed what was demonstrated by the insights of Einstein and Bohr, when they found that changing the form of our informal language radically altered the content of physics.

Bohr was in fact the first one explicitly to call attention to the relevance of the question of form and content for physics. His insight shows itself especially clearly in connection with the question of the observer and the observed. Now, these have generally been treated in scientific discussions as separate or existentially disjoint entities in some kind of interaction. This means that the relationship between observer and observed has in effect been regarded as part of the content of scientific knowledge (just as is the relationship between one particle and another). But what Bohr said implicitly is that these are merely two terms of discourse, having to do with the form of our communications, rather than with the content.

One should add here that generally speaking, the term "observer" or "instrument" corresponds to the form of the quantum description (e.g. the general context or experimental conditions), while the "observed result" corresponds similarly to the content of the description. But since form and content of a description are an unanalyzable whole, there can be no relevance to efforts to analyze the "interaction" of "observing instrument and observed object". Therefore, in Bohr's language, there is, and can be, no "measurement problem." It is significant here that Bohr never talked in terms of von Neumann's "cut" between observer and observed, neither explicitly nor implicitly. For example, he did not even refer to the famous Heisenberg "microscope experiment", which is usually discussed with the aid of a diagram showing electron, microscope, and photographic plate all together. This diagram tacitly and informally implies the separation of observer and observed because of the way they appear in the drawing. Nevertheless, it explicitly seeks to communicate Bohr's notions on the subject and thus to deny this seperation. Such a mode of discussion is a disharmony between form (disjunction of observer and observed) and content (wholeness of observer and observed). It is in many ways similar to the statement: "Never use a preposition to end a sentence with," in which likewise form is not in harmony with content.

This disharmony of form and content has confused the communication of what Bohr tried to say, and thus led to a very general misunderstanding in the whole question of whether statements such as the principle of complementarity refer to forms of the language of physics or to the content of physical knowledge. Thus, Heisenberg's approach was such as tacitly and informally to suggest that Bohr's notions have to do mainly with the content of physics. Other physicists, such as von Neumann, went even further in this direction. Indeed, by introducing the notion of disjoint sets of "quantum" and "classical" observables (or, similarly, the disjunction of "physical" and "psychological" domains), the latter adopted a mode of discourse that tacitly implied the irrelevance of the entire general form of Bohr's language. Therefore, with regard to the terms of informal description, there was no way for Bohr and von Neumann even to meet and argue about which point of view is better. The "measurement problem" which is of basic interest in von Neumann's approach, cannot even be discussed in Bohr's language, while Bohr's notion of the unanalyzable wholeness of the quantum description likewise cannot be discussed in von Neumann's language.

The following is an imaginary dialogue between Bohr and von Neumann, with regard to the terms of the informal description:–

N: "A clear distinction between the object and the apparatus is essential to me, and I can distinguish between them in a way which is internally consistent and purely descriptive. The 'object' is that part of the world which is considered to exhibit quantum effects, and the 'apparatus' is the part which is not."

B: "The clear distinction between the 'object' and the 'apparatus', the specification of the experimental conditions, is no less important to me, but the very word 'quantum', or 'quantum effects', can refer only to the whole description, including both the experimental outcome and the specification of the experimental conditions."

N: "All right, I can call the 'object' that part of the world which is affected by the mutual object-apparatus interaction, and the 'apparatus', the part which is not affected by it."

B: "What do you mean, 'affected by interaction'? The apparatus is affected in the sense that there is an irreversible registration of a recording mark in it. From the trend of your arguments it is clear that you wish to regard that as part of the content cf physical description."

N: "So I do, but the 'object' and the 'apparatus' can still be consistently distinguished analytically. The 'object' is dynamically affected by the interaction of the two, and the 'apparatus' is not dynamically affected by it."

B: "In using or implying a phrase like 'dynamical influence of the apparatus on the object', you are ultimately denying the principal point that quantum theory demonstrates, that the content of the quantum mechanical description is not separable from the specification of the experimental conditions." (...)
N: "But are not both the 'object' and the 'apparatus' parts of our world in the most general sense? If so, we certainly ought to be able to include them both in the content of an internally consistent description."
B: "But we were talking about some other type of description, to which the term 'quantum' cannot be applied. You have made the common mistake of using the term 'world' as if it had some meaning which is entirely independent of the content of the physical description. These terms 'object' and 'apparatus' do not denote 'parts of the world', one part obeying one sort of law and the other, another. The principle of complementarity is ... not itself part of content of the physical description.., (it is a form of that description.)"

The informal language forms of leading physicists working in this field had thus split into mutually irrelevant fragments. Yet, it was widely believed that physicists were in general communication on the quantum theory. Such a belief was possible because of the common notion that the "essence" of the theory is its formalism, which all physicists shared in common. Thus, it would appear that the formalism plays a role analogous to a perceived object which each person can see and touch and then describe verbally. That is to say, physicists could tacitly and informally treat the formalism as a perceived "mental object", so that they could be thought to communicate about this "object", as it were, by "pointing" to various features of the formalism. Thus differences of informal language (such as that of Bohr from that of von Neumann) could be regarded as having little more significance than differences in different people's ways of describing a perceived object.

In an indirect manner, the sharp distinction between formal and informal languages described above leads to a correspondingly sharp distinction between form and content. For the formalism is then being treated as a kind of content which is being "talked about" in the informal form of ordinary language.

It is noteworthy that even Bohr, who was sensitive to the significance of the wholeness of form and content in other contexts (e.g. observer and observed), also tended to make a sharp distinction between the formal algorithm of the quantum theory and the informal description of the experiment in ordinary language. To be sure, he showed that these two were related. He pointed out that the informal description in terms of two mutually incompatible but complementary kinds of general experimental conditions corresponds in the formal algorithm to a pair of operators that do not commute. But in this mode of discussion, there is a sharp tacit separation between formal and informal languages. This separation is evidently the necessary precondition for the possibility of a relationship between the two kinds of language. (Indeed, the notion of a "problem of measurement" in the context is such a relationship). Thus, Bohr saw the unanalyzable wholeness of the description of experimental conditions and experimental results, but he did not seem similarly to regard formal and informal descriptions as a whole. As a result, the way was left open for a consideration of the formal description as "representing a separate physical reality" which it would make sense to "measure". Therefore, it was not made clear that "the problem of measurement" is irrelevant.

If one thinks about this question carefully, he will see that formal and informal languages are indeed inseparable. For the essence of the formal language is that it gives a set of well-defined rules. But these rules have no meaning or relevance without a discussion of the informal situations in which they would either apply or not apply. For example, consider the various formal rules discussed in the field of spectroscopy. These rules would be nothing but empty verbal expressions if there were not a wide range of instances where the rules are relevant, so that one could see in each case whether they applied or whether there was an "exception". Thus, in a certain sense, the formal rules are the instances in which they are relevant either as exceptions or as non-exceptions. But the description of such instances is evidently basically informal. Thus, the formal and informal descriptions are (like that of experimental conditions and experimental results) a whole, in which what each aspect is reveals itself in and through the other aspect. It would not be appropriate to say that they are different but related through some kind of disjunction and correspondence which is in turn an object of further relevant discourse.

Because a formal description is the informal situation in which it is relevant, in the sense described above, confusion in the informal language will inevitably lead ultimately to a corresponding confusion in the formal language. In areas that are well-known and familiar, we can learn informally to "adapt" and "adjust", so as to tend to avoid this kind of confusion (e.g. we can judiciously alternate between different and mutually incompatible "models", none of which is actually fully relevant to the situations under investigation). But when we enter new and unknown domains we can no longer do this; and then, confusion in the informal language can easily confuse what is done with the formalism.

It is generally accepted that physics is now entering new and unknown domains (e.g. the study of elementary particles, cosmology, and possibly other fields). Indeed, there is a growing feeling that older forms of thought are (or soon will be) exhausted. Of course, no one has yet been able to call attention to broad ranges of phenomena where current notions are clearly revealed as irrelevant (e.g. as Bohr did in spectroscopy and Einstein did in connection with notions of time and space). Nevertheless, there are growing indications of such irrelevance. First of all, there are the infinities and other formal inconsistencies of field theories. And then there is a general informal situation, characterized by the continual appearance of new kinds of particles, and the ultimate inadequacy of current theories to deal with the corresponding facts fully and coherently, in spite of frequent adaptations and modifications (e.g. the introduction of S matrices, symmetry groups, current algebras, etc.)

In all of these developments, the general informal language of physics has not been given very much attention. But one may well ask whether the time is not ripe for a consideration of these informal forms, in which, as has been seen, the formal forms have to obtain their meanings and to show their relevance or irrelevance. A very important indication of the irrelevance of older language forms can even now be seen by asking how quantum theory and relativity are to be related. Thus far, such inquiry has been based largely on attempts to develop "relativistic quantum" formalisms, which still work in terms of the older informal forms. Little attention has however been given to the question of whether the informal language forms of relativity and quantum theory are actually capable of being combined in any consistent way. Rather, it seems to have been tacitly taken for granted that such a combination is possible. But as will be seen, a more careful inquiry into this question shows a kind of mutual irrelevance of these two informal forms that might well serve to indicate in more detail just how one could introduce a new general informal form (rather as happened in the work of Einstein and Bohr).

Although Einstein played a key part in early developments of the quantum theory, it is well know that he was never able to accept the informal form that this theory ultimately took. Generally speaking, most physicists tacitly ignored Einstein's difficulties with the quantum theory as irrelevant. It is significant, however, that Bohr did not do this. Rather, he engaged in extensive and serious discussions with Einstein, in which he tried, apparently without success, to communicate the novel character of his informal language forms. This failure of communication was unfortunate in its consequences for the development of physics, because it meant that contact was never really made between the informal language forms of relativity and of quantum theory. Thus the language of physics was effectively split into two disjoint parts between which a real interchange or dialogue has ceased to to be possible.

In the discussion between Bohr and Einstein, a crucial turning point was the interchange that took place concerning the hypothetical experiment of Einstein, Podolsky and Rosen.[14] This experiment was suggested as part of an extensive and thorough criticism of Bohr's views concerning the "quantum" mode of description. Starting from the non-relativistic quantum formalism, EPR were able by means of informal discussions about the significance of the formal equations to obtain certain results that they felt to be paradoxical. For example, consider a molecule of "total spin zero", which disintegrates into two atoms, each of "spin one-half". The atoms are allowed to separate, so that the distance between them is ultimately very large, at which time they may be presumed to have ceased to interact significantly. Then, the component of the "spin" of one of them can be "measured" in any desired direction. One deduces from the formalism that the component of the spin of the other atom in that same direction is always opposite. But from the uncertainty principle, one also deduces that the remaining components of the spin of each atom are undefined, (so that they are described by many physicists as "fluctuating at random").

Now, the difficulty noted by EPR arises from the circumstance that while the atoms are still in flight, we can re-orient the apparatus for "measuring" the spin of the first atom in any desired direction. It would seem then that immediately the second atom "knows" that its "spin component" must be well defined in that particular direction, and undefined (or "fluctuating') in other directions. Thus one would have to suppose that one atom had instantaneously "signalled" some kind of information concerning "spin direction" to the other atom. This contradicts the generally accepted notion that such atoms have ceased to interact significantly after they are far apart.

[14]Einstein et al. (1983).

EPR therefore argued that in some sense there must belong to the second atom certain "elements of reality" which were already defined before the "spin" of the first atom was measured and that the "quantum" language is therefore incapable of providing a "complete description of reality". This meant in fact that Bohr's notion of unanalyzable wholeness of the experimental results and the experimental conditions would have to be inadequate and indeed self-contradictory.

Bohr's answer to EPR[15] was in a very carefully worded article in which he showed essentially that EPR's criticisms were based on a certain informal language (i.e. an extension of the language of classical dynamics) which was not compatible with the language of "wholeness of experimental results and experimental conditions". More specifically, Bohr introduced a restricted application of the term "measurement" to the quantities defined as "classical" and said that it is not formally correct in quantum theory to employ a description of "measurement of the spin", such as that used by EPR. Thus Bohr implied that EPR had not actually met his views.

Bohr's answer to EPR was in a certain limited sense an adequate and consistent one as far as its content was concerned. But in a broader sense, it overlooked a key point and thus contributed greatly to the breakdown of communication which followed. What Bohr overlooked was the unspoken background of the whole discussion of EPR, in which there was implied the relevance of the informal language of relativity and therefore the irrelevance of Bohr's informal language of "wholeness".

To indicate the significance of tacit and informal language forms explicitly, it is useful to begin by asking why EPR did not conclude from their argument that there should be a novel kind of interaction or interconnection which would in some way transmit "information" concerning the "spin direction" of the first atom to the second. Such a proposal would for example have been in a general kind of harmony with the Newtonian conception of "action at a distance". However, EPR apparently did not regard this possibility to be worthy of mention. Indeed, in the informal language form of relativity, in terms of which these authors did all of their work, the question of "action at a distance" could not even arise, because all "communication" between different regions had to be by way of "signals" transmitted continuously across the intervening space. It was moreover necessary for the consistency of their language form that no "signals" faster than light are possible. It would follow then that even if one proposed some new and unknown means of producing interactions, this would still not be acceptable because the hypothetical experiment of EPR implied that the instantaneous transfer of "information" from one atom to the other. It is clear then that EPR's rejection of Bohr's informal "quantum description" tacitly presupposed the context of the general language form of relativity.

Now Bohr himself had frequently emphasized that his own informal language is relevant only in a non-relativistic context. Thus, he recognized that it can be used only on condition of the smallness of the fine structure constant, $e^2/c\hbar$. Moreover, in relativity a basic notion is that of separate and distinct "point events" which can then be related by "signals". However, in Bohr's language, the notion of "unanalyzable wholeness" leaves no place for a basic term like "point event" and therefore no place

[15] Bohr (1983).

for the corresponding term "signal" whose essential meaning is just the relationship between such events.

EPR had in fact tried to meet this point by starting with a non-relativistic quantum formalism. But tacitly, the notion of a limiting signal velocity was not separable from their whole informal language form. In view of the fact that the general language form could not be criticised explicitly in their particular mode of discussion, it probably did not seem significant either to Bohr or to EPR to note that it had not proved possible to extend Bohr's informal discussion in terms of unanalyzable wholeness to the relativistic context. But this implies, of course, that the whole informal language form of EPR, which took the notion of a "signal" for granted, made Bohr's informal form completly irrelevant. On the other hand, Bohr's language made Einstein's informal language involving the (tacit) notion of a "signal" completely irrelevant.

The attempt of these two men to communicate was evidently confused, because the essence of the EPR paper, insofar as it was relevant, was entirely unstated, whereas Bohr answered only the explicit content of the paper. Indeed, Bohr's appropriate response would have been to say that the discussion of EPR had from the very outset, nothing to do with the meaning of "quantum" in Bohr's language, so that no further answer was really required. However, by nevertheless giving a detailed "answer", Bohr tacitly implied that the informal form of the language of EPR was in fact relevant. Thus, in his "informal form" of responding to EPR, Bohr was not in harmony with the formal content of his reply, which latter explicitly and in great detail brought out the irrelevance of the discussion of EPR. But evidently, it was precisely this very detailed and serious form of an "answer" that informally implied the relevance of the "questions" raised by EPR. Thus, the whole discussion became confused. And subsequent efforts to "clear up" this confusion tended to make it worse, because tacitly and informally they continued to imply that the language of EPR was relevant to that of Bohr and vice versa, with each other.

In connection with the apparently natural wish to "clear up" confusion of the kind indicated above, it must be emphasized that such efforts will tend to maintain the confusion rather than to clear it up. For example, in order to express the entire description of the general situation constituting the illustrative example about EPR, it was necessary for us to employ incompatible informal language forms of "wholeness" and of "separate events and objects". Any discussion which attempts to relate such forms must in some way be confused. And here it should be noted that merely in order to criticize the statements of someone who uses these incompatible forms, we cannot avoid the informal use of essentially the same forms. Thus, even the attempt to "define the confusion" in detail will tend to add to it informally. For this reason, even our own discussion of the arguments between EPR and Bohr must still show a certain residue of confusion if they are very carefully analyzed. The only way out of such confusion of form is, at a certain point, just to drop the issue and to go on to something new. We need simply to be aware here, as elsewhere, of the wholeness of form and content.

If one thus perceives that confusion inevitably results from trying to use the mutually incompatible informal languages of "quantum" and "relativity" one will thus stop trying to accommodate these two languages to each other. But one tends to feel that the situation in physics cannot be left as it is, i.e. a "state of fragmentation" into branches that actually have almost nothing to do with each other. Indeed, this feeling, although concealed, is doubtless behind the many efforts that have been made to develop formalisms that would unite relativity and quantum theory. But now, one sees that these efforts are irrelevant, as long as the informal languages are so different. What then is to be done?

What is called for is evidently an awareness that both relativity and quantum theory are forms of communication, each of which is its own informal context that does not include the other. To drop the effort to accommodate these language forms to each other, and to develop a language that is relevant in a broader context requires that we cease to start with the older forms in a basic role. That is to say, by talking "about" relativity or quantum theory, we cannot get beyond them. We have therefore to begin anew, with novel forms, in which terms like "quantum" or "signal" are no longer taken to be primal. Thus, we can come to a new infomal language with a broader context of relevance. The older "quantum" and "relativistic" forms will then be abstracted as simplifications, limiting cases, or approximations. But the new descriptions and the new laws will be as different from those of relativity and quantum theory as these latter are from classical Newtonian theory.

In further papers that are to follow this one, we shall suggest specific new forms of informal language that go beyond the contexts of relevance of "quantum" and "relativistic" languages. These lead not only to novel formal theories, but also to the indication of new kinds of experiments. Thus, in experimental arrangements similar to those considered by EPR, it can be seen with the aid of such novel language forms that it is possible to ask new kinds of experimental questions that would have no relevance in terms of the older language forms. In particular, one can see the importance of studying, for example, how "spin measurements" are actually related when, according to relativistic notions, there is not enough time for a "signal" to pass from one atom to the other. One can anticipate the possibility of entirely new kinds of relationships in these observations which, would indicate the irrelevance of the whole informal language form of quantum theory, (rather as spectroscopic observations similarly indicate the general irrelevance of the informal language form of classical dynamics). And likewise, yet other observations are possible that would indicate the irrelevance of the relativistic notion of "signal". Indeed, as will be seen in later articles, the irrelevance of the notion of "signal" may perhaps also be significant in other fields such as cosmology, where one is already confronted with a wide range of phenomena (in connection with "quasars", etc.) that appear to be very confused, when discussed in terms of current language forms. Thus, new developments in language of the kind indicated above may well bring about a different approach to experiment in general which would help to provide the clear indications which have so far been lacking, of irrelevance of older forms of physical description.

It must be emphasized that the informal language of physics plays just as important a role with regard to the form of experiment as it does with regard to the formal theories. In a certain sense, it may properly be said that experiment is one extension of the general informal language of physics while theory is another. For example, in terms of the Ptolemaic language form, an indicated experiment for astronomy was observation of the epicycles needed for the description of planetary motions. After the work of Kepler, Copernicus, Galileo and Newton, it was clear that epicycles constituted an irrelevant descriptive form. Instead, what is relevant was seen to be a description in terms of orbits involving positions, velocities, curvatures, etc. All of these would have been regarded as totally irrelevant in ancient times. The orbit does not even appear in direct perception; it is an abstraction those relevance is indicated largely with the aid of the formal language forms of algebra, analytic geometry, and the calculus.

As has already been pointed out, however, Einstein later saw the irrelevance of Newton's informal description of space and time as absolute, while Bohr saw the irrelevance of the precisely defined orbits and of the separation of "observer and observed". The new "relativistic" and "quantum" language forms that emerged then led to yet newer kinds of experimental observations (e.g. variation of mass in nuclear reactions, the symbolic classification of spectral lines in terms of orbital angular momentum quantum numbers, spin, etc, etc.) The formal elaboration of this informal language then elicited the development of "high energy machines", which are evidently designed to answer mainly the sort of questions that can be raised in terms of these forms.

It is clear that the situation described above can lead to a certain kind of trap, of which it is difficult for us to be aware. The machines, which are an extension of the current language forms of physics, provide us almost entirely with the kinds of data that can readily be described in terms of these very forms. It is therefore extremely difficult to get out of our present theoretical language forms, because the "need" for continuing them is apparently constantly being indicated by the kind of experimental data that these machines are able to provide (so that it may thus appear that "nature itself" is such as to demand just these forms and no others). Indeed, the existence of technical means to raise new kinds of experimental questions (such as those involved in probing into the relevance of current "quantum" and "relativistic" languages) now depends on largely fortuitous and accidental developments, because these are, of course, not "planned for" at all, when the time comes to design laboratories and their equipment. So our present way of doing experiments in itself constitutes a very strong commitment to continue the current informal language forms of physics (and of science in general) into the indefinite future.

The possibility of such a self-stabilizing commitment may be seen, at least implicitly, in a recent article of L. Rosenfeld,[16] who mentions the following about Bohr. "... In one of his last conversations, he observed that the reason why no progress was being made in the theory of the transformations of matter occurring at very high energies is that we have not so far found among these processes any one exhibiting a sufficiently violent contradiction with what could be expected from current ideas to give a clear and unambiguous indication of how we have to modify these ideas." In view of what has been pointed out here, the possibility is emerging that the content of the explicit "experimental results" can never present the "contradiction" indicated by Bohr's remark. Rather, the relevant cue for modifying current ideas may have to

[16]Rosenfeld (1967).

come from the whole informal situation which involves the indefinite development of apparatus and techniques that enable us to look only for the general kind of result that we already expect to find. Indeed, even the search for an explicit "contradiction" between theory and experiment may well be an elaboration of this same informal situation, in the sense that we are still seeking something similar to the "contradictions" that developed in the past, i.e. contradiction between already accepted theories and the experimental facts that result from the effort to test or apply these theories. But now, we may be faced with an essentially new significance of "wholeness", that of the highly-developed apparatus with the results it provides, so that the relevant "contradiction" is of a very implicit character, and inseparable from the over-all informal situation indicated in this article.

What has to be emphasized especially in connection with this informal situation is that theories play a key role in determining what kind of experiment is judged to be worth doing. Indeed, as has already been indicated, the form of the theoretical language tends to call attention to certain kinds of questions (such as those connected with epicycles, orbits, quantum numbers, or scattering cross-sections), thus making those questions "stand out" as being interesting for experimental investigation. Other questions are, in the very same forms, tacitly relegated to the role of being unimportant or not worthy of notice at all.

For this reason, it would perhaps be helpful here to revive the old word "to relevate", which has dropped out of common usage. This means "to lift up" and has the same root as the word "relief", so that to "relevate something" is to make it "stand out in relief" in the foreground of attention, while to "irrelevate" it is to push it into the unnoticed background. Each form of physical description relevates certain sorts of content and irrelevates others.

What is meant by the word "relevant"? Evidently, this is one of those words (like "hot") which can have no explicit and formal definition (or at least none that would be relevant). Rather, its meaning always appears informally in the way the word is used. Indeed, the notion of relevance is extremely subtle, much more so, for example, than that of formal truth or falsity. Thus, before one can meaningfully discuss the truth or falsity of a statement, one has to judge its relevance. And there is no way to describe how this is done. It is basically an act of perception which is an unanalyzable whole. Relevance and judgment <u>are</u> this whole, so even the verbal disjunction between relevance and judgment is a contingency of the language.

The tendency of our language to "relevate the irrelevant" has been satirized by Lewis Carroll.[17] For example, the March Hare says that the Mad Hatter's watch doesn't run properly although he used the <u>best</u> butter. Thus there could be a long subsequent discussion as to whether the butter was first or second grade, with theoretical arguments and experimental observations, attempting to ascertain what the grade of butter actually was and to study how variations in this grade influence the running of watches. But this irrelevant kind of discussion as a whole is a form of language of which we are generally unaware. The perception of this form would mean the ending of such forms. But in complex scientific inquiries, it is not so easy

---

[17]See C134, p. 205, n 3—CT.

to see that a somewhat similar irrelevance is often at the root of what is most puzzling and confusing in our work. Thus, when Einstein objected to Bohr's form of description tacitly because it implied a signal velocity greater than that of light, it escaped the notice of both of them that the speed of signals has no more relevance to the "quantum" context than the grade of butter has to the running of watches. And this happened in spite of the fact that both of these men had been unusually sensitive in other contexts to similar questions of the relevance of terms in informal description. This shows that all of us can very easily become confused by arguing "about" irrelevant questions in great detail. This activity is just what occupies our attention so completely that we are *informally* prevented from perceiving the irrelevance of what is being "talked about". In other words, the informal form determined by the general academic situation in which we work can lead to an unbalanced kind of interest in narrow questions, whose irrelevance thus escapes the awareness of all those who participate in this situation. (It should be added, however, that no value-judgment of the situation, e.g. as "confused", will change that situation. Thus, relevance is not the same as value.)

If one carefully considers the kind of situation indicated above, he will see that such reactions are actually playing a key role in determining our general way of looking at the relationship of theory and experiment. For example, there is currently a great deal of discussion as to whether or not there is adequate experimental evidence to make possible a decision concerning the validity of the general theory of relativity. In all this discussion, it seems to be tacitly agreed that the main requirement for clearing up the situation is to get more experimental results. From what has been pointed out earlier in this article, hovever, it is clear that an even more significant question is whether the informal language of relativity (special and general) does not have some context of relevance whose limitations may be demonstrated, not only in the context of the "quantum", but also, perhaps, in that of cosmology (with all the confusion in this subject concerning "quasars", "pulsars", etc.)

However, because of the common "consensus" that theories are "about the experimental facts", there tends to be a very heavy emphasis on the effort to "get more facts". In this effort, it is not noticed that such facts are actually *elicited* by the general language form of physics. Or as has been indicated earlier, a particular theoretical form, such as that of general relativity, may be said to *relevate* certain kinds of data (such as the careful measurement of planetary precessions and the deflection of light rays by the sun), while it tacitly *irrelevates* a wide and unspecifiable range of further kinds of experimental observations, to which new informal language forms could call attention. Because interest is so heavily focussed on the questions that are thus relevated, we cannot see their limitad relevance.

The notion that "facts come first" and that theories are relevant in sofar as they correlate the facts mathematically or give knowledge about them in some other way is by now very deeply ingrained in the whole informal situation in which scientific research is carried out. To some extent, this may have begun as a kind of reaction against Medieval scholastic tendencies to engage in elaborate discussions of purely verbal questions. But it does not seem to have been noticed that this reaction goes too far. It is necessary to question this notion as a whole and in each detail in order to help bring about a more balanced approach to scientific research.

First of all, we may ask what is meant by the term "experimental fact". Consider, for example, what actually happens in a laboratory experiment. A geiger counter may click a certain number of times, and this "fact" is noted and recorded as an item of data. But at the same time, a truck may pass by, the telephone may ring, the light may be switched on, and some of the laboratory technicians may be engaged in an argument. Although all of this (and much more) is part of the fact of what happened at that moment, none of it is regarded as relevant. And this relevance judgment is evidently determined by the language form of physics, which relevates a very small part of the fact in the content of its particular kind of description. Without such judgments of relevance (and irrelevance), the "fact" would be of unlimited verbal extent; it could not be recorded as such, and could have no definite meaning or content. The notion of a "fact" is thus seen to be a form of our current mode of expression.

In a certain way, it can be said that the language of physics "shapes" or "makes" the fact. Indeed, the very word "fact" means "to make" (e.g. as in "manufacture"). So each language form determines the making of a corresponding kind of fact. A person using psychological language forms would thus probably regard the argument between the laboratory technicians as the relevant fact. It is therefore clear that "facts" are not to be confused with objects that one would pick up in the laboratory, to be preserved in notebooks or in other ways, and then published as objects of discourse. Rather, the experimental fact is an extension of the language forms determining the relevance judgments that are to be operative in a particular kind of situation which is termed a "scientific experiment". Thus, as with form and content, and formal and informal forms, the fact and the language forms needed in its description are an unanalyzable whole. And this means, of course, that it is not appropriate to regard theory and experiment as disjoint (but related) activities.

A more subtle form of the notion of intrinsic separateness of theory and experiment is that theories form a "hypothetical-deductive system" or "knowledge" that is either to be confirmed or falsified by experiment. In this point of view, it is recognized that the theory is not a simple correlation of already given facts. Rather, it is regarded as a proposed form of knowledge (so that theoretical statements could be termed "propositions"). Such proposals might be inspired by almost anything, so that their origin may have no obvious relationship to experimental fact. But whatever their source, it is argued that they constitute a kind of "knowledge" that has to be tested experimentally.

This view tends to give too much emphasis to the question of truth and falsity of particular theories and too little to that of the relevance of the general language forms of these theories. It is well known that when a particular theory is falsified by experient, it is always possible by adapting or modifying some secondary hypothesis to "save the theory", i.e. to retain the same general form of description. So no experiment can ever lead definitely to the need to drop a particular general language form. Indeed, even when there is a great deal of empirical evidence falsifying particular theories that are expressed in terms of such a form, it is still appropriate to hold onto this form, if one has reason to believe it to be relevant. For example, Galileo very persistently held on to his new dynamical forms of description, in spite of a wealth of fairly detailed empirical evidence tending to falsify particular theoretical expressions of this form.

The key question is therefore that of relevance, rather than that of truth or falsity. And the judgment of relevance is not primarily based on experiments aimed at confirming or falsifying predictions of a theory. Rather, it depends on a kind of perception that cannot be described or analyzed in detail. Bohr's perception of the irrelevance of classical language forms in the context of spectroscopy did not involve confirmation or falsification in any clearly definable way. Yet it was the basic step that led ultimately to the dropping of classical forms of theory. And after this step was taken, what followed was the "making" of entirely new kinds of facts (e.g. quantum numbers of spectral lines, etc.)

It may be perhaps be helpful here to point out that the Greek word "theoria" has the same basic root as the word "theater". It means "to view". So a theory could be regarded mainly as giving a kind of insight, and not a kind of knowledge. In a similar way, a drama can give insight into human character without giving "factual news" about what actually happened or "knowledge" about particular individuals who may be represented as characters in the play. Thus, it would have no meaning to try experimentally either to confirm or to refute the insights afforded by a drama. Rather, one would have to perceive their relevance and irrelevance. Similarly, if a theory is mainly a form of insight, then what is called for is a perception of its relevance and irrelevance, while the attempt to regard it as "proposed knowledge" subject to confirmation and falsification could be a source of confusion.

Our whole scientific tradition has however developed in such a way that we find it very hard not to think that a theory aims to "give true knowledge about how things are". On the basis of this tradition, the continual succession of theories having radically different forms nakes the actual situation in science almost incomprehensible. But if we regard theories primarily as language forms that give insight, we need not be surprised by the continual appearance of radically different language forms. The main role of experiment is then to give us nonverbal indications of the relevance of that content which is relevated by these forms (rather than to add to "the total of knowledge about nature").

In a very significant way, even this degree of disjunction between theory and experiment can be seen to be largely a formal and verbal one, without genuine content. For when a theorist changes the general language form, what he is doing is an <u>experiment with theories</u>. That is, when he considers a new description, his activity is also to perceive whether its content is actually relevant. In doing this, he first sees the mutual relevance of his new theoretical language and a series of older and less abstract theoretical languages that gradually extend onward to the language of experiment. In this "two way" flow of relevance judgments there emerges an overall perception of the relevance or irrelevance of his "experiment with theory", and perhaps of his activity as a whole. Evidently, such an approach is not possible if one regards theory as potentially true or false knowledge. What significance could be attached to an "experiment with truth and falsity"?

The above discussion indicates a very thoroughgoing unity of theory and experiment. Both are actually experimental in their informal form and both are aimed mainly at insight into what is relevant, rather than at knowledge of what is true and false.

The tendency to continue the disjunction between theory and experiment is however rooted in certain very pervasive aspects of our civilization, some of which can be formally and explicitly described, and some of which are so tacit and informal that it is very difficult even to indicate how they operate.

What can be said formally about this question is that our attitude to theory and experiment is, to a certain extent, an outgrowth of the effort of Kant and others to solve the problem implied by the Cartesian disjunction between "mind" and "matter". As indicated earlier, Kant was thus led to assume "things in themselves", which were distinct from the "Ego" or "observing subject" which only "has knowledge about them". But this still did not make clear what the status of this "knowledge" is. It was perhaps the "Zeitgeist" (i.e. the prevailing informal form of thought) to try to solve this problem by giving heavy emphasis to the formal aspects of knowledge (mathematics, logic, etc.). In the formalism, one seems to have something which is simultaneously a definable and describable object, and which is yet a form of knowledge, and therefore part of the "knowing Ego". So in the formalism one appears in some sense to be able to unite both subject and object. But, of course, this knowledge would be merely "empty formalism" unless it were "about" something external to it. This is where the experimental fact comes in. Tho experimental fact is what the theoretical formalism is supposed to be "about".

We have seen, however, that this way of thinking does not stand up to a close scrutiny. Actually, formal and informal languages are an unanalyzable whole in which *the formal rules* ***are*** *the informal situations in which they are* relevant, (either in the role of exceptions or of non-exceptions.) Similarly, the formal and informal languages of science *are* the factual situations in which they are relevant, either in the role of statements that are falsified or of statements that are confirmed. Indeed, it is only in and through these languages that what is observed can give rise to a scientific fact. And as has already been indicated earlier, even the instruments and techniques of experimentation are extensions of the general informal language forms of science (e.g. the "high energy machines").

So there is no disjunction or separation between scientific discourse and anything else that is an essential aspect of science. The approach that treats such aspects as separate is tho result of a certain contingent language form that disjoins thought from action, knowledge from what is known, the description from what is described, etc. The contingent character of this language form has been able to escape general awareness largely because the difficulties in the Cartesian duality of mind and matter were regarded as "problems" by those who followed Descartes. As a result, all attention (which was judged to be relevant to the question) tended to be focussed on efforts, tacit or explicit, to "solve the problem". It was thus not noticed that this way of using language is actually irrelevant as a whole, so that there *is* no problem. It was rather like taking a torch to go and look for a fire, or indeed investigating the influence of the grade of butter on the running of watches. It is thus appropriate simply to drop the informal language form which assigns basic relevance to the distinction of knowledge from what is known, observer from observed, and which tacitly dismisses the wholeness of form and content of physical description and discourse.

By continuing the old and familiar tradition however, we tend to go on with the form in which we regard the content of our discourse (both theoretical and experimental) as quite distinct and separate from the act of discourse. The former is said to belong to the object under investigation, while the latter is said to be one of the activities or manifestations of the observing subject. Hitherto, it has been generally accepted that this subject (i.e. the scientist himself) is a separate entity who is only observing and doing experiments in whatever it is that enters into the content of the discourse, but who is not himself an essential part of the object under investigation. But clearly this separation itself is merely an irrelevant form of discourse; when the irrelevance of this form is perceived, it is possible to do scientific research in a new way, i.e. by experimenting with the informal forms of discourse along with their extensions as formal theories and as experimental instruments and techniques. What is perceived non-verbally in such an "experiment" thus gives insight, not only into the content of the discourse, but also into the form (both formal and informal). But since the form of our communications is what we are, such a perception is as relevant to ourselves as to the world we live in. To perceive in this manner is evidently to see the irrelevance of the traditional disjunction of self from nature or the rost of the world. For the form of our communication, which is us, is now involved in every inquiry in a way that cannot be separated from the rest of the inquiry (including for example the specific content that is under investigation).

The traditional disjunction of self from nature implies that in an experiment, it is only the content of the discourse which is being tested with the aid of laboratory instruments, while the theory (which includes the general form of discourse) is tacitly regarded as a kind of ineffable" subjective" factor that could never be "part of the experiment". The recognition of the inseparability of form and content means, however, that our language and mode of thinking are just as much in question in each "experiment" as is "nature" or the "material world". Nevertheless, the indication of the unanalyzable wholeness of form and content does not attribute a unique relevance to language, or to associated forms of thought (i.e. ideas). Rather, the relevance of both form and content is to be indicated by perception going essentially beyond words.

The "problem" of disjunction of self from nature, or equivalently, of theory from experiment, has hovever been very deeply ingrained into the informal form of human life (and thus into every aspect of the teaching of successive generations) over the centuries of development of modern society. It was therefore generally speaking not at all easy for those in this philosophical tradition simply to set this disjunction aside as irrelevant. Even those like Wittgenstein and Bohr, who were able to see the irrelevance of such forms in their academic contexts, fell into a certain kind of disharmony when they tacitly relevated the content of formal language (logical or mathematical) to the level of some kind of disjoint and commonly perceivable set of objects. For example, in the Tractatus Logico-Philosophicus, Wittgenstein spoke about "atomic propositions", thus informally implying that the content of formal statements could in some way be regarded as constituting such objects, disjoined from the formal language. Later he naturally saw the limited relevance of this disjunction since the formal language of Boolean algebra did not cover all "formal systems" that could arise in mathematics as objects of discourse. As a further example which has already been mentioned, Bohr was inclined to make a rather sharp disjunction between the formal language (the algorithm) and the language of everyday life. He emphasized that experiment had to be described in terms of the language of everyday life (suitably refined, where necessary, to that of classical physics). However, the formal algorithm had a very different language, with basic terms like "operator", "comutator", "square root of minus one", etc. etc. In this way, he was led tacitly to make a correspondingly sharp distinction between the language of theory and the language of experiment. But since, as has been seen, the language relevates both the kinds of experimental fact and the kinds of theoretical questions that will be regarded as worthy of consideration, Bohr's distinction between these languages therefore implies that experiment and theory are correspondingly distinct and separate kinds of activities.

To understand why it was so difficult to go out of the involvement of the "Zeitgeist" in the question of formalism and the closely related one of the separation of theory and experiment, it is necessary to consider a wide context of informal factors within which the formal factors have their rolavance. It may perhaps be useful hore to try to give a kind of insight into the whole informal situation in which our discussion is taking place by saying that the disjunction of theory and experiment (along with that of formal and informal languages) is informally a kind of game which developed gradually, that was not generally noticed by those who took part in it. As a result, whoever now studies physics (or other sciences) tacitly "picks up the rules of the game", which include the notion (mainly informal) that "experiment is one thing" and "theory is another". And whoever wants to work in the field will evidently have to "play according to the rules", not only because those who do not may bo "penalized", but even more because it would evidently seem futile to any reasonable person to enter the game and then to refuse to follow the rules of that game.

Even those few who do not follow the rules are still generally giving relevance to these rules by regarding their behavior as "exceptional" or "heretical". The notion that these rules may not have unlimited relevance thus seems not worthy of serious consideration. Indeed, as experiments are done within the rules, a form of data is produced which informally and tacitly demands that the theorist regard his own work as a separate effort to develop theoretical knowledge that explains or correlates the data. Vice versa, the theorist produces abstract conclusions which tacitly and informally demand that the experimenter regard his work as an effort to confirm or falsify the prediction of the theory. Thus, it appears in cach detailed aspect of the physicist's life that "experience itself" is demonstrating the inescapable necessity of these rules and these only.

Nevertheless, it is evidently quite possible that while these rules may have been relevant in certain ways in the past, and while they may still be relevant today in certain limited contexts, they may actually be getting in the way of progress on fundamental questions of new kinds that are now emerging. As has been indicated in connection with relativity, quantum theory and cosmology, it may well be a more fruitful approach to inquire into the informal language forms, which relevate questions that are common both to experiment and to theory. As has already been indicated, such an inquiry is at once an experiment with language forms and an experiment with the instruments and techniques which are extensions of those forms. The overall aim of such experimental activity is insight, (rather than knowledge). But this kind of insight is not actually possible as long as scientists generally accept the relevance of the traditional game, in which "the ball", i.e. knowledge, is continually "tossed from the theoretical side to the experimental side and back again".

It is evidently not easy for us to set these rules aside and to begin something very different. The great difficulty of such an approach is just that which is informal and hard to define or indicate, such as a vague fear of becoming lost in a very wide context where one would have no assurance of doing meaningful work. The game then is strengthened by seemingly negligible and irrelevant feelings which we are conditioned not to express completely. Thus the game is not completely separable from a personal security in which we will be able to get the satisfaction of solving well defined problems (as well as that of "knowing what nature is really like", etc.) In addition, such activities will of course be helpful for advancement of one's reputation or status. Thus, in uncountable ways, one's energies tend to become directed back into the game, even if one may have seen that it could be worthwhile to question its relevance.

The notion of a game calls attention to the wholeness of form and contont, and also, of the formal and informal forms. Thus, in each move (or transaction) one is expressing or embodying the rules; the form of the game is a whole with the content of each transaction. One may be aware of the rules "in the back of the mind", but in the "foreground" there is the need to meet the changing situation in terms of the rules. This challenge is felt as a frustration about success, fear of failure, pleasure about a winning move, excitement about the prospect of success, etc. The rules, which by themselves would be empty verbal forms, take on a "living quality" as they infuse sensations, emotions, thoughts, and directed actions, which are one's response to the informal game situation. Indeed, it can be said that the rules are the total informal situation, as described above. And whoever is playing the game is the game at that moment, for his whole "character" is that mode of response which is aimed at meeting the game situation. ("Externalizing" the game from the player as objects of discourse would be continuing the disjunction of subject and object.) If a person however feels the game to be irrelevant, he loses interest in it, and thus ceases to be the game.

The recognition of a game in human society is essentially a new recognition about the "object" of such a game. The "object" and the game are a whole. Clearly, without such recognition, the game, regardless of the detailed human behavior, will show changes of content rather than of form.

The description of human activities in terms of games has evidently a broad significance going beyond natural science. Wittgenstein regarded language in a dialectical sense to be a kind of game in which there is a wholeness of the form of discourse with its "living content". Von Neumann developed a description of formalized situations in terms of games, and developed a mathematical theory of games which is applicable to the subject of computing machines as well as to the determination of political and military strategies. More recently there has appeared a book by Berne entitled Games People Play.[18] Implicitly, this title is a disjunction equivalent to that of subject and object. Explicitly, the content is an application in the context of psychology of certain identifiable game forms. From the informal point of view, this work involves value-judgments about certain of these identifiable games. The implication is that people are in ordinary circumstances not generally serious in their relationships, nor even really honest. However, no value-judgment is implied here as we wish only to call attention to the wholeness of the identifiable game with the content of each detailed "game transaction" and to theory and experiment as forms rather than content of game transactions. The wholeness of form and content in human activities such as physics is clearly expressible in these terms. (The disjunction of subject and object, implicit in the form in which the "people" are separate from the "game", as an object of discourse, has a further formal expression in the work cited above through the metaphysical notion of a "stimulus hunger", i.e. of a need to "structure time" with content of some kind for the proper functioning of the nervous system.)

We do tend quite frequently to go on with games that have ceased to be relevant, but any value-judgment of the whole game is not appropriate in the context of interest here and would indeed have no clear basis. Thus rather than judging the game to be good or bad, we might simply be aware that there is a general tendency for the form of an indentifiable game to enter informally into human relationships in an irrelevant way. And, of course, this kind of irrelevance as regards the forms of theory and experiment is suggested in the present context. To see the wholeness of theory and experiment as extension of common language is therefore not only to open up the possibility of changing science in a fundamental way, but it also opens up the possibility of corresponding changes in the scientist. (e.g., as indicated earlier, in the discussion of how a change of language form is as relevant to ourselves as to the world we live in).

We may thus be able to see that it may no longer be relevant to continue responding to the challenge of the "old game" in which theory and experiment are separated, and begin a new kind of inquiry perceiving the wholeness of theory and experiment, formal and informal, description and described, knowledge and what is known.

Such a new kind of inquiry could have far-reaching consequences in many contexts other than those which have been discussed in this article. For example, the notion of "order" in present physical science is confused precisely because of a sharp distinction between form and content. Thus such notions of order as "information", "genetic codes", "entropy", "randomness", etc. are now commonly regarded as objects of discourse. Nevertheless, they are clearly inseparable from the form of our communications. The separation of ourselves from our communications is a game not basically different from that of theory and experiment. Thus to end this kind of game can lead to quite novel forms of discourse which may well imply the irrelevance of many of the questions that are now considered central in these contexts (e.g. the distinction of order and disorder).

---

[18] Berne (1966).

In conclusion, it might be emphasized that this paper is its implications. That is, it would not be appropriate to introduce a further "game" by distinguishing between the contents of the present paper and further discussions "about" it in the form of interpretation of a critical character.

Generally, if we "play the game" of talking "about" any object or human activity, then rather than producing anything essentially new, we will produce only a further "display" of what we call that object or activity. And going beyond the "game" of talking "about" also goes beyond purely intellectual criteria of judgment (which may have aspects of emotional character defense). Thus, to speak "about" Bohr, Einstein, or Wittgenstein is not to understand them. This is a purely intellectual function, broadly speaking, without essentially novel perception.

Often when wide cultural implications are involved, we tend to retreat from them, as they seem too pervasive and too "deeply" ingrained to be changed easily by any individual. This is also an "externalizing" of personal emotional defenses that are pervasive in human life for just the reason that they are not fully perceived in wide varieties of informal situations. If we seriously ask why we do what we do, something essentially new may happen.

References

Berne, E. (1966). *Games people play: The psychology of human relationships*. Andre Deutsch.
Bohr, N. (1963). *Atomic physics and human knowledge*. New Jersey: Wiley Interscience.
Bohr, N. (1983). Can qauntum-mechanical description of physical reality be considered complete? In J. A. Wheeler, & W. H. Zurek, (Eds.), *Quantum theory and measurement*, (pp. 145–151). Princeton: Princeton University Press. Originally published in Physical Review, *48*, 696–702 (1935).
Einstein, A., Podolsky, B., & Rosen, N. (1983). Can quantum-mechanical description of physical reality be considered complete? In J. A. Wheeler, & W. H. Zurek, (Eds.), *Quantum theory and measurement*, (pp. 138–141). Princeton: Princeton University Press. Originally published in Physical Review, *47*, 777–780 (1935).
Rosenfeld, L. (1967). Niels Bohr in the thirties: Consolidation and extension of the conception of complementarity. In Rozental, S., (Ed.), *Niels Bohr: His life and work as seen by his friends and colleagues*, (pp. 114–136). North-Holland.
von Neumann, J. (1955). *Mathematical foundations of quantum mechanics*. Princeton: Princeton University Press.

Printed in the United States
by Baker & Taylor Publisher Services